Analog Communications

Kasturi Vasudevan

Analog Communications

Problems and Solutions

Ane Books
Pvt. Ltd.

 Springer

Kasturi Vasudevan
Electrical Engineering
Indian Institute of Technology Kanpur
Kanpur, Uttar Pradesh, India

ISBN 978-3-030-50339-0   ISBN 978-3-030-50337-6   (eBook)
https://doi.org/10.1007/978-3-030-50337-6

Jointly published with ANE Books Pvt. Ltd.
In addition to this printed edition, there is a local printed edition of this work available via Ane Books in
South Asia (India, Pakistan, Sri Lanka, Bangladesh, Nepal and Bhutan) and Africa (all countries in the
African subcontinent).
ISBN of the Co-Publisher's edition: 9789386761811

This Springer imprint is published by the registered company Springer Nature Switzerland AG
The registered company address is: Gewerbestrasse 11, 6330 Cham, Switzerland

To my family

Preface

Analog Communications: Problems and Solutions is suitable for third-year undergraduates taking a first course in communications. Although most of the present-day communication systems are digital, they continue to use many basic concepts like Fourier transform, Hilbert transform, modulation, synchronization, signal-to-noise ratio analysis, and so on that have roots in analog communications. The book is richly illustrated with figures and easy to read.

Chapter 1 covers some of the basic concepts like the Fourier and Hilbert transforms, Fourier series, and the canonical representation of bandpass signals. Random variables and random processes are covered in Chap. 2. Amplitude modulation (AM), envelope detection, Costas loop, squaring loop, single sideband modulation, vestigial sideband modulation, and quadrature amplitude modulation are covered in Chap. 3. In particular, different implementations of the Costas loop is considered in Chap. 3. Chapter 4 deals with frequency modulation (FM) and various methods of generation and demodulation of FM signals. The figure-of-merit analysis of AM and FM receivers is dealt with in Chap. 5. The principles of quantization are covered in Chap. 6. Finally, topics on digital communications are covered in Chap. 7.

I would like to express my gratitude to some of my instructors at IIT Kharagpur (where I had completed my undergraduate)—Dr. S. L. Maskara (Emeritus faculty), Dr. T. S. Lamba (Emeritus faculty), Dr. R. V. Rajkumar and Dr. S. Shanmugavel, Dr. D. Dutta, and Dr. C. K. Maiti.

During the early stages of my career (1991–1992), I was associated with the CAD-VLSI Group, Indian Telephone Industries Ltd., Bangalore. I would like to express my gratitude to Mr. K. S. Raghunathan (formerly a Deputy Chief Engineer at the CAD-VLSI Group) for his supervision of the implementation of a statistical fault analyzer for digital circuits. It was from him that I learnt the concepts of good programming, which I cherish and use to this day.

During the course of my master's degree and Ph.D. at IIT Madras, I had the opportunity to learn the fundamental concepts of digital communications from my instructors, Dr. V. G. K. Murthy, Dr. V. V. Rao, Dr. K. Radhakrishna Rao, Dr. Bhaskar Ramamurthi, and Dr. Ashok Jhunjhunwalla. It is a pleasure to acknowledge their teaching. I also gratefully acknowledge the guidance of Dr. K. Giridhar and Dr. Bhaskar Ramamurthi who were jointly my Doctoral supervisors. I also wish to thank Dr. Devendra Jalihal for introducing me to the *LATEX* document processing system, without which this book would not have been complete.

Special mention is also due to Dr. Bixio Rimoldi of the Mobile Communications Lab, EPFL Switzerland, and Dr. Raymond Knopp, now with Institute Eurecom, Sophia Antipolis France, for providing me the opportunity to implement some of the signal processing algorithms in real time, for their software radio platform.

I would like to thank many of my students for their valuable feedback. I thank my colleagues at IIT Kanpur, in particular Dr. S. C. Srivastava, Dr. V. Sinha (Emeritus faculty), Dr. Govind Sharma, Dr. Pradip Sircar, Dr. R. K. Bansal, Dr. K. S. Venkatesh, Dr. A. K. Chaturvedi, Dr. Y. N. Singh, Dr. Ketan Rajawat, Dr. Abhishek Gupta, and Dr. Rohit Budhiraja for their support and encouragement.

I would also like to thank the following people for encouraging me to write this book:

- Dr. Surendra Prasad, IIT Delhi, India
- Dr. P. Y. Kam, NUS Singapore
- Dr. John M. Cioffi, Emeritus faculty, Stanford University, USA
- Dr. Lazos Hanzo, University of Southampton, UK
- Dr. Prakash Narayan, University of Maryland, College Park, USA
- Dr. P. P. Vaidyanathan, Caltech, USA
- Dr. Vincent Poor, Princeton, USA
- Dr. W. C. Lindsey, University of Southern California, USA
- Dr. Bella Bose, Oregon State University, USA
- Dr. S. Pal, former President IETE, India
- Dr. G. Panda, IIT Bhubaneswar, India
- Dr. Arne Svensson, Chalmers University of Technology, Sweden
- Dr. Lev B. Levitin, Boston University, USA
- Dr. Lillikutty Jacob, NIT Calicut, India
- Dr. Khoa N. Le, University of Western Sydney, Australia
- Dr. Hamid Jafarkhani, University of California Irvine, USA
- Dr. Aarne Mämmelä, VTT Technical Research Centre, Finland
- Dr. Behnaam Aazhang, Rice University, USA
- Dr. Thomas Kailath, Emeritus faculty, Stanford University, USA
- Dr. Stephen Boyd, Stanford University, USA
- Dr. Rama Chellappa, University of Maryland, College Park, USA

Thanks are also due to the open source community for providing operating systems like Linux and software like Scilab, *LATEX*, Xfig, and Gnuplot, without which this book would not have been complete. I also wish to thank Mr. Jai Raj Kapoor and his team for their skill and dedication in bringing out this book.

In spite of my best efforts, some errors might have gone unnoticed. Suggestions for improving the book are welcome.

Kanpur, India   Kasturi Vasudevan

Contents

About the Author

Kasturi Vasudevan completed his Bachelor of Technology (Honours) from the Department of Electronics and Electrical Communication Engineering, IIT Kharagpur, India, in 1991, and his M.S. and Ph.D. from the Department of Electrical Engineering, IIT Madras, in 1996 and 2000, respectively. During 1991–1992, he was employed with Indian Telephone Industries Ltd., Bangalore, India. He was a Postdoctoral Fellow at Mobile Communications Lab, EPFL, Switzerland, and then an Engineer at Texas Instruments, Bangalore. Since July 2001, he has been a faculty member at the Electrical Department at IIT Kanpur, where he is now an Professor. His interests lie in the area of communication.

Notation

$a \wedge b$	Logical AND of a and b
$a \vee b$	Logical OR of a and b
$a \stackrel{?}{=} b$	a may or may not be equal to b
$\forall$	For all
$\lfloor x \rfloor$	Largest integer less than or equal to x
$\lceil x \rceil$	Smallest integer greater than or equal to x
j	$\sqrt{-1}$
$\stackrel{\Delta}{=}$	Equal to by definition
$*$	Convolution
$\delta_D(\cdot)$	Dirac delta function
$\delta_K(\cdot)$	Kronecker delta function
$\tilde{x}$	A complex quantity
$\hat{x}$	Estimate of x
$\mathbf{x}$	A vector or matrix
$\mathbf{I}_M$	An $M \times M$ identity matrix
S	Complex symbol (note the absence of tilde)
$\Re\{\cdot\}$	Real part
$\Im\{\cdot\}$	Imaginary part
x_I	Real or in-phase part of $\tilde{x}$
x_Q	Imaginary or quadrature part of $\tilde{x}$
$E[\cdot]$	Expectation
$\mathrm{erfc}(\cdot)$	Complementary error function
$[x_1, x_2]$	Closed interval, inclusive of x_1 and x_2
$[x_1, x_2)$	Open interval, inclusive of x_1 and exclusive of x_2
(x_1, x_2)	Open interval, exclusive of x_1 and x_2
$P(\cdot)$	Probability
$f_X(\cdot)$	Probability density function of X
Hz	Frequency in Hertz
wrt	With respect to

Chapter 1
Signals and Systems

1. We know that the Fourier transform is given by

$$G(f) = \int_{t=-\infty}^{\infty} g(t) \exp\left(-j 2\pi f t\right) dt. \tag{1.1}$$

Using the properties of the Dirac-delta function, prove the inverse Fourier transform, that is,

$$g(t) = \int_{f=-\infty}^{\infty} G(f) \exp\left(j 2\pi f t\right) df. \tag{1.2}$$

Hint: Substitute for $G(f)$ from (1.1).

- *Solution*: Substituting for $G(f)$ in the right-hand side of (1.2), we get

$$
\begin{aligned}
&\int_{f=-\infty}^{\infty} G(f) \exp\left(j 2\pi f t\right) df \\
&= \int_{f=-\infty}^{\infty} \int_{x=-\infty}^{\infty} g(x) \exp\left(-j 2\pi f x\right) dx \exp\left(j 2\pi f t\right) df \\
&= \int_{x=-\infty}^{\infty} g(x) dx \int_{f=-\infty}^{\infty} \exp\left(j 2\pi f (t-x)\right) df. \tag{1.3}
\end{aligned}
$$

However, we know that

$$\delta(t) \rightleftharpoons 1$$

$$\Rightarrow \int_{f=-\infty}^{\infty} \exp\left(j 2\pi f t\right) df = \delta(t)$$

$$\Rightarrow \int_{f=-\infty}^{\infty} \exp\left(j 2\pi f (t-x)\right) df = \delta(t-x). \tag{1.4}$$

© The Editor(s) (if applicable) and The Author(s), under exclusive license
to Springer Nature Switzerland AG 2021
K. Vasudevan, *Analog Communications*,
https://doi.org/10.1007/978-3-030-50337-6_1

Therefore (1.3) becomes

$$\int_{f=-\infty}^{\infty} G(f) \exp\left(j\,2\pi f t\right) df = \int_{x=-\infty}^{\infty} \delta(t-x)g(x)\,dx$$

$$= g(t). \tag{1.5}$$

Thus proved.

2. (Simon Haykin 1983) Consider a pulse-like function $g(t)$ that consists of a small number of straight-line segments. Suppose that this function is differentiated with respect to time twice so as to generate a sequence of weighted delta functions as shown by

$$\frac{d^2 g(t)}{dt^2} = \sum_i k_i \delta(t - t_i), \tag{1.6}$$

where k_i are related to the slopes of the straight-line segments.

(a) Given the values of k_i and t_i show that the Fourier transform of $g(t)$ is given by

$$G(f) = -\frac{1}{4\pi^2 f^2} \sum_i k_i \exp\left(-j\,2\pi f t_i\right). \tag{1.7}$$

(b) Using the above procedure find the Fourier transform of the pulse in Fig. 1.1.

• *Solution*: To solve part (a) we note that

$$\frac{d^2 g(t)}{dt^2} = \int_{f=-\infty}^{\infty} (j\,2\pi f)^2\, G(f) e^{j\,2\pi f t}\, df, \tag{1.8}$$

where $G(f)$ is the Fourier transform of $g(t)$. In other words

$$\frac{d^2 g(t)}{dt^2} = g_1(t) \rightleftharpoons (j\,2\pi f)^2\, G(f) = G_1(f) \quad \text{(say)}. \tag{1.9}$$

Consequently

$$G(f) = -\frac{G_1(f)}{4\pi^2 f^2}. \tag{1.10}$$

Fig. 1.1 $g(t)$

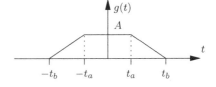

In part (a), it is clear that

$$G_1(f) = \sum_i k_i e^{-j 2\pi f t_i}. \tag{1.11}$$

Therefore

$$G(f) = -\frac{1}{4\pi^2 f^2} \sum_i k_i e^{-j 2\pi f t_i}. \tag{1.12}$$

To solve part (b), we note that

$$\frac{d^2 g(t)}{dt^2} = \frac{A}{t_b - t_a} \left(\delta(t + t_b) - \delta(t + t_a) - \delta(t - t_a) + \delta(t - t_b) \right). \tag{1.13}$$

This is illustrated in Fig. 1.2. Hence

$$G(f) = -\frac{A}{(t_b - t_a) 4\pi^2 f^2} \left(\exp(j 2\pi f t_b) - \exp(j 2\pi f t_a) \right.$$
$$\left. - \exp(-j 2\pi f t_a) + \exp(-j 2\pi f t_b) \right)$$
$$\Rightarrow G(f) = -\frac{A}{(t_b - t_a) 2\pi^2 f^2} \left(\cos(2\pi f t_b) - \cos(2\pi f t_a) \right)$$
$$\Rightarrow G(f) = \frac{A}{(t_b - t_a) \pi^2 f^2} \sin(\pi f (t_b + t_a)) \sin(\pi f (t_b - t_a)), \tag{1.14}$$

Fig. 1.2 Illustrating the various derivatives of $g(t)$

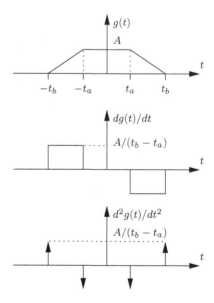

Fig. 1.3 $g_p(t)$

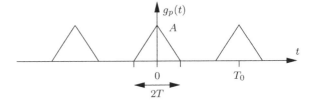

where we have used the formula:

$$\cos(B) - \cos(A) = 2\sin\left(\frac{A+B}{2}\right)\sin\left(\frac{A-B}{2}\right). \tag{1.15}$$

3. (Simon Haykin 1983) Using the above procedure compute the complex Fourier series representation of the periodic train of triangular pulses shown in Fig. 1.3. Here $T < T_0/2$. Hence also find the Fourier transform of $g_p(t)$.

- *Solution*: Consider the pulse:

$$g(t) = \begin{cases} g_p(t) \text{ for } -T_0/2 < t < T_0/2 \\ 0 \qquad \text{elsewhere.} \end{cases} \tag{1.16}$$

Clearly

$$\frac{d^2 g(t)}{dt^2} = \frac{A}{T} \left(\delta(t+T) - 2\delta(t) + \delta(t-T) \right). \tag{1.17}$$

Hence

$$G(f) = -\frac{A}{4\pi^2 f^2 T} \left(\exp(\mathrm{j}\, 2\pi f T) - 2 + \exp(-\mathrm{j}\, 2\pi f T) \right)$$

$$= \frac{AT}{\pi^2 f^2 T^2} \sin^2(\pi f T)$$

$$= AT \operatorname{sinc}^2(fT). \tag{1.18}$$

We also know that

$$g_p(t) = \sum_{n=-\infty}^{\infty} c_n \exp\left(\mathrm{j}\, 2\pi n t / T_0\right), \tag{1.19}$$

where

$$c_n = \frac{1}{T_0} \int_{t=-\infty}^{\infty} g(t) \exp\left(-j\,2\pi nt / T_0\right) dt$$

$$= \frac{1}{T_0} G(n/T_0).$$ (1.20)

Substituting for $G(n/T_0)$ we have

$$g_p(t) = \frac{AT}{T_0} \sum_{n=-\infty}^{\infty} \operatorname{sinc}^2(nT/T_0) \exp\left(j\,2\pi nt / T_0\right).$$ (1.21)

Using the fact that

$$\exp\left(j\,2\pi f_c t\right) \rightleftharpoons \delta(f - f_c)$$ (1.22)

the Fourier transform of $g_p(t)$ is given by

$$G_p(f) = \frac{AT}{T_0} \sum_{n=-\infty}^{\infty} \operatorname{sinc}^2(nT/T_0)\delta\left(f - \frac{n}{T_0}\right).$$ (1.23)

4. (Simon Haykin 1983) Consider a signal $g_p(t)$ defined by

$$g_p(t) = A_0 + A_1 \cos(2\pi f_1 t + \theta) + A_2 \cos(2\pi f_2 t + \theta),$$ (1.24)

which is given to be periodic with a period of T_0.

(a) Determine the autocorrelation $R_{g_p}(\tau)$.
(b) Is any information lost in obtaining $R_{g_p}(\tau)$?

• *Solution*: Since $g_p(t)$ is periodic with a period T_0, the autocorrelation of $g_p(t)$ is also periodic with period T_0. We also have

$$T_0 = \frac{n}{f_1} = \frac{m}{f_2}$$

$$\Rightarrow T_0 = \frac{n-m}{f_1 - f_2} = \frac{m+n}{f_1 + f_2} = \frac{2n}{2f_1} = \frac{2m}{2f_2},$$ (1.25)

where m and n are integers. Thus, frequencies $f_1 - f_2$, $f_1 + f_2$, $2f_1$, and $2f_2$ are also periodic with period T_0. Now, the autocorrelation of $g_p(t)$ is equal to (note that $g_p(t)$ is real-valued)

$$R_{g_p}(\tau) \triangleq \frac{1}{T_0} \int_{t=0}^{T_0} g_p(t)g_p(t - \tau)\, dt$$

$$= A_0^2 + \frac{A_1^2}{2} \cos(2\pi f_1 \tau) + \frac{A_2^2}{2} \cos(2\pi f_2 \tau),$$ (1.26)

where we have made use of the fact that sinusoids integrated over one period is equal to zero. In computing the autocorrelation, we have lost the phase information.

5. Consider a signal $g_p(t)$ defined by

$$g_p(t) = A_0 + A_1 \sin(2\pi f_1 t + \theta) - A_2 \cos(2\pi f_2 t + \alpha), \qquad (1.27)$$

which is given to be periodic with a period of T_0.

(a) Determine the autocorrelation $R_{g_p}(\tau)$.
(b) Is any information lost in obtaining $R_{g_p}(\tau)$?

- *Solution*: Since $g_p(t)$ is periodic with a period T_0, the autocorrelation of $g_p(t)$ is also periodic with period T_0. We also have

$$T_0 = \frac{n}{f_1} = \frac{m}{f_2}$$
$$\Rightarrow T_0 = \frac{n-m}{f_1 - f_2} = \frac{m+n}{f_1 + f_2} = \frac{2n}{2f_1} = \frac{2m}{2f_2}, \qquad (1.28)$$

where m and n are integers. Thus, frequencies $f_1 - f_2$, $f_1 + f_2$, $2f_1$, and $2f_2$ are also periodic with period T_0. Now, the autocorrelation of $g_p(t)$ is equal to (note that $g_p(t)$ is real-valued)

$$R_{g_p}(\tau) \triangleq \frac{1}{T_0} \int_{t=0}^{T_0} g_p(t) g_p(t - \tau) \, dt$$
$$= A_0^2 + \frac{A_1^2}{2} \cos(2\pi f_1 \tau) + \frac{A_2^2}{2} \cos(2\pi f_2 \tau), \qquad (1.29)$$

where we have made use of the fact that sinusoids integrated over one period is equal to zero. In computing the autocorrelation, we have lost the phase information.

6. (Simon Haykin 1983) Let $G(f)$ denote the Fourier transform of a real-valued signal $g(t)$ and $R_g(\tau)$ its autocorrelation function. Show that

$$\int_{\tau=-\infty}^{\infty} \left| \frac{d R_g(\tau)}{d\tau} \right|^2 d\tau = 4\pi^2 \int_{f=-\infty}^{\infty} f^2 |G(f)|^4 \, df. \qquad (1.30)$$

- *Solution*: We know that

$$R_g(\tau) \rightleftharpoons |G(f)|^2$$
$$\Rightarrow \frac{d R_g(\tau)}{d\tau} \rightleftharpoons j 2\pi f \, |G(f)|^2. \qquad (1.31)$$

Let

$$\frac{dR_g(\tau)}{d\tau} = g_1(\tau).$$ (1.32)

Let $G_1(f)$ denote the Fourier transform of $g_1(t)$. Thus

$$G_1(f) = j2\pi f \, |G(f)|^2.$$ (1.33)

By Rayleigh's energy theorem we have

$$\int_{\tau=-\infty}^{\infty} |g_1(\tau)|^2 \, d\tau = \int_{f=-\infty}^{\infty} |G_1(f)|^2 \, df.$$ (1.34)

Substituting for $g_1(t)$ and $G_1(f)$ in the above equation we get the required result.

7. A sinusoidal signal of the form $A\cos(2\pi f_0 t)$, $A > 0$, is full wave rectified to obtain $g_p(t)$.

(a) Using the complex Fourier series representation of $g_p(t)$ and the Parseval's power theorem, compute

$$S = 9\left(\frac{2}{\pi}\right)^2 \left[1 + \frac{2}{3^2} + \cdots + \frac{2}{(1-4n^2)^2} + \cdots\right],$$ (1.35)

where n is an integer greater than or equal to one.

(b) The full wave rectified signal is passed through an ideal bandpass filter having a gain of 2 in the passband frequency range of $f_0 < |f| < 5f_0$. Compute the power of the output signal.

• *Solution*: The complex Fourier series representation of a periodic signal $g_p(t)$, having a period T_1, is

$$g_p(t) = \sum_{n=-\infty}^{\infty} c_n \exp(j2\pi nt/T_1),$$ (1.36)

where

$$c_n = \frac{1}{T_1} \int_{-T_1/2}^{T_1/2} g_p(t) \exp(-j2\pi nt/T_1) \, dt$$ (1.37)

denotes the complex Fourier series coefficient. In the given problem:

$$g_p(t) = A|\cos(2\pi f_0 t)|$$
$$T_1 = 1/(2f_0).$$ (1.38)

Hence

$$
\begin{aligned}
c_n &= \frac{A}{T_1} \int_{-T_1/2}^{T_1/2} |\cos(2\pi f_0 t)| \exp(-j\,2\pi nt/T_1)\, dt \\
&= \frac{2A}{2T_1} \int_0^{T_1/2} 2\cos(2\pi f_0 t)\cos(2\pi nt/T_1)\, dt \\
&= \frac{A}{T_1} \int_0^{T_1/2} [\cos(2\pi(f_0 - n/T_1)t) + \cos(2\pi(f_0 + n/T_1)t)]\, dt \\
&= \frac{A}{T_1} [\sin(2\pi(f_0 - n/T_1)T_1/2)/(2\pi(f_0 - n/T_1)) \\
&\quad + \sin(2\pi(f_0 + n/T_1)T_1/2)/(2\pi(f_0 + n/T_1))] \\
&= \frac{AT_1}{T_1} [\cos(n\pi)/(2\pi(f_0 T_1 - n)) \\
&\quad + \cos(n\pi)/(2\pi(f_0 T_1 + n))] \\
&= \frac{A(-1)^n}{\pi} [1/(1-2n) + 1/(1+2n)] \\
&= \frac{2A(-1)^n}{\pi(1-4n^2)}.
\end{aligned}
\tag{1.39}
$$

Observe that in (1.39)

$$
c_{-n} = c_n.
\tag{1.40}
$$

From Parseval's power theorem, we have

$$
\begin{aligned}
\frac{1}{T_1} \int_{t=-T_1/2}^{T_1/2} |g_p(t)|^2\, dt &= \sum_{n=-\infty}^{\infty} |c_n|^2 \\
&= |c_0|^2 + 2\sum_{n=1}^{\infty} |c_n|^2 \\
&= \left(\frac{2A}{\pi}\right)^2 \left[1 + \frac{2}{3^2} + \cdots + \frac{2}{(1-4n^2)^2} + \cdots\right],
\end{aligned}
\tag{1.41}
$$

where we have used (1.40). Comparing (1.35) with the last equation of (1.41), we obtain

$$
A = 3.
\tag{1.42}
$$

Now, the left-hand side of (1.41) is equal to

$$\frac{1}{T_1} \int_{t=-T_1/2}^{T_1/2} |g_p(t)|^2 \, dt = \frac{A^2}{T_1} \int_{t=-T_1/2}^{T_1/2} \cos^2(2\pi f_0 t) \, dt$$

$$= \frac{A^2}{2}. \qquad (1.43)$$

Substituting for A from (1.42), the sum in (1.35) is equal to

$$S = 4.5. \qquad (1.44)$$

The (periodic) signal at the bandpass filter output is

$$g_{p1}(t) = \sum_{\substack{n=-2 \\ n \neq 0}}^{2} 2c_n \exp(j \, 2\pi n t / T_1), \qquad (1.45)$$

where we have assumed that the gain of the bandpass filter is 2 and c_n is given by (1.39). The signal power at the bandpass filter output is

$$P = \sum_{\substack{n=-2 \\ n \neq 0}}^{2} 4 \, |c_n|^2$$

$$= 3.372. \qquad (1.46)$$

8. (Simon Haykin 1983) Determine the autocorrelation of the Gaussian pulse given by

$$g(t) = \frac{1}{t_0} \exp\left(-\frac{\pi t^2}{t_0^2}\right). \qquad (1.47)$$

- *Solution*: We know that

$$e^{-\pi t^2} \rightleftharpoons e^{-\pi f^2}$$

$$\Rightarrow \frac{1}{t_0} e^{-\pi t^2} \rightleftharpoons \frac{1}{t_0} e^{-\pi f^2}. \qquad (1.48)$$

Using time scaling with $a = 1/t_0$ we get

$$g(t) = \frac{1}{t_0} e^{-\pi t^2 / t_0^2} \rightleftharpoons \frac{|t_0|}{t_0} e^{-\pi f^2 t_0^2} = G(f). \qquad (1.49)$$

Since $G(f)$ is real-valued, the Fourier transform of the autocorrelation of $g(t)$ is simply $G^2(f)$. Thus

$$G^2(f) = e^{-2\pi f^2 t_0^2} = \psi_g(f) \qquad \text{(say)}. \qquad (1.50)$$

Fig. 1.4 $X(f)$

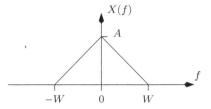

Once again we use the time scaling property of the Fourier transform, namely,

$$|a|g(at) \rightleftharpoons G(f/a) \tag{1.51}$$

with $1/a = \sqrt{2}$ in (1.49) to get

$$R_g(t) = \frac{1}{|t_0|\sqrt{2}} e^{-\pi t^2/(2t_0^2)}, \tag{1.52}$$

which is the required autocorrelation of the Gaussian pulse. Observe the $|t_0|$ in the denominator of (1.52), since $R_g(0)$ must be positive.

9. (Rodger and William 2002) Assume that the Fourier transform of $x(t)$ has the shape as shown in Fig. 1.4. Determine and plot the spectrum of each of the following signals:

   (a) $x_1(t) = (3/4)x(t) + (1/4)\mathrm{j}\,\hat{x}(t)$
   (b) $x_2(t) = \left[(3/4)x(t) + (3/4)\mathrm{j}\,\hat{x}(t)\right] e^{\mathrm{j}2\pi f_0 t}$
   (c) $x_3(t) = \left[(3/4)x(t) + (1/4)\mathrm{j}\,\hat{x}(t)\right] e^{\mathrm{j}2\pi W t}$,

   where $f_0 \gg W$ and $\hat{x}(t)$ denotes the Hilbert transform of $x(t)$.

   • *Solution*: We know that

$$\hat{x}(t) \rightleftharpoons -\mathrm{j}\,\mathrm{sgn}\,(f)X(f). \tag{1.53}$$

   Therefore

$$X_1(f) = (3/4)X(f) + (1/4)\mathrm{sgn}\,(f)X(f)$$
$$= \begin{cases} X(f) & \text{for } f > 0 \\ (3/4)A & \text{for } f = 0. \\ (1/2)X(f) & \text{for } f < 0 \end{cases} \tag{1.54}$$

   The plot of $X_1(f)$ is given in Fig. 1.5.
   To solve part (b) let

$$m(t) = (3/4)x(t) + (3/4)\mathrm{j}\,\hat{x}(t). \tag{1.55}$$

Fig. 1.5 $X_1(f)$

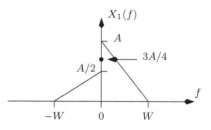

Fig. 1.6 $X_2(f)$

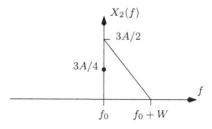

Then

$$M(f) = (3/4)X(f) + (3/4)\text{sgn}\,(f)X(f)$$
$$= \begin{cases} (3/2)X(f) & \text{for } f > 0 \\ (3/4)A & \text{for } f = 0 \,. \\ 0 & \text{for } f < 0 \end{cases} \tag{1.56}$$

Then $X_2(f) = M(f - f_0)$ and is given by

$$X_2(f) = \begin{cases} (3/2)X(f - f_0) & \text{for } f > f_0 \\ (3/4)A & \text{for } f = f_0 \,. \\ 0 & \text{for } f < f_0 \end{cases} \tag{1.57}$$

The plot of $X_2(f)$ is shown in Fig. 1.6.
To solve part (c) let

$$m(t) = (3/4)x(t) + (1/4)\text{j}\,\hat{x}(t). \tag{1.58}$$

Then

$$M(f) = (3/4)X(f) + (1/4)\text{sgn}(f)\,X(f)$$
$$= \begin{cases} X(f) & \text{for } f > 0 \\ (3/4)A & \text{for } f = 0 \,. \\ (1/2)X(f) & \text{for } f < 0 \end{cases} \tag{1.59}$$

Fig. 1.7 $X_3(f)$

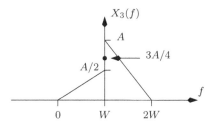

Thus $X_3(f) = M(f - W)$ and is given by

$$X_3(f) = \begin{cases} X(f - W) & \text{for } f > W \\ (3/4)A & \text{for } f = W \\ (1/2)X(f - W) & \text{for } f < W \end{cases} \tag{1.60}$$

The plot of $X_3(f)$ is given in Fig. 1.7.

10. Consider the input

$$x(t) = \text{rect}\,(t/T)\cos(2\pi(f_0 + \Delta f)t) \tag{1.61}$$

for $\Delta f \ll f_0$. When $x(t)$ is input to a filter with impulse response

$$h(t) = \alpha e^{-\alpha t}\cos(2\pi f_0 t)u(t) \tag{1.62}$$

find the output by convolving the complex envelopes. Assume that the impulse response of the filter is of the form:

$$h(t) = 2h_c(t)\cos(2\pi f_0 t) - 2h_s(t)\sin(2\pi f_0 t). \tag{1.63}$$

- *Solution*: The canonical representation of a bandpass signal $g(t)$ centered at frequency f_c is

$$g(t) = g_c(t)\cos(2\pi f_c t) - g_s(t)\sin(2\pi f_c t), \tag{1.64}$$

where $g_c(t)$ and $g_s(t)$ extend over the frequency range $[-W, W]$, where $W \ll f_c$. The complex envelope of $g(t)$ is defined as

$$\tilde{g}(t) = g_c(t) + j\,g_s(t). \tag{1.65}$$

The given signal $x(t)$ can be written as

$$x(t) = \text{rect}\,(t/T)\left[\cos(2\pi f_0 t)\cos(2\pi\Delta f t) - \sin(2\pi f_0 t)\sin(2\pi\Delta f t)\right], \tag{1.66}$$

where $\Delta f \ll f_0$. By inspection, we conclude that the complex envelope of $x(t)$ is given by

$$\tilde{x}(t) = \text{rect } (t/T) \left[\cos(2\pi \Delta f t) + j \sin(2\pi \Delta f t) \right]$$
$$= \text{rect } (t/T) e^{j 2\pi \Delta f t}. \tag{1.67}$$

Thus

$$x(t) = \Re \left\{ \tilde{x}(t) e^{j 2\pi f_0 t} \right\}. \tag{1.68}$$

In order to facilitate computation of the filter output using the complex envelopes, it is given that the filter impulse response is to be represented by

$$h(t) = 2h_c(t) \cos(2\pi f_0 t) - 2h_s(t) \sin(2\pi f_0 t), \tag{1.69}$$

where the complex envelope of the filter is

$$\tilde{h}(t) = h_c(t) + j h_s(t). \tag{1.70}$$

Again by inspection, the complex envelope of the channel is

$$\tilde{h}(t) = \frac{1}{2} \alpha e^{-\alpha t} u(t). \tag{1.71}$$

Thus

$$h(t) = \Re \left\{ 2\tilde{h}(t) e^{j 2\pi f_0 t} \right\}. \tag{1.72}$$

Now, the complex envelope of the output is given by

$$\tilde{y}(t) = \tilde{x}(t) \star \tilde{h}(t)$$
$$= \int_{\tau=-\infty}^{\infty} \tilde{x}(\tau) \tilde{h}(t - \tau) \, d\tau. \tag{1.73}$$

Substituting for $\tilde{x}(\cdot)$ and $\tilde{h}(\tau)$ in the above equation, we get

$$\tilde{y}(t) = \frac{\alpha}{2} \int_{\tau=-T/2}^{T/2} e^{j 2\pi \Delta f \tau} e^{-\alpha(t-\tau)} u(t - \tau) \, d\tau. \tag{1.74}$$

Since $u(t - \tau) = 0$ for $t < \tau$, and since $-T/2 < \tau < T/2$, it is clear that $\tilde{y}(t) = 0$, and hence $y(t) = 0$ for $t < -T/2$.
Now, for $-T/2 < t < T/2$, we have

$$\tilde{y}(t) = \frac{\alpha}{2} \int_{\tau=-T/2}^{t} e^{j 2\pi \Delta f \tau} e^{-\alpha(t-\tau)} \, d\tau$$

$$= \frac{\alpha}{2} e^{-\alpha t} \int_{\tau=-T/2}^{t} e^{\tau(\alpha+j 2\pi \Delta f)} \, d\tau$$

$$= \frac{\alpha}{2(\alpha + j 2\pi \Delta f)} e^{-\alpha t} \left[e^{t(\alpha+j 2\pi \Delta f)} - e^{-T/2(\alpha+j 2\pi \Delta f)} \right]$$

$$= \frac{\alpha}{2(\alpha + j 2\pi \Delta f)} \left[e^{j 2\pi \Delta f t} - e^{-\alpha(t+T/2)-j\pi \Delta f T} \right]. \tag{1.75}$$

Let

$$\theta_1 = \tan^{-1}\left(\frac{2\pi \Delta f}{\alpha}\right)$$

$$\theta_2 = \pi \Delta f T$$

$$r = \sqrt{\alpha^2 + (2\pi \Delta f)^2}. \tag{1.76}$$

Then for $-T/2 < t < T/2$ the output is

$$y(t) = \Re\left\{ \tilde{y}(t) e^{j 2\pi f_0 t} \right\}$$

$$= \frac{\alpha}{2r} \left[\cos(2\pi(f_0 + \Delta f)t - \theta_1) \right.$$

$$\left. - e^{-\alpha(t+T/2)} \cos(2\pi f_0 t - \theta_1 - \theta_2) \right]. \tag{1.77}$$

Similarly, for $t > T/2$ we have

$$\tilde{y}(t) = \frac{\alpha}{2} \int_{\tau=-T/2}^{T/2} e^{j 2\pi \Delta f \tau} e^{-\alpha(t-\tau)} \, d\tau$$

$$= \frac{\alpha}{2} e^{-\alpha t} \int_{\tau=-T/2}^{T/2} e^{\tau(\alpha+j 2\pi \Delta f)} \, d\tau$$

$$= \frac{\alpha}{2(\alpha + j 2\pi \Delta f)} e^{-\alpha t} \left[e^{T/2(\alpha+j 2\pi \Delta f)} - e^{-T/2(\alpha+j 2\pi \Delta f)} \right]$$

$$= \frac{\alpha}{2(\alpha + j 2\pi \Delta f)} \left[e^{-\alpha(t-T/2)} e^{j\pi \Delta f T} - e^{-\alpha(t+T/2)} e^{-j\pi \Delta f T} \right].$$

$$\tag{1.78}$$

Therefore for $t > T/2$ the output is

$$y(t) = \Re\left\{ \tilde{y}(t) e^{j 2\pi f_0 t} \right\}$$

$$= \frac{\alpha}{2r} \left[e^{-\alpha(t-T/2)} \cos(2\pi f_0 t - \theta_1 + \theta_2) \right.$$

$$\left. - e^{-\alpha(t+T/2)} \cos(2\pi f_0 t - \theta_1 - \theta_2) \right]. \tag{1.79}$$

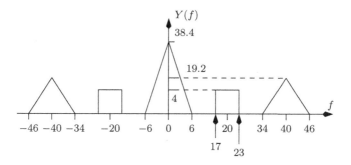

Fig. 1.8 $Y(f)$

11. A nonlinear system defined by

$$y(t) = x(t) + 0.2x^2(t) \tag{1.80}$$

has an input signal with a bandpass spectrum given by

$$X(f) = 4\,\mathrm{rect}\left(\frac{f-20}{6}\right) + 4\,\mathrm{rect}\left(\frac{f+20}{6}\right). \tag{1.81}$$

Sketch the spectrum at the output labeling all the important frequencies and amplitudes.

- *Solution*: The Fourier transform of the output is

$$Y(f) = X(f) + 0.2X(f) \star X(f). \tag{1.82}$$

The spectrum of $Y(f)$ can be found out by inspection and is shown in Fig. 1.8.

12. (Simon Haykin 1983) Let $R_{12}(\tau)$ denote the cross-correlation function of two energy signals $g_1(t)$ and $g_2(t)$.

(a) Using Fourier transforms, show that

$$R_{12}^{(m+n)}(\tau) = (-1)^n \int_{t=-\infty}^{\infty} h_1(t)h_2^*(t-\tau)\,dt, \tag{1.83}$$

where

$$h_1(t) = g_1^{(m)}(t)$$
$$h_2(t) = g_2^{(n)}(t) \tag{1.84}$$

denote the mth and nth derivatives of $g_1(t)$ and $g_2(t)$, respectively.

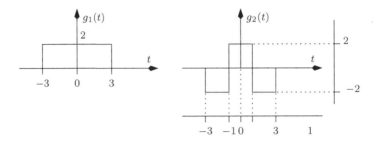

Fig. 1.9 Pulses $g_1(t)$ and $g_2(t)$

(b) Use the above relation with $m = 1$ and $n = 0$ to evaluate and sketch the cross-correlation function $R_{12}(\tau)$ of the pulses $g_1(t)$ and $g_2(t)$ shown in Fig. 1.9.

- *Solution*: We know that

$$R_{12}(t) = g_1(t) \star g_2^*(-t) \rightleftharpoons G_1(f) G_2^*(f)$$
$$\Rightarrow R_{12}^{(m+n)}(t) \rightleftharpoons (j2\pi f)^m \, G_1(f) \, (j2\pi f)^n \, G_2^*(f)$$
$$\Rightarrow R_{12}^{(m+n)}(t) \rightleftharpoons (j2\pi f)^m \, G_1(f) \, (-1)^n \, (-j2\pi f)^n \, G_2^*(f). \quad (1.85)$$

Let

$$h_2(t) = g_2^{(n)}(t) \rightleftharpoons (j2\pi f)^n \, G_2(f)$$
$$\Rightarrow h_2^*(-t) \rightleftharpoons (-j2\pi f)^n \, G_2^*(f). \quad (1.86)$$

Similarly

$$h_1(t) = g^{(m)}(t) \rightleftharpoons (j2\pi f)^m \, G(f). \quad (1.87)$$

Using the fact that multiplication in the frequency domain is equivalent to convolution in the time domain, we get

$$R_{12}^{(m+n)}(t) = (-1)^n \left(h_1(t) \star h_2^*(-t) \right)$$
$$\Rightarrow R_{12}^{(m+n)}(\tau) = (-1)^n \int_{t=-\infty}^{\infty} h_1(t) h_2^*(t - \tau) \, dt. \quad (1.88)$$

Hence proved. For the signal in Fig. 1.9, using $m = 1$ and $n = 0$, we get

$$h_1(t) = 2\delta(t + 3) - 2\delta(t - 3)$$
$$h_2(t) = g_2(t). \quad (1.89)$$

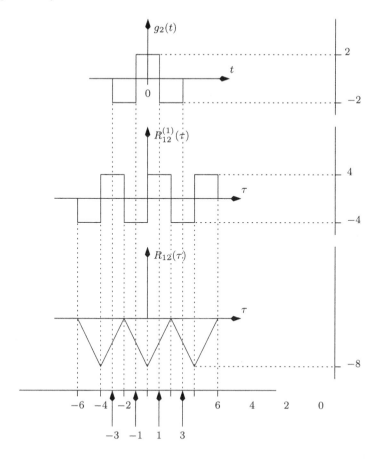

Fig. 1.10 Cross-correlation $R_{12}(\tau)$

Thus

$$R_{12}^{(1)}(\tau) = \int_{t=-\infty}^{\infty} h_1(t)h_2(t-\tau)\,dt$$
$$= 2g_2(-\tau-3) - 2g_2(3-\tau). \tag{1.90}$$

$R_{12}^{(1)}(\tau)$ and $R_{12}(\tau)$ are plotted in Fig. 1.10.

13. Let $R_{12}(\tau)$ denote the cross-correlation function of two energy signals $g_1(t)$ and $g_2(t)$.

   (a) Using Fourier transforms, show that

$$R_{12}^{(m+n)}(\tau) = (-1)^n \int_{t=-\infty}^{\infty} h_1(t)h_2^*(t-\tau)\,dt, \tag{1.91}$$

Fig. 1.11 Pulses $g_1(t)$ and $g_2(t)$

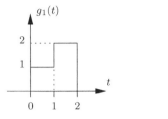

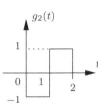

where

$$h_1(t) = g_1^{(m)}(t)$$
$$h_2(t) = g_2^{(n)}(t) \tag{1.92}$$

denote the mth and nth derivatives of $g_1(t)$ and $g_2(t)$, respectively.

(b) Use the above relation with $m = 1$ and $n = 0$ to evaluate and sketch the cross-correlation function $R_{12}(\tau)$ of the pulses $g_1(t)$ and $g_2(t)$ shown in Fig. 1.11.

- *Solution*: We know that

$$R_{12}(t) = g_1(t) \star g_2^*(-t) \rightleftharpoons G_1(f)G_2^*(f)$$
$$\Rightarrow R_{12}^{(m+n)}(t) \rightleftharpoons (\mathrm{j}2\pi f)^m\, G_1(f)\, (\mathrm{j}2\pi f)^n\, G_2^*(f)$$
$$\Rightarrow R_{12}^{(m+n)}(t) \rightleftharpoons (\mathrm{j}2\pi f)^m\, G_1(f)\, (-1)^n\, (-\mathrm{j}2\pi f)^n\, G_2^*(f). \tag{1.93}$$

Let

$$h_2(t) = g_2^{(n)}(t) \rightleftharpoons (\mathrm{j}2\pi f)^n\, G_2(f)$$
$$\Rightarrow h_2^*(-t) \rightleftharpoons (-\mathrm{j}2\pi f)^n\, G_2^*(f). \tag{1.94}$$

Similarly

$$h_1(t) = g^{(m)}(t) \rightleftharpoons (\mathrm{j}2\pi f)^m\, G(f). \tag{1.95}$$

Using the fact that multiplication in the frequency domain is equivalent to convolution in the time domain, we get

$$R_{12}^{(m+n)}(t) = (-1)^n \left(h_1(t) \star h_2^*(-t) \right)$$
$$\Rightarrow R_{12}^{(m+n)}(\tau) = (-1)^n \int_{t=-\infty}^{\infty} h_1(t) h_2^*(t - \tau)\, dt. \tag{1.96}$$

Hence proved. For the signal in Fig. 1.11, using $m = 1$ and $n = 0$, we get

$$h_1(t) = \delta(t) + \delta(t-1) - 2\delta(t-2)$$
$$h_2(t) = g_2(t). \tag{1.97}$$

Thus

$$
\begin{aligned}
R_{12}^{(1)}(\tau) &= \int_{t=-\infty}^{\infty} h_1(t) h_2(t-\tau)\, dt \\
&= \int_{t=-\infty}^{\infty} [\delta(t) + \delta(t-1) - 2\delta(t-2)]\, g_2(t-\tau) \\
&= g_2(-\tau) + g_2(1-\tau) - 2g_2(2-\tau). \tag{1.98}
\end{aligned}
$$

$R_{12}^{(1)}(\tau)$ and $R_{12}(\tau)$ are plotted in Fig. 1.12.

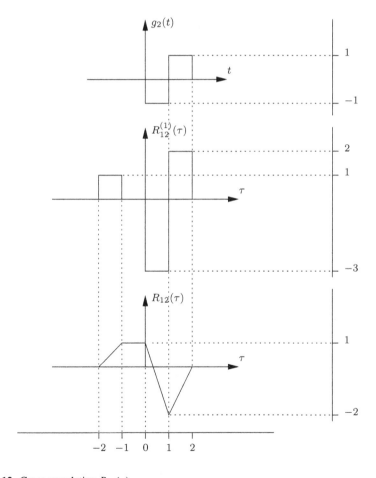

Fig. 1.12 Cross-correlation $R_{12}(\tau)$

14. (Simon Haykin 1983) Let $x(t)$ and $y(t)$ be the input and output signals of a linear time-invariant filter. Using Rayleigh's energy theorem, show that if the filter is stable and the input signal $x(t)$ has finite energy, then the output signal $y(t)$ also has finite energy.

- *Solution*: Let $H(f)$ denote the Fourier transform of $h(t)$. We have

$$H(f) = \int_{t=-\infty}^{\infty} h(t)e^{-j2\pi ft}\, dt$$

$$\Rightarrow |H(f)| = \left| \int_{t=-\infty}^{\infty} h(t)e^{-j2\pi ft}\, dt \right|$$

$$\leq \int_{t=-\infty}^{\infty} \left| h(t)e^{-j2\pi ft}\, dt \right|$$

$$\Rightarrow |H(f)| \leq \int_{t=-\infty}^{\infty} |h(t)|\, dt, \tag{1.99}$$

where we have used Schwarz's inequality. Since $h(t)$ is stable

$$\int_{t=-\infty}^{\infty} |h(t)|\, dt < \infty$$

$$\Rightarrow |H(f)| < \infty. \tag{1.100}$$

Let

$$A = \max |H(f)|. \tag{1.101}$$

The energy of the output signal $y(t)$ is

$$\int_{t=-\infty}^{\infty} |y(t)|^2\, dt = \int_{f=-\infty}^{\infty} |Y(f)|^2\, df \tag{1.102}$$

where we have used the Rayleigh's energy theorem. Using the fact that

$$Y(f) = H(f)X(f) \tag{1.103}$$

the energy of the output signal is

$$\int_{f=-\infty}^{\infty} |H(f)|^2 |X(f)|^2\, df \leq A^2 \int_{f=-\infty}^{\infty} |X(f)|^2\, df$$

$$\Rightarrow \int_{t=-\infty}^{\infty} |y(t)|^2\, dt \leq A^2 \int_{f=-\infty}^{\infty} |x(t)|^2\, dt. \tag{1.104}$$

Since the input signal has finite energy, so does the output signal.

15. (Simon Haykin 1983) Prove the following properties of the complex exponential Fourier series representation, for a real-valued periodic signal $g_p(t)$:

(a) If the periodic function $g_p(t)$ is even, that is, $g_p(-t) = g_p(t)$, then the Fourier coefficient c_n is purely real and an even function of n.

(b) If $g_p(t)$ is odd, that is, $g_p(-t) = -g_p(t)$, then c_n is purely imaginary and an odd function of n.

(c) If $g_p(t)$ has half-wave symmetry, that is, $g_p(t \pm T_0/2) = -g_p(t)$, where T_0 is the period of $g_p(t)$, then c_n consists of only odd order terms.

- *Solution*: The Fourier series for any periodic signal $g_p(t)$ is given by

$$g_p(t) = \sum_{n=-\infty}^{\infty} c_n e^{j 2\pi n t / T_0}$$

$$= a_0 + 2 \sum_{n=1}^{\infty} a_n \cos(2\pi n t / T_0) + b_n \sin(2\pi n t / T_0), \quad (1.105)$$

where

$$c_n = \begin{cases} a_n - j b_n & \text{for } n > 0 \\ a_0 & \text{for } n = 0 \\ a_{-n} + j b_{-n} & \text{for } n < 0. \end{cases} \quad (1.106)$$

Note that a_n and b_n are real-valued, since $g_p(t)$ is real-valued. To prove the first part, we note that

$$g_p(-t) = \sum_{n=-\infty}^{\infty} c_n e^{-j 2\pi n t / T_0}. \quad (1.107)$$

Substituting $n = -m$ in the above equation we get

$$g_p(-t) = \sum_{m=\infty}^{-\infty} c_{-m} e^{j 2\pi m t / T_0}$$

$$= \sum_{m=-\infty}^{\infty} c_{-m} e^{j 2\pi m t / T_0}. \quad (1.108)$$

Since $g_p(-t) = g_p(t)$, comparing (1.105) and (1.108) we must have $c_{-m} = c_m$ (even function of m). Moreover, from (1.106), c_m must be purely real.
To prove the second part, we note from (1.105) and (1.108) that $c_{-m} = -c_m$ (odd function of m) and moreover from (1.106) it is clear that c_m must be purely imaginary.
To prove the third part, we note that

$$g_p(t \pm T_0/2) = \sum_{n=-\infty}^{\infty} c_n e^{j 2\pi n(t \pm T_0/2)/T_0}$$

$$= \sum_{n=-\infty}^{\infty} c_n (-1)^n e^{j 2\pi n t/T_0}. \qquad (1.109)$$

Since it is given that $g_p(t \pm T_0/2) = -g_p(t)$, comparing (1.105) and (1.109) we conclude that $c_n = 0$ for $n = 2m$.

16. (Simon Haykin 1983) A signal $x(t)$ of finite energy is applied to a square-law device whose output $y(t)$ is given by

$$y(t) = x^2(t). \qquad (1.110)$$

The spectrum of $x(t)$ is limited to the frequency interval $-W \le f \le W$. Show that the spectrum of $y(t)$ is limited to $-2W \le f \le 2W$.

- *Solution*: We know that multiplication of signals in the time domain is equivalent to convolution in the frequency domain. Let

$$x(t) \rightleftharpoons X(f). \qquad (1.111)$$

Therefore

$$Y(f) = \int_{\alpha=-\infty}^{\infty} X(\alpha)X(f - \alpha)\, d\alpha$$

$$= \int_{\alpha=-W}^{W} X(\alpha)X(f - \alpha)\, d\alpha, \qquad (1.112)$$

where we have used the fact that $X(\alpha)$ is bandlimited to $-W \le \alpha \le W$. Consequently, we must also have

$$-W \le f - \alpha \le W \qquad (1.113)$$

so that the product $X(\alpha)X(f - \alpha)$ is non-zero and contributes to the integral. The plot of $X(f - \alpha)$ as a function of α and f is shown in Fig. 1.13. We see that the convolution integral is non-zero (for this particular example) only for $-2W \le f \le 2W$.

The important point to note here is that the spectrum of $Y(f)$ *cannot exceed* $-2W \le f \le 2W$.

17. A signal $x(t)$ has the Fourier transform shown in Fig. 1.14.

   (a) Compute its Hilbert transform $\hat{x}(t)$.
   (b) Compare the area under $\hat{x}(t)$ and the value of $\hat{X}(0)$. Comment on your answer.

Fig. 1.13 Plot of $X(\alpha)$ and $X(f - \alpha)$ for different values of f

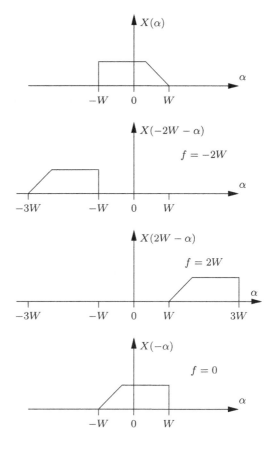

Fig. 1.14 Fourier transform of $x(t)$

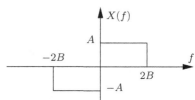

- *Solution*: We know that

$$A \, \text{rect}(t/T) \rightleftharpoons AT \, \text{sinc}(fT). \qquad (1.114)$$

Using duality and substituting $2B$ for T we have

$$A \, \text{rect}\left(\frac{f}{2B}\right) \rightleftharpoons 2AB \, \text{sinc}(2Bt). \qquad (1.115)$$

Now

$$X(f) = A \operatorname{rect}\left(\frac{f-B}{2B}\right) - A \operatorname{rect}\left(\frac{f+B}{2B}\right)$$
$$\rightleftharpoons 2AB \operatorname{sinc}(2Bt)e^{\mathrm{j}2\pi Bt} - 2AB \operatorname{sinc}(2Bt)e^{-\mathrm{j}2\pi Bt}$$
$$\Rightarrow X(f) \rightleftharpoons \mathrm{j}4AB \operatorname{sinc}(2Bt)\sin(2\pi Bt). \tag{1.116}$$

The Fourier transform of $\hat{x}(t)$ is

$$\hat{X}(f) = \begin{cases} -\mathrm{j}\, A \operatorname{rect}\left(\frac{f-B}{2B}\right) - \mathrm{j}\, A \operatorname{rect}\left(\frac{f+B}{2B}\right) & \text{for } f \neq 0 \\ 0 & \text{for } f = 0. \end{cases} \tag{1.117}$$

From (1.115) we see that the inverse Fourier transform of $\hat{X}(f)$ is

$$\hat{X}(f) \rightleftharpoons \hat{x}(t) = -\mathrm{j}\,2AB \operatorname{sinc}(2tB)e^{\mathrm{j}2\pi Bt} - \mathrm{j}\,2AB \operatorname{sinc}(2tB)e^{-\mathrm{j}2\pi Bt}$$
$$= -\mathrm{j}\,4AB \operatorname{sinc}(2tB)\cos(2\pi Bt)$$
$$= -\mathrm{j}\,4AB \operatorname{sinc}(4tB). \tag{1.118}$$

Note that

$$\int_{t=-\infty}^{\infty} \hat{x}(t)\,dt = -\mathrm{j}\,A \neq \hat{X}(0) = 0. \tag{1.119}$$

This is because the property

$$\int_{t=-\infty}^{\infty} g(t)\,dt = G(0) \tag{1.120}$$

is valid only when $G(f)$ is continuous at $f = 0$.

18. Consider the system shown in Fig. 1.15. Assume that the current in branch AB is zero. The voltage at point A is $v_1(t)$.

   (a) Find out the time-domain expression that relates $v_1(t)$ and $v_i(t)$.

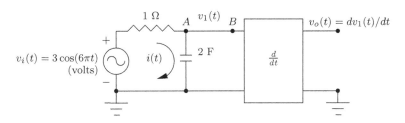

Fig. 1.15 Block diagram of the system

(b) Using the Fourier transform, find out the relation between $V_1(f)$ and $V_i(f)$.

(c) Find the relation between $V_o(f)$ and $V_i(f)$.

(d) Compute the power dissipated when the output voltage $v_o(t)$ is applied across a 1 Ω resistor.

- *Solution*: We know that

$$v_i(t) = i(t)R + v_1(t). \tag{1.121}$$

However

$$i(t) = C\frac{dv_1(t)}{dt}. \tag{1.122}$$

Thus

$$v_i(t) = RC\frac{dv_1(t)}{dt} + v_1(t). \tag{1.123}$$

Taking the Fourier transform of both sides, we get

$$V_i(f) = RC\,\mathrm{j}\,2\pi f V_1(f) + V_1(f)$$
$$\Rightarrow V_1(f) = \frac{V_i(f)}{1 + \mathrm{j}\,2\pi f RC}. \tag{1.124}$$

It is given that

$$v_o(t) = \frac{dv_1(t)}{dt}$$
$$\Rightarrow V_o(f) = \mathrm{j}\,2\pi f V_1(f)$$
$$= \frac{\mathrm{j}\,2\pi f V_i(f)}{1 + \mathrm{j}\,2\pi f RC}. \tag{1.125}$$

Now

$$V_i(f) = \frac{3}{2}[\delta(f-3) + \delta(f+3)]. \tag{1.126}$$

Therefore

$$V_o(f) = \frac{3}{2}\delta(f-3)\frac{\mathrm{j}\,6\pi}{1 + \mathrm{j}\,12\pi} + \frac{3}{2}\delta(f+3)\frac{-\mathrm{j}\,6\pi}{1 - \mathrm{j}\,12\pi}. \tag{1.127}$$

Taking the inverse Fourier transform we get

$$v_o(t) = c_1 \mathrm{e}^{\mathrm{j}\,6\pi t} + c_2 \mathrm{e}^{-\mathrm{j}\,6\pi t}. \tag{1.128}$$

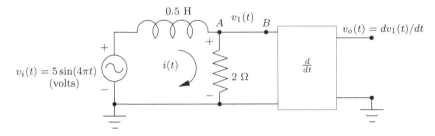

Fig. 1.16 Block diagram of the system

Using Parseval's power theorem we get the power of $v_o(t)$ as

$$P = |c_1|^2 + |c_2|^2 = \frac{2 \times 9}{4} \frac{36\pi^2}{1 + 144\pi^2} \approx 9/8\,\text{W}. \qquad (1.129)$$

19. Consider the system shown in Fig. 1.16. Assume that the current in branch AB is zero. The voltage at point A is $v_1(t)$.

(a) Find out the time-domain expression that relates $v_i(t)$ and $i(t)$.
(b) Using the Fourier transform, find out the relation between $V_1(f)$ and $V_i(f)$.
(c) Find the relation between $V_o(f)$ and $V_i(f)$.
(d) Compute $v_o(t)$.
(e) Compute the power dissipated when the output voltage $v_o(t)$ is applied across a $1/2\,\Omega$ resistor.

• *Solution*: We know that

$$v_i(t) = v_1(t) + L\,di(t)/dt$$
$$= i(t)R + L\,di(t)/dt. \qquad (1.130)$$

Taking the Fourier transform of both sides we get

$$V_i(f) = RI(f) + j2\pi fLI(f)$$
$$\Rightarrow I(f) = \frac{V_i(f)}{R + j2\pi fL}. \qquad (1.131)$$

Therefore

$$V_1(f) = RI(f)$$
$$= \frac{V_i(f)}{1 + j2\pi fL/R}$$
$$= \frac{V_i(f)}{1 + j\pi f/2}, \qquad (1.132)$$

where we have substituted $L = 0.5$ H and $R = 2\,\Omega$. Now

$$
\begin{aligned}
v_o(t) &= dv_1(t)/dt \\
\Rightarrow V_o(f) &= \mathrm{j}\,2\pi f V_1(f) \\
&= \frac{\mathrm{j}\,2\pi f}{1 + \mathrm{j}\,\pi f/2} V_i(f).
\end{aligned}
\tag{1.133}
$$

Since

$$
V_i(f) = \frac{5}{2\mathrm{j}}\,[\delta(f - 2) - \delta(f + 2)]
\tag{1.134}
$$

we get

$$
V_o(f) = 5\delta(f - 2)\frac{2\pi}{1 + \mathrm{j}\,\pi} - 5\delta(f + 2)\frac{(-2\pi)}{1 - \mathrm{j}\,\pi}.
\tag{1.135}
$$

Taking the inverse Fourier transform we get

$$
v_o(t) = c_1 \mathrm{e}^{\mathrm{j}\,4\pi t} + c_2 \mathrm{e}^{-\mathrm{j}\,4\pi t},
\tag{1.136}
$$

where

$$
\begin{aligned}
c_1 &= \frac{10\pi}{1 + \mathrm{j}\,\pi} \\
&= A\mathrm{e}^{\mathrm{j}\phi} \\
c_2 &= \frac{10\pi}{1 - \mathrm{j}\,\pi} \\
&= A\mathrm{e}^{-\mathrm{j}\phi},
\end{aligned}
\tag{1.137}
$$

where

$$
\begin{aligned}
A &= \frac{10\pi}{\sqrt{1 + \pi^2}} \\
\phi &= -\tan^{-1}(\pi).
\end{aligned}
\tag{1.138}
$$

Substituting (1.137) and (1.138) in (1.136), we obtain

$$
v_o(t) = 2A\cos(4\pi t + \phi).
\tag{1.139}
$$

Using Parseval's power theorem we get the power of $v_o(t)$ as

$$P = |c_1|^2 + |c_2|^2$$
$$= 2A^2$$
$$= \frac{200\pi^2}{1 + \pi^2} \, \text{W}. \tag{1.140}$$

The power dissipated across the $1/2 \, \Omega$ resistor is $2P$.

20. Consider a complex-valued signal $g(t)$. Let $g_1(t) = g^*(-t)$. Let $g_1^{(n)}(t)$ denote the nth derivative of $g_1(t)$. Consider another signal $g_2(t) = g^{(n)}(t)$. Is $g_2^*(-t) = g_1^{(n)}(t)$? Justify your answer using Fourier transforms.

   • *Solution*: Let $G(f)$ denote the Fourier transform of $g(t)$. Then we have

$$g_1(t) = g^*(-t) \rightleftharpoons G^*(f) = G_1(f) \quad \text{(say)}. \tag{1.141}$$

   Therefore

$$g_1^{(n)}(t) \rightleftharpoons (j \, 2\pi f)^n G_1(f)$$
$$\Rightarrow g_1^{(n)}(t) \rightleftharpoons (j \, 2\pi f)^n G^*(f). \tag{1.142}$$

   Next consider $g_2(t)$. We have

$$g_2(t) = g^{(n)}(t) \rightleftharpoons (j \, 2\pi f)^n G(f) = G_2(f) \quad \text{(say)}$$
$$\Rightarrow g_2^*(-t) \rightleftharpoons G_2^*(f) = (-j \, 2\pi f)^n G^*(f). \tag{1.143}$$

   Comparing (1.142) and (1.143) we see that

$$g_2^*(-t) = g_1^{(n)}(t) \quad \text{when } n \text{ is even}$$
$$g_2^*(-t) \neq g_1^{(n)}(t) \quad \text{when } n \text{ is odd}. \tag{1.144}$$

21. (Simon Haykin 1983) Consider N stages of the RC-lowpass filter as illustrated in Fig. 1.17.

   (a) Compute the magnitude response $|V_o(f)/V_i(f)|$ of the overall cascade connection. The current drawn by the buffers is zero.

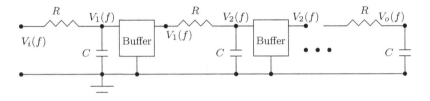

Fig. 1.17 N stages of the RC-lowpass filter

(b) Show that the magnitude response in the vicinity of $f = 0$ approaches a Gaussian function given by $\exp(-(N/2)4\pi^2 f^2 R^2 C^2)$, for large values of N.

- *Solution*: For a single stage, the input and output voltages are related as follows:

$$v_1(t) = i(t)R + v_2(t). \tag{1.145}$$

However,

$$i(t) = C\frac{dv_2(t)}{dt}. \tag{1.146}$$

Therefore (1.145) can be rewritten as

$$v_1(t) = RC\frac{dv_2(t)}{dt} + v_2(t). \tag{1.147}$$

Taking the Fourier transform of both sides, we get

$$V_1(f) = (j2\pi f RC + 1)V_2(f)$$
$$\Rightarrow \frac{V_2(f)}{V_1(f)} = \frac{1}{1 + j2\pi f RC}. \tag{1.148}$$

The magnitude response of the overall cascade connection is

$$\left|\frac{V_o(f)}{V_i(f)}\right| = \frac{1}{(1 + 4\pi^2 f^2 \tau_0^2)^{N/2}}, \tag{1.149}$$

where $\tau_0 = RC$. Note that

$$\lim_{f \to 0}\left|\frac{V_o(f)}{V_i(f)}\right| = 1 \tag{1.150}$$

and for large values of N

$$\left|\frac{V_o(f)}{V_i(f)}\right| \to 0 \quad \text{for } 4\pi^2 f^2 \tau_0^2 > 1. \tag{1.151}$$

Therefore, we are interested in the magnitude response in the vicinity of $f = 0$, for large values of N.

We make use of the Maclaurin series expansion of a function $f(x)$ about $x = 0$ as follows (assuming that all the derivatives exist):

$$f(x) = f(0) + \frac{f^{(1)}(0)}{1!}x + \frac{f^{(2)}(0)}{2!}x^2 + \cdots \tag{1.152}$$

where

$$f^{(n)}(0) = \frac{d^n f(x)}{dx^n}\bigg|_{x=0}.$$

(1.153)

In the present context

$$f(x) = \frac{1}{(1 + 4\pi^2 \tau_0^2 x)^{N/2}}$$

(1.154)

with $x = f^2$. We have

$$f(0) = 1$$

$$f^{(1)}(0) = \frac{-N}{2} \frac{4\pi^2 \tau_0^2}{(1 + 4\pi^2 \tau_0^2 x)^{N/2+1}}\bigg|_{x=0}$$

$$= \frac{-N}{2}(4\pi^2 \tau_0^2)$$

$$f^{(2)}(0) = \frac{N}{2}\left(\frac{N}{2}+1\right)\frac{\left(4\pi^2 \tau_0^2\right)^2}{(1 + 4\pi^2 \tau_0^2 x)^{N/2+2}}\bigg|_{x=0}$$

$$\approx \left(\frac{N}{2}4\pi^2 \tau_0^2\right)^2,$$

(1.155)

where we have assumed that for large N

$$\frac{N}{2} + 1 \approx \frac{N}{2}.$$

(1.156)

Generalizing (1.155), we can obtain the nth derivative of the Maclaurin series as

$$f^{(n)}(0) = (-1)^n \frac{N}{2}\left(\frac{N}{2}+1\right)\cdots\left(\frac{N}{2}+n-1\right)$$

$$\times \frac{\left(4\pi^2 \tau_0^2\right)^n}{(1 + 4\pi^2 \tau_0^2 x)^{N/2+n}}\bigg|_{x=0}$$

$$\approx (-1)^n \left(\frac{N}{2}4\pi^2 \tau_0^2\right)^n,$$

(1.157)

where we have used the fact that for any finite $n \ll N$

$$\frac{N}{2} + n - 1 \approx \frac{N}{2}.$$

(1.158)

Thus $f(x)$ in (1.154) can be written as

Fig. 1.18 A periodic waveform

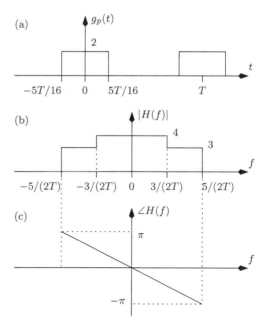

$$f(x) = 1 - \frac{y}{1!}x + \frac{y^2}{2!}x^2 - \cdots$$
$$= \exp(-xy), \tag{1.159}$$

where

$$y = (N/2)4\pi^2\tau_0^2 = (N/2)4\pi^2 R^2 C^2. \tag{1.160}$$

Finally, substituting for x we get the desired result which is Gaussian. Observe that, in order to satisfy the condition $n \ll N$, the nth term of the series in (1.159) must tend to zero for $n \ll N$. This can happen only when

$$xy < 1$$
$$\Rightarrow (N/2)4\pi^2 R^2 C^2 f^2 < 1$$
$$\Rightarrow |f| < \frac{1}{\pi RC\sqrt{2N}}, \tag{1.161}$$

which is in the vicinity of $f = 0$ for large N.

22. Consider the periodic waveform $g_p(t)$ in Fig. 1.18a.

   (a) Compute the real Fourier series representation of $g_p(t)$. Give the expression for the nth term in the Fourier series.
   (b) Hence compute the Fourier transform of $g_p(t)$.

(c) Express the autocorrelation of $g_p(t)$ ($R_{g_p}(\tau)$) in terms of the autocorrelation of the generating function $g(t)$ ($R_g(\tau)$). There is no need to derive the expression.

(d) Sketch $R_g(\tau)$. Label all the important points.
Assume that $g(t)$ is defined from $-T/2$ to $T/2$.

(e) Sketch $R_{g_p}(\tau)$ for $0 < \tau < T$. Label all the important points.

(f) The pulse width of $g_p(t)$ in Fig. 1.18a is $5T/8$. What is the maximum possible pulse width, so that no aliasing occurs when $R_{g_p}(\tau)$ is expressed in terms of $R_g(\tau)$.

(g) If $g_p(t)$ is passed through a filter having magnitude and phase response as shown in Fig. 1.18b, c, determine the filter output.

● *Solution*: The Fourier series representation for a periodic signal is given by

$$g_p(t) = a_0 + 2 \sum_{n=1}^{\infty} a_n \cos(2\pi nt/T) + b_n \sin(2\pi nt/T). \quad (1.162)$$

For the given problem $g_p(t) = g_p(-t)$, therefore $b_n = 0$. We have

$$a_0 = \frac{1}{T} \int_{t=-T/2}^{T/2} g_p(t)\, dt$$

$$= \frac{1}{T} \int_{t=-5T/16}^{5T/16} 2\, dt$$

$$= \frac{5}{4}. \quad (1.163)$$

Similarly,

$$a_n = \frac{1}{T} \int_{t=-T/2}^{T/2} g_p(t) \cos(2\pi nt/T)\, dt$$

$$= \frac{1}{T} \int_{t=-5T/16}^{5T/16} 2 \cos(2\pi nt/T)\, dt$$

$$= \frac{2}{n\pi} \sin(5n\pi/8) \quad \text{for } n \geq 1. \quad (1.164)$$

The Fourier transform of $g_p(t)$ is given by

$$g_p(t) \rightleftharpoons a_0 \delta(f) + \sum_{n=1}^{\infty} a_n \left[\delta(f - n/T) + \delta(f + n/T) \right], \quad (1.165)$$

where a_n for $n \geq 0$ is given by (1.163) and (1.164).
The generating function $g(t)$ is defined as follows:

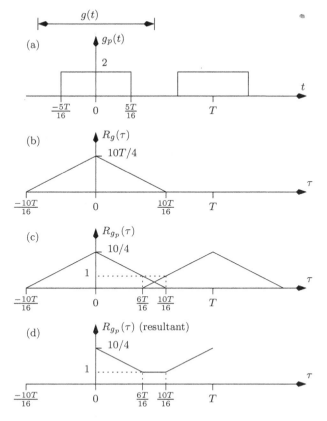

Fig. 1.19 a $g_p(t)$. **b** $R_g(\tau)$. **c** Superposition of $R_g(\tau)/T$ and $R_g(\tau - T)/T$. **d** Resultant $R_{g_p}(\tau)$ for $0 < \tau < T$

$$g(t) = \begin{cases} g_p(t) & \text{for } -T/2 < t < T/2 \\ 0 & \text{elsewhere.} \end{cases} \quad (1.166)$$

We know that the autocorrelation of $g_p(t)$ $(R_{g_p}(\tau))$ is given by

$$R_{g_p}(\tau) = \frac{1}{T} \sum_{m=-\infty}^{\infty} R_g(\tau + mT), \quad (1.167)$$

where $R_g(\tau)$ is the autocorrelation of the generating function $g(t)$. $R_g(\tau)$ is plotted in Fig. 1.19b and $R_{g_p}(\tau)$ is plotted in Fig. 1.19d.
The maximum pulse width for no aliasing is $T/2$.

Clearly, the filter output is a dc component plus the first and second harmonics. The gain of the dc component is 4 and the phase shift is 0. Hence, the output dc signal is $4a_0 = 5$.

Fig. 1.20 A periodic
waveform $g_p(t)$

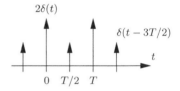

Similarly, the gain of the first harmonic is also 4 and the phase shift is $-2\pi/5$.
Finally, the gain of the second harmonic is 3 and the phase shift is $-4\pi/5$.
Thus, the filter output is

$$x(t) = 5 + \frac{16}{\pi} \sin(5\pi/8) \cos(2\pi t/T - 2\pi/5)$$

$$+ \frac{6}{\pi} \sin(10\pi/8) \cos(4\pi t/T - 4\pi/5). \qquad (1.168)$$

23. Consider the periodic waveform $g_p(t)$ given in Fig. 1.20, where T denotes the
period.

   (a) Compute the real Fourier series representation of $g_p(t)$. Give the expression
       for the coefficient of the nth term.
   (b) Compute the Fourier transform of $g_p(t)$.

   • *Solution*: The Fourier series representation for a periodic signal is given by

$$g_p(t) = a_0 + 2 \sum_{n=1}^{\infty} a_n \cos(2\pi nt/T) + b_n \sin(2\pi nt/T). \qquad (1.169)$$

For the given problem $g_p(t) = g_p(-t)$, therefore $b_n = 0$. We have

$$a_0 = \frac{1}{T} \int_{t=0^-}^{T^-} g_p(t)\, dt$$

$$= \frac{1}{T} \int_{t=0^-}^{T^-} [2\delta(t) + \delta(t - T/2)]\, dt$$

$$= \frac{3}{T}. \qquad (1.170)$$

Similarly,

$$a_n = \frac{1}{T} \int_{t=0^-}^{T^-} g_p(t) \cos(2\pi n t/T) \, dt$$

$$= \frac{1}{T} \int_{t=0^-}^{T^-} [2\delta(t) + \delta(t - T/2)] \cos(2\pi n t/T) \, dt$$

$$= \frac{1}{T} [2 + \cos(n\pi)] \qquad \text{for } n \geq 1. \tag{1.171}$$

The Fourier transform of $g_p(t)$ is given by

$$g_p(t) \rightleftharpoons a_0 \delta(f) + \sum_{n=1}^{\infty} a_n \left[\delta(f - n/T) + \delta(f + n/T)\right], \tag{1.172}$$

where a_n for $n \geq 0$ is given by (1.170) and (1.171).

24. If $g(t)$ has the Fourier transform $G(f)$, compute the inverse Fourier transform of $g(af - f_0)$ for

(a) $a \neq 0$,
(b) $a = 0$.

- *Solution*: From the duality property of the Fourier transform, we know that

$$g(f) \rightleftharpoons G(-t). \tag{1.173}$$

In other words

$$G(-t) = \int_{f=-\infty}^{\infty} g(f) e^{j 2\pi f t} \, df. \tag{1.174}$$

Let $G_1(t)$ be the required inverse Fourier transform. Then

$$G_1(t) = \int_{f=-\infty}^{\infty} g(af - f_0) e^{j 2\pi f t} \, df. \tag{1.175}$$

Let

$$af - f_0 = x$$
$$\Rightarrow a \, df = dx. \tag{1.176}$$

Assuming $a < 0$, we have

$$|a| \, df = -dx. \tag{1.177}$$

Therefore, (1.175) becomes

$$
\begin{aligned}
G_1(t) &= \frac{1}{|a|} \int_{x=-\infty}^{\infty} g(x)\, \mathrm{e}^{\mathrm{j}\, 2\pi(x+f_0)t/a}\, dx \\
&= \frac{1}{|a|} G(-t/a) \mathrm{e}^{\mathrm{j}\, 2\pi f_0 t/a}.
\end{aligned}
\tag{1.178}
$$

The above result is also valid for $a > 0$.

When $a = 0$, the function reduces to a constant $g(-f_0)$ whose inverse Fourier transform is

$$
\begin{aligned}
G_1(t) &= \int_{f=-\infty}^{\infty} g(-f_0)\, \mathrm{e}^{\mathrm{j}\, 2\pi f t}\, df \\
&= g(-f_0)\delta(t).
\end{aligned}
\tag{1.179}
$$

The alternate solution is as follows:

$$
g(t) \rightleftharpoons G(f)
$$
$$
\Rightarrow h(t) = g(at) \rightleftharpoons H(f) = \frac{1}{|a|} G(f/a)
$$
$$
\Rightarrow h(t - t_0/a) = g(at - t_0) \rightleftharpoons H(f)\mathrm{e}^{-\mathrm{j}\, 2\pi f t_0/a} = \frac{1}{|a|} G(f/a)\mathrm{e}^{-\mathrm{j}\, 2\pi f t_0/a}.
\tag{1.180}
$$

Since

$$
p(f) \rightleftharpoons P(-t)
$$
$$
\Rightarrow g(af - f_0) \rightleftharpoons \frac{1}{|a|} G(-t/a)\mathrm{e}^{\mathrm{j}\, 2\pi t f_0/a}.
\tag{1.181}
$$

25. (Simon Haykin 1983) Let $R_g(t)$ denote the autocorrelation of an energy signal $g(t)$.

   (a) Using Fourier transforms, show that

$$
R_g^{(m+n)}(\tau) = (-1)^n \int_{t=-\infty}^{\infty} g_2(t)g_1^*(t - \tau)\, dt,
\tag{1.182}
$$

   where the asterisk denotes the complex conjugate and

$$
\begin{aligned}
g_1(t) &= g^{(n)}(t), \\
g_2(t) &= g^{(m)}(t),
\end{aligned}
\tag{1.183}
$$

   where $g^{(m)}(t)$ denotes the mth derivative of $g(t)$.

Fig. 1.21 Plot of $g(t)$

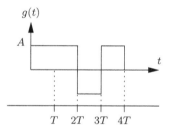

(b) Using the above relation, evaluate and sketch the autocorrelation of the signal in Fig. 1.21. You can use $m = 1$ and $n = 0$.

- *Solution*: We know that

$$R_g(t) = g(t) \star g^*(-t) \rightleftharpoons G(f)G^*(f)$$
$$\Rightarrow R^{(m+n)}(t) \rightleftharpoons (j\,2\pi f)^m \, G(f) \, (j\,2\pi f)^n \, G^*(f)$$
$$\Rightarrow R^{(m+n)}(t) \rightleftharpoons (j\,2\pi f)^m \, G(f) \, (-1)^n \, (-j\,2\pi f)^n \, G^*(f), \quad (1.184)$$

where "$\star$" denotes convolution. Let

$$g_1(t) = g^{(n)}(t) \rightleftharpoons (j\,2\pi f)^n \, G(f)$$
$$\Rightarrow g_1^*(-t) \rightleftharpoons (-j\,2\pi f)^n \, G^*(f). \quad (1.185)$$

Similarly let

$$g_2(t) = g^{(m)}(t) \rightleftharpoons (j\,2\pi f)^m \, G(f). \quad (1.186)$$

Thus, using the fact that multiplication in the frequency domain is equivalent to convolution in the time domain, we get

$$R_g^{(m+n)}(t) = (-1)^n \left(g_2(t) \star g_1^*(-t) \right)$$
$$\Rightarrow R_g^{(m+n)}(\tau) = (-1)^n \int_{t=-\infty}^{\infty} g_2(t) g_1^*(t - \tau) \, dt. \quad (1.187)$$

Hence proved. For the signal in Fig. 1.21, using $m = 1$ and $n = 0$ we get

$$g_1(t) = g(t)$$
$$g_2(t) = g^{(1)}(t)$$
$$= A \left(\delta(t) - 2\delta(t - 2T) + 2\delta(t - 3T) - \delta(t - 4T) \right). \quad (1.188)$$

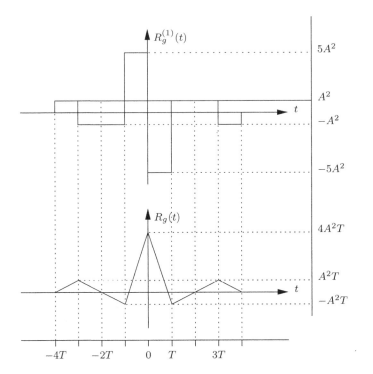

Fig. 1.22 Plot of $R_g^{(1)}(t)$ and $R_g(t)$

Hence

$$R_g^{(1)}(t) = g_2(t) \star g^*(-t). \tag{1.189}$$

Using the fact that for any signal $x(t)$

$$x(t) \star \delta(t) = x(t), \tag{1.190}$$

we get

$$R_g^{(1)}(t) = A\left(g(-t) - 2g(-t + 2T) + 2g(-t + 3T) - g(-t + 4T)\right). \tag{1.191}$$

$R_g^{(1)}(t)$ and $R_g(t)$ are plotted in Fig. 1.22.

26. Let $R_g(t)$ denote the autocorrelation of an energy signal $g(t)$ shown in Fig. 1.23.

   (a) Express the first derivative of $R_g(t)$ in terms of the derivative of $g(t)$.
   (b) Draw the first derivative of $R_g(t)$. Show all the steps.

Fig. 1.23 An energy signal

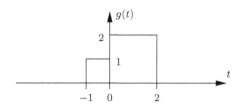

(c) Hence sketch $R_g(t)$.

- *Solution*: We know that

$$R_g(t) = g(t) \star g(-t) \rightleftharpoons G(f)G^*(f)$$
$$\Rightarrow dR_g(t)/dt \rightleftharpoons (j\,2\pi f G(f))\,G^*(f)$$
$$\Rightarrow dR_g(t)/dt = dg(t)/dt \star g(-t), \tag{1.192}$$

where "$\star$" denotes convolution. Now

$$dg(t)/dt = \delta(t+1) + \delta(t) - 2\delta(t-2). \tag{1.193}$$

Therefore, from (1.192), we get

$$dR_g(t)/dt = s(t+1) + s(t) - 2s(t-2), \tag{1.194}$$

where

$$s(t) = g(-t). \tag{1.195}$$

The function $dR_g(t)/dt$ is plotted in Fig. 1.24. Finally, the autocorrelation of $g(t)$ is presented in Fig. 1.25.

27. (Simon Haykin 1983) Let $R_{12}(\tau)$ and $R_{21}(\tau)$ denote the cross-correlation functions of two energy signals $g_1(t)$ and $g_2(t)$

(a) Show that the total area under $R_{12}(\tau)$ is defined by

$$\int_{\tau=-\infty}^{\infty} R_{12}(\tau)\,d\tau = \left[\int_{t=-\infty}^{\infty} g_1(t)\,dt\right]\left[\int_{t=-\infty}^{\infty} g_2(t)\,dt\right]^*. \tag{1.196}$$

(b) Show that

$$R_{12}(\tau) = R_{21}^*(-\tau). \tag{1.197}$$

- *Solution*: We know that

Fig. 1.24 Obtaining the first
derivative of the
autocorrelation of $g(t)$ in
Fig. 1.23

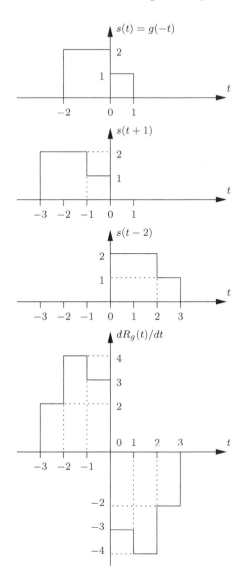

$$R_{12}(\tau) = \int_{t=-\infty}^{\infty} g_1(t)g_2^*(t-\tau)\,dt$$
$$= g_1(t) \star g_2^*(-t)$$
$$\rightleftharpoons G_1(f)G_2^*(f), \tag{1.198}$$

where "$\star$" denotes convolution and we have used the fact that convolution in
the time domain is equivalent to multiplication in the frequency domain. Thus,
we have

Fig. 1.25 Obtaining the autocorrelation of $g(t)$ in Fig. 1.23

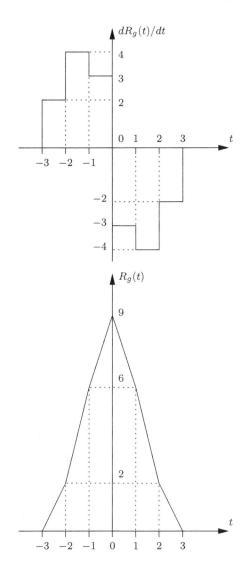

$$\int_{\tau=-\infty}^{\infty} R_{12}(\tau) \exp\left(-j\,2\pi f \tau\right)\,d\tau = G_1(f)G_2^*(f)$$

$$\Rightarrow \int_{\tau=-\infty}^{\infty} R_{12}(\tau) = G_1(0)G_2^*(0)$$

$$= \left[\int_{t=-\infty}^{\infty} g_1(t)\,dt\right]$$

$$\left[\int_{t=-\infty}^{\infty} g_2(t)\,dt\right]^*. \quad (1.199)$$

Thus proved. To prove the second part we note that

$$R_{21}(\tau) = \int_{t=-\infty}^{\infty} g_2(t)g_1^*(t-\tau)\,dt$$

$$\Rightarrow R_{21}^*(\tau) = \int_{t=-\infty}^{\infty} g_2^*(t)g_1(t-\tau)\,dt. \qquad (1.200)$$

Now we substitute $t - \tau = x$ to get

$$R_{21}^*(\tau) = \int_{x=-\infty}^{\infty} g_2^*(x+\tau)g_1(x)\,dx.$$

$$\Rightarrow R_{21}^*(-\tau) = \int_{x=-\infty}^{\infty} g_2^*(x-\tau)g_1(x)\,dx.$$

$$= R_{12}(\tau). \qquad (1.201)$$

Thus proved.

28. (Simon Haykin 1983) Consider two periodic signals $g_{p1}(t)$ and $g_{p2}(t)$, both of period T_0. Show that the cross-correlation function $R_{12}(\tau)$ satisfies the Fourier transform pair:

$$R_{12}(\tau) \rightleftharpoons \frac{1}{T_0^2} \sum_{n=-\infty}^{\infty} G_1(n/T_0)G_2^*(n/T_0)\delta(f-n/T_0), \qquad (1.202)$$

where $G_1(n/T_0)$ and $G_2(n/T_0)$ are the Fourier transforms of the generating functions for the periodic functions $g_{p1}(t)$ and $g_{p2}(t)$, respectively.

• *Solution*: For periodic signals we know that

$$R_{12}(\tau) = \frac{1}{T_0} \int_{t=-T_0/2}^{T_0/2} g_{p1}(t)g_{p2}^*(t-\tau)\,dt. \qquad (1.203)$$

Define

$$g_1(t) = \begin{cases} g_{p1}(t) \text{ for } -T_0/2 \leq t < T_0/2 \\ 0 \qquad \text{elsewhere} \end{cases}$$

$$g_2(t) = \begin{cases} g_{p2}(t) \text{ for } -T_0/2 \leq t < T_0/2 \\ 0 \qquad \text{elsewhere.} \end{cases} \qquad (1.204)$$

Note that $g_1(t)$ and $g_2(t)$ are the generating functions of $g_{p1}(t)$ and $g_{p2}(t)$, respectively. Hence, we have the following relationships:

$$g_{p1}(t) = \sum_{m=-\infty}^{\infty} g_1(t - mT_0)$$

$$g_{p2}(t) = \sum_{m=-\infty}^{\infty} g_2(t - mT_0). \tag{1.205}$$

Using the first equation of (1.204) and the second equation of (1.205), we get

$$R_{12}(\tau) = \frac{1}{T_0} \int_{t=-T_0/2}^{T_0/2} g_1(t) \sum_{m=-\infty}^{\infty} g_2^*(t - \tau - mT_0)\, dt$$

$$\Rightarrow R_{12}(\tau) = \frac{1}{T_0} \int_{t=-\infty}^{\infty} g_1(t) \sum_{m=-\infty}^{\infty} g_2^*(t - \tau - mT_0)\, dt. \tag{1.206}$$

Interchanging the order of the summation and the integration, we get

$$R_{12}(\tau) = \frac{1}{T_0} \sum_{m=-\infty}^{\infty} R_{g_1 g_2}(\tau + mT_0). \tag{1.207}$$

Note that

(a) $R_{12}(\tau)$ is periodic with period T_0.
(b) The span of $R_{g_1 g_2}(\tau)$ may exceed T_0, hence the summation in (1.207) may result in aliasing (overlapping). Therefore

$$R_{12}(\tau) \neq \frac{1}{T_0} R_{g_1 g_2}(\tau) \quad \text{for } -T_0/2 \leq \tau < T_0/2. \tag{1.208}$$

Since $R_{12}(\tau)$ is periodic with period T_0, it can be expressed as a complex Fourier series as follows:

$$R_{12}(\tau) = \sum_{n=-\infty}^{\infty} c_n \exp\left(j\, 2\pi n\tau/T_0\right), \tag{1.209}$$

where

$$c_n = \frac{1}{T_0} \int_{\tau=-T_0/2}^{T_0/2} R_{12}(\tau) \exp\left(-j\, 2\pi n\tau/T_0\right) d\tau. \tag{1.210}$$

Substituting for $R_{12}(\tau)$ in the above equation, we get

$$c_n = \frac{1}{T_0^2} \int_{\tau=-T_0/2}^{T_0/2} \sum_{m=-\infty}^{\infty} R_{g_1 g_2}(\tau + mT_0) \exp\left(-j\, 2\pi n\tau/T_0\right) d\tau. \tag{1.211}$$

Substituting for $R_{g_1 g_2}(\tau)$ in the above equation, we get

$$c_n = \frac{1}{T_0^2} \int_{\tau=-T_0/2}^{T_0/2} \sum_{m=-\infty}^{\infty} \int_{t=-\infty}^{\infty} g_1(t) g_2^*(t - \tau - mT_0) \, dt$$
$$\exp(-j 2\pi n\tau/T_0) \, d\tau. \tag{1.212}$$

Substituting

$$t - \tau - mT_0 = x \tag{1.213}$$

and interchanging the order of the integrals, we get

$$c_n = \frac{1}{T_0^2} \int_{t=-\infty}^{\infty} g_1(t) \, dt \sum_{m=-\infty}^{\infty} \int_{x=t-T_0/2-mT_0}^{t+T_0/2-mT_0} g_2^*(x)$$
$$\exp(-j 2\pi n(t - x - mT_0)/T_0) \, dx. \tag{1.214}$$

Combining the summation and the second integral and rearranging terms, we get

$$c_n = \frac{1}{T_0^2} \int_{t=-\infty}^{\infty} g_1(t) \exp(-j 2\pi nt/T_0) \, dt$$
$$\int_{x=-\infty}^{\infty} g_2^*(x) \exp(j 2\pi nx/T_0) \, dx$$
$$= G_1(n/T_0) G_2^*(n/T_0). \tag{1.215}$$

Hence

$$R_{12}(\tau) = \frac{1}{T_0^2} \sum_{n=-\infty}^{\infty} G_1(n/T_0) G_2^*(n/T_0) \exp(j 2\pi n\tau/T_0)$$
$$\rightleftharpoons \frac{1}{T_0^2} \sum_{n=-\infty}^{\infty} G_1(n/T_0) G_2^*(n/T_0) \delta(f - n/T_0). \tag{1.216}$$

Thus proved.

29. Compute the Fourier transform of

$$\widehat{\frac{dx(t)}{dt}}, \tag{1.217}$$

where the "hat" denotes the Hilbert transform. Assume that the Fourier transform of $x(t)$ is $X(f)$.

- *Solution*: We have

Fig. 1.26 Plot of $g(t)$

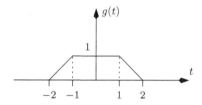

$$\frac{dx(t)}{dt} \rightleftharpoons j2\pi f X(f)$$

$$\Rightarrow \frac{\widehat{dx(t)}}{dt} \rightleftharpoons -j\,\mathrm{sgn}\,(f)\,(j2\pi f)\,X(f)$$

$$= 2\pi|f|X(f). \tag{1.218}$$

30. Prove the following using Schwarz's inequality:

$$|G(f)| \leq \int_{t=-\infty}^{\infty} |g(t)|\,dt$$

$$|j2\pi f G(f)| \leq \int_{t=-\infty}^{\infty} \left|\frac{dg(t)}{dt}\right|\,dt$$

$$\left|(j2\pi f)^2 G(f)\right| \leq \int_{t=-\infty}^{\infty} \left|\frac{d^2 g(t)}{dt^2}\right|\,dt. \tag{1.219}$$

Schwarz's inequality states that

$$\left|\int_{x=x_1}^{x_2} f(x)\,dx\right| \leq \int_{x=x_1}^{x_2} |f(x)|\,dx. \tag{1.220}$$

Evaluate the three bounds on $|G(f)|$ for the pulse shown in Fig. 1.26.

- *Solution*: We start from the Fourier transform relation:

$$G(f) = \int_{f=-\infty}^{\infty} g(t)e^{-j2\pi ft}\,dt. \tag{1.221}$$

Invoking Schwarz's inequality, we have

$$|G(f)| \leq \int_{f=-\infty}^{\infty} \left|g(t)e^{-j2\pi ft}\right|\,dt$$

$$\Rightarrow |G(f)| \leq \int_{f=-\infty}^{\infty} |g(t)|\,dt. \tag{1.222}$$

Similarly, we have

$$j 2\pi f G(f) = \int_{f=-\infty}^{\infty} \frac{dg(t)}{dt} e^{-j2\pi ft}\, dt$$

$$\Rightarrow |j 2\pi f G(f)| \leq \int_{f=-\infty}^{\infty} \left|\frac{dg(t)}{dt}\right|\, dt \tag{1.223}$$

and

$$(j 2\pi f)^2\, G(f) = \int_{f=-\infty}^{\infty} \frac{d^2 g(t)}{dt^2} e^{-j2\pi ft}\, dt$$

$$\Rightarrow \left|(j 2\pi f)^2\, G(f)\right| \leq \int_{f=-\infty}^{\infty} \left|\frac{d^2 g(t)}{dt^2}\right|\, dt. \tag{1.224}$$

The various derivatives of $g(t)$ are shown in Fig. 1.27. For the given pulse

$$\int_{t=-2}^{2} |g(t)|\, dt = 3$$

$$\int_{t=-2}^{2} \left|\frac{dg(t)}{dt}\right|\, dt = 2$$

$$\int_{t=-2^-}^{2^+} \left|\frac{d^2 g(t)}{dt^2}\right|\, dt = 4. \tag{1.225}$$

Fig. 1.27 Various derivatives of $g(t)$

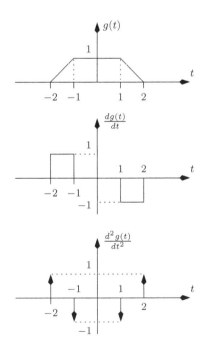

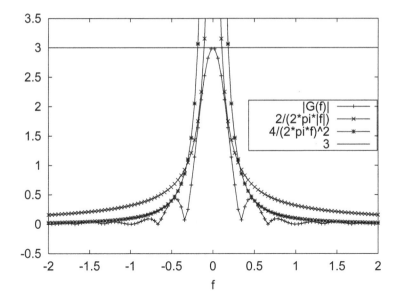

Fig. 1.28 Various bounds for $|G(f)|$

Therefore, the three bounds are

$$|G(f)| \leq 3$$

$$|G(f)| \leq \frac{2}{2\pi|f|}$$

$$|G(f)| \leq \frac{4}{(2\pi f)^2}. \qquad (1.226)$$

These bounds on $|G(f)|$ are shown plotted in Fig. 1.28. It can be shown that

$$G(f) = \frac{\cos(2\pi f) - \cos(4\pi f)}{2\pi^2 f^2}. \qquad (1.227)$$

31. (Simon Haykin 1983) A signal that is popularly used in communication systems
    is the raised cosine pulse. Consider a periodic sequence of these pulses as shown
    in Fig. 1.29. Compute the first three terms ($n = 0, 1, 2$) of the (real) Fourier
    series expansion.

   • *Solution*: The (real) Fourier series expansion of a periodic signal $g_p(t)$ can be
     written as follows:

   $$g_p(t) = a_0 + 2 \sum_{n=1}^{\infty} a_n \cos(2\pi nt/T_0) + b_n \sin(2\pi nt/T_0). \qquad (1.228)$$

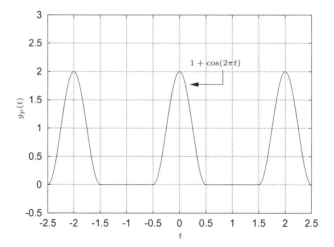

Fig. 1.29 Plot of $g_p(t)$

In the given problem $T_0 = 2$. Moreover, since the given function is even, $b_n = 0$. Thus, we have

$$
\begin{aligned}
a_0 &= \frac{1}{2} \int_{t=-1}^{1} g_p(t)\, dt \\
&= \frac{1}{2} \int_{t=-1/2}^{1/2} (1 + \cos(2\pi t))\, dt \\
&= \frac{1}{2}.
\end{aligned}
\tag{1.229}
$$

Similarly

$$
\begin{aligned}
a_n &= \frac{1}{2} \int_{t=-1}^{1} g_p(t) \cos(2\pi n t/2)\, dt \\
&= \frac{1}{2} \int_{t=-1/2}^{1/2} (1 + \cos(2\pi t)) \cos(n\pi t)\, dt \\
&= \frac{1}{n\pi} \sin(n\pi/2) + \frac{1}{2} \left[\frac{\sin(n-2)\pi/2}{(n-2)\pi} + \frac{\sin(n+2)\pi/2}{(n+2)\pi} \right].
\end{aligned}
\tag{1.230}
$$

Substituting $n = 1, 2$ in (1.230), we have

$$
a_1 = \frac{4}{3\pi}
$$
$$
a_2 = \frac{1}{4}.
\tag{1.231}
$$

32. (Simon Haykin 1983) Any function $g(t)$ can be expressed as the sum of $g_e(t)$ and $g_o(t)$, where

$$g_e(t) = \frac{1}{2}[g(t) + g(-t)]$$
$$g_o(t) = \frac{1}{2}[g(t) - g(-t)]. \qquad (1.232)$$

(a) Sketch the even and odd parts of

$$g(t) = A \operatorname{rect}\left(\frac{t}{T} - \frac{1}{2}\right). \qquad (1.233)$$

(b) Find the Fourier transform of these two parts.

- *Solution*: Note that

$$A \operatorname{rect}\left(\frac{t}{T} - \frac{1}{2}\right) = A \operatorname{rect}\left(\frac{t - (T/2)}{T}\right), \qquad (1.234)$$

which is nothing but rect (t/T) shifted by $T/2$. This is illustrated in Fig. 1.30, along with the even and odd parts of $g(t)$. We know that

Fig. 1.30 Plot of $g(t)$ and the corresponding even and odd functions

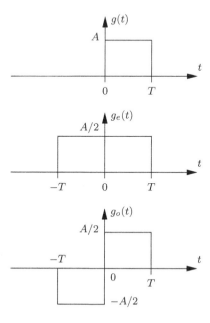

$$A \operatorname{rect}\left(\frac{t}{T}\right) \rightleftharpoons AT \operatorname{sinc}\,(fT). \tag{1.235}$$

Therefore by inspection, we have

$$g_e(t) \rightleftharpoons \frac{A}{2}(2T)\operatorname{sinc}\,(f(2T)) = AT \operatorname{sinc}\,(2fT). \tag{1.236}$$

Similarly

$$g_o(t) \rightleftharpoons \frac{A}{2}T \operatorname{sinc}\,(fT)\,\mathrm{e}^{-\mathrm{j}2\pi fT/2} - \frac{A}{2}T \operatorname{sinc}\,(fT)\,\mathrm{e}^{\mathrm{j}2\pi fT/2}$$
$$= -\mathrm{j}\,AT \operatorname{sinc}\,(fT)\sin(\pi fT). \tag{1.237}$$

33. Let $g(t) = \mathrm{e}^{-\pi t^2}$. Let

$$y(t) = g(at) \star \delta(t - t_0), \tag{1.238}$$

where "$\star$" denotes convolution, a and t_0 are real-valued constants.

(a) Write down the Fourier transform of $g(t)$. No derivation is required.
(b) Derive the Fourier transform of $y(t)$.
(c) Derive the Fourier transform of the autocorrelation of $y(t)$.
(d) Find the autocorrelation of $y(t)$.

Show all the steps. Use Fourier transform properties. Derivation of any Fourier transform property is not required.

- *Solution*: We know that

$$g(t) = \mathrm{e}^{-\pi t^2} \rightleftharpoons \mathrm{e}^{-\pi f^2} = G(f). \tag{1.239}$$

Now

$$y(t) = g(at) \star \delta(t - t_0)$$
$$= \int_{\tau=-\infty}^{\infty} \delta(\tau - t_0)g[a(t - \tau)]\,d\tau$$
$$= g[a(t - t_0)]$$
$$= \mathrm{e}^{-\pi a^2 (t-t_0)^2}. \tag{1.240}$$

Using the Fourier transform properties of time scaling and time shifting, we have

$$x_1(t) = g(at) \rightleftharpoons \frac{1}{|a|} G(f/a) = X_1(f)$$

$$\Rightarrow x_1(t - t_0) = g[a(t - t_0)] = y(t) \rightleftharpoons X_1(f) e^{-j 2\pi f t_0}$$

$$= \frac{1}{|a|} G(f/a) e^{-j 2\pi f t_0}$$

$$= Y(f)$$

$$= \frac{1}{|a|} e^{-\pi f^2/a^2} e^{-j 2\pi f t_0}. \quad (1.241)$$

The Fourier transform of the autocorrelation of $y(t)$ is

$$|Y(f)|^2 = \frac{1}{a^2} e^{-2\pi f^2/a^2} = H(f) \quad \text{(say)}. \quad (1.242)$$

The autocorrelation of $y(t)$ is the inverse Fourier transform of $H(f)$. Therefore

$$H(f) = \frac{1}{|a|\sqrt{2}} \frac{\sqrt{2}}{|a|} e^{-\pi (f\sqrt{2}/a)^2}$$

$$\rightleftharpoons \frac{1}{|a|\sqrt{2}} g(ct), \quad (1.243)$$

where

$$c = \frac{a}{\sqrt{2}}. \quad (1.244)$$

Thus, the autocorrelation of $y(t)$ is

$$h(t) = \frac{1}{|a|\sqrt{2}} e^{-\pi a^2 t^2/2}. \quad (1.245)$$

34. Consider the periodic pulse train in Fig. 1.31a. Here $T \leq T_0/2$.

   (a) Using the Fourier transform property of differentiation in the time domain, compute the complex Fourier series representation of $g_p(t)$ in Fig. 1.31a. Hence, also find the Fourier transform of $g_p(t)$.
   (b) If $g_p(t)$ is passed through a filter having the magnitude and phase response as depicted in Fig. 1.31b, c, find the output $y(t)$.

   • *Solution*: Consider the pulse:

$$g(t) = \begin{cases} g_p(t) & \text{for } -T_0/2 < t < T_0/2 \\ 0 & \text{elsewhere.} \end{cases} \quad (1.246)$$

Fig. 1.31 $g_p(t).g_p(t)$

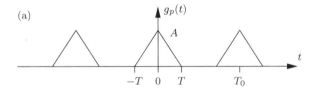

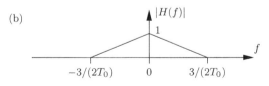

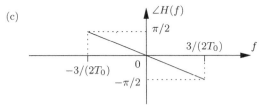

Clearly

$$\frac{d^2 g(t)}{dt^2} = \frac{A}{T} \left(\delta(t + T) - 2\delta(t) + \delta(t - T) \right). \tag{1.247}$$

Hence

$$\begin{aligned} G(f) &= -\frac{A}{4\pi^2 f^2 T} \left(\exp(\mathrm{j} 2\pi f T) - 2 + \exp(-\mathrm{j} 2\pi f T) \right) \\ &= \frac{AT}{\pi^2 f^2 T^2} \sin^2(\pi f T) \\ &= AT \operatorname{sinc}^2(f T). \end{aligned} \tag{1.248}$$

We also know that

$$g_p(t) = \sum_{n=-\infty}^{\infty} c_n \exp\left(\mathrm{j} 2\pi n t / T_0\right), \tag{1.249}$$

where

$$\begin{aligned} c_n &= \frac{1}{T_0} \int_{t=-\infty}^{\infty} g(t) \exp\left(-\mathrm{j} 2\pi n t / T_0\right) dt \\ &= \frac{1}{T_0} G(n / T_0). \end{aligned} \tag{1.250}$$

Substituting for $G(n/T_0)$ we have

$$g_p(t) = \frac{AT}{T_0} \sum_{n=-\infty}^{\infty} \text{sinc}^2(nT/T_0) \exp(j\,2\pi nt/T_0) . \tag{1.251}$$

Using the fact that

$$\exp(j\,2\pi f_c t) \rightleftharpoons \delta(f - f_c) \tag{1.252}$$

the Fourier transform of $g_p(t)$ is given by

$$G_p(f) = \frac{AT}{T_0} \sum_{n=-\infty}^{\infty} \text{sinc}^2(nT/T_0)\delta\left(f - \frac{n}{T_0}\right) . \tag{1.253}$$

The filter output will have the frequency components 0, $\pm 1/T_0$. From Fig. 1.31b, c, we note that

$$H(0) = 1$$
$$H(1/T_0) = \frac{1}{3}e^{-j\pi/3}$$
$$H(-1/T_0) = \frac{1}{3}e^{j\pi/3}. \tag{1.254}$$

Hence

$$y(t) = \frac{AT}{T_0} + \frac{AT}{3T_0} \text{sinc}^2(T/T_0)e^{j\,(2\pi t/T_0 - \pi/3)}$$
$$\quad + \frac{AT}{3T_0} \text{sinc}^2(T/T_0)e^{j\,(-2\pi t/T_0 + \pi/3)}$$
$$\quad = \frac{AT}{T_0} + \frac{2AT}{3T_0} \text{sinc}^2(T/T_0) \cos(2\pi t/T_0 - \pi/3). \tag{1.255}$$

35. (Simon Haykin 1983) Prove the following Hilbert transform:

$$\frac{\sin(t)}{t} \xrightarrow{HT} \frac{1 - \cos(t)}{t}. \tag{1.256}$$

- *Solution*: We start with the familiar Fourier transform pair:

$$A \text{ rect}(t/T) \rightleftharpoons AT \text{ sinc}(fT). \tag{1.257}$$

Applying duality and replacing T by B, we get

$$A \text{ rect}(f/B) \rightleftharpoons AB \text{ sinc}(tB). \tag{1.258}$$

Substituting $A = B = 1$, we get

$$\text{rect}(f) \rightleftharpoons \text{sinc}(t) = \sin(\pi t)/(\pi t). \tag{1.259}$$

Time scaling by $1/\pi$, we get

$$\pi \, \text{rect}(\pi f) \rightleftharpoons \sin(t)/t = g(t) \quad \text{(say).} \tag{1.260}$$

Now, the Hilbert transform of $\sin(t)/t$ is best computed in the frequency domain. Thus

$$\hat{G}(f) = -\text{j} \, \text{sgn}(f) \pi \, \text{rect}(\pi f)$$

$$\Rightarrow \hat{g}(t) = \int_{f=-\infty}^{\infty} \hat{G}(f) \exp(\text{j} 2\pi f t) \, df$$

$$\Rightarrow \hat{g}(t) = -\text{j} \pi \int_{f=0}^{1/(2\pi)} \exp(\text{j} 2\pi f t) \, df$$

$$+ \text{j} \pi \int_{-1/(2\pi)}^{0} \exp(\text{j} 2\pi f t) \, df$$

$$= -\text{j} \pi \frac{\exp(\text{j} t) - 1}{\text{j} 2\pi t} + \text{j} \pi \frac{1 - \exp(-\text{j} t)}{\text{j} 2\pi t}$$

$$= \frac{1 - \cos(t)}{t}. \tag{1.261}$$

Thus proved.

36. (Simon Haykin 1983) Prove the following Hilbert transform:

$$\frac{1}{1 + t^2} \xrightarrow{HT} \frac{t}{1 + t^2}. \tag{1.262}$$

- *Solution*: We start with the Fourier transform pair:

$$\frac{1}{2} \exp(-|t|) \rightleftharpoons \frac{1}{1 + (2\pi f)^2}. \tag{1.263}$$

Time scaling by 2π, we get

$$\frac{1}{2} \exp(-|2\pi t|) \rightleftharpoons \frac{1}{(2\pi)(1 + f^2)}. \tag{1.264}$$

Multiplying both sides by 2π and applying duality, we get

$$\pi \exp(-|2\pi f|) \rightleftharpoons \frac{1}{1 + t^2} = g(t) \quad \text{(say).} \tag{1.265}$$

Now the Hilbert transform of $g(t)$ in the above equation is best obtained in the frequency domain. Thus

$$
\begin{aligned}
\hat{g}(t) &= -j \int_{f=-\infty}^{\infty} \operatorname{sgn}(f)\pi \exp\left(-|2\pi f|\right) \exp\left(j\,2\pi f t\right) df \\
&= -j\pi \int_{f=0}^{\infty} \exp\left(-2\pi f(1 - jt)\right) df \\
&\quad + j\pi \int_{f=-\infty}^{0} \exp\left(2\pi f(1 + jt)\right) df \\
&= -j\pi \frac{-1}{-2\pi(1 - jt)} + j\pi \frac{1}{2\pi(1 + jt)} \\
&= \frac{t}{1+t^2}.
\end{aligned}
\tag{1.266}
$$

Thus proved.

37. (Simon Haykin 1983) Prove the following Hilbert transform:

$$
\operatorname{rect}(t) \xrightarrow{HT} -\frac{1}{\pi}\ln\left|\frac{t - 1/2}{t + 1/2}\right|.
\tag{1.267}
$$

- *Solution*: This problem can be easily solved in the time domain. From the basic definition of the Hilbert transform, we have

$$
\begin{aligned}
\hat{g}(t) &= \frac{1}{\pi} \int_{\tau=-\infty}^{\infty} \frac{g(\tau)}{t - \tau} d\tau \\
&= \frac{1}{\pi} \int_{\tau=-1/2}^{1/2} \frac{1}{t - \tau} d\tau \\
&= \frac{-1}{\pi}\ln\left|\frac{t - 1/2}{t + 1/2}\right|.
\end{aligned}
\tag{1.268}
$$

Thus proved.

38. (Simon Haykin 1983) Let $\hat{g}(t)$ denote the Hilbert transform of a real-valued energy signal $g(t)$. Show that the cross-correlation functions of $g(t)$ and $\hat{g}(t)$ are given by

$$
\begin{aligned}
R_{g\hat{g}}(\tau) &= -\hat{R}_g(\tau) \\
R_{\hat{g}g}(\tau) &= \hat{R}_g(\tau),
\end{aligned}
\tag{1.269}
$$

where $\hat{R}_g(\tau)$ denotes the Hilbert transform of $R_g(\tau)$.

- *Solution*: We start from the basic definition of the cross-correlation function:

$$R_{g_1 g_2}(\tau) = \int_{t=-\infty}^{\infty} g_1(t) g_2^*(t-\tau)\, dt$$
$$= g_1(\tau) \star g_2^*(-\tau)$$
$$= \int_{f=-\infty}^{\infty} G_1(f) G_2^*(f) \exp(j\, 2\pi f \tau)\, df. \qquad (1.270)$$

Here

$$g_1(t) = g(t) \rightleftharpoons G(f)$$
$$g_2(t) = \hat{g}(t) \rightleftharpoons -j\, \mathrm{sgn}\,(f) G(f)$$
$$\Rightarrow g_2^*(-t) = \hat{g}^*(-t) = \hat{g}(-t) \rightleftharpoons j\, \mathrm{sgn}\,(f) G^*(f), \qquad (1.271)$$

since $g(t)$ and hence $\hat{g}(t)$ are real-valued. Hence

$$R_{g\hat{g}}(\tau) = j \int_{f=-\infty}^{\infty} \mathrm{sgn}\,(f) G(f) G^*(f) \exp(j\, 2\pi f \tau)\, df$$
$$= j \int_{f=-\infty}^{\infty} \mathrm{sgn}\,(f) |G(f)|^2 \exp(j\, 2\pi f \tau)\, df. \qquad (1.272)$$

However, we know that

$$R_g(\tau) \rightleftharpoons |G(f)|^2$$
$$\Rightarrow \hat{R}_g(\tau) \rightleftharpoons -j\, \mathrm{sgn}\,(f) |G(f)|^2. \qquad (1.273)$$

From (1.272) and (1.273), it is clear that

$$R_{g\hat{g}}(\tau) = -\hat{R}_g(\tau). \qquad (1.274)$$

The second part can be similarly proved.

39. (Simon Haykin 1983) Consider two bandpass signals $g_1(t)$ and $g_2(t)$ whose pre-envelopes are denoted by $g_{1+}(t)$ and $g_{2+}(t)$, respectively.

(a) Show that

$$\int_{t=-\infty}^{\infty} \Re\{g_{1+}(t)\}\, \Re\{g_{2+}(t)\}\, dt = \frac{1}{2}\Re\left\{\int_{t=-\infty}^{\infty} g_{1+}(t) g_{2+}^*(t)\, dt\right\}.$$
$$(1.275)$$

How is this relation modified if $g_2(t)$ is replaced by $g_2(-t)$?

(b) Assuming that $g(t)$ is a narrowband signal with complex envelope $\tilde{g}(t)$ and carrier frequency f_c, use the result of part (a) to show that

$$\int_{t=-\infty}^{\infty} g^2(t)\, dt = \frac{1}{2} \int_{t=-\infty}^{\infty} |\tilde{g}(t)|^2\, dt. \qquad (1.276)$$

- *Solution*: From the basic definition of the pre-envelope, we have

$$g_{1+}(t) = g_1(t) + j\,\hat{g}_1(t),$$
$$g_{2+}(t) = g_2(t) + j\,\hat{g}_2(t), \qquad (1.277)$$

where $g_1(t)$ and $g_2(t)$ are real-valued signals. Hence, $\hat{g}_1(t)$ and $\hat{g}_2(t)$ are also real-valued. Next, we note that the real part of the integral is equal to the integral of the real part. Hence, the right-hand side of (1.275) becomes

$$\frac{1}{2}\int_{t=-\infty}^{\infty} \big(g_1(t)g_2(t) + \hat{g}_1(t)\hat{g}_2(t)\big)\,dt. \qquad (1.278)$$

Next, we note that

$$\int_{t=-\infty}^{\infty} g_1(t)g_2(t)\,dt = g_1(t) \star g_2(-t)\big|_{t=0} = \int_{f=-\infty}^{\infty} G_1(f)G_2^*(f)\,df, \qquad (1.279)$$

where "$\star$" denotes convolution. Similarly

$$\int_{t=-\infty}^{\infty} \hat{g}_1(t)\hat{g}_2(t)\,dt = \hat{g}_1(t) \star \hat{g}_2(-t)\big|_{t=0} = \int_{f=-\infty}^{\infty} (-j\,\mathrm{sgn}\,(f))\,G_1(f)$$
$$(j\,\mathrm{sgn}\,(f))\,G_2^*(f)\,df$$
$$= \int_{f=-\infty}^{\infty} G_1(f)G_2^*(f)\,df$$
$$= \int_{t=-\infty}^{\infty} g_1(t)g_2(t)\,dt, \quad (1.280)$$

where we have used the fact that

$$\mathrm{sgn}^2(f) = 1 \qquad \text{for } f \neq 0. \qquad (1.281)$$

Now, the left-hand side of (1.275) is nothing but

$$\int_{t=-\infty}^{\infty} g_1(t)g_2(t)\,dt. \qquad (1.282)$$

Thus (1.275) is proved. Now, let us see what happens when $g_2(t)$ is replaced by $g_2(-t)$. Let

$$g_3(t) = g_2(-t)$$
$$\Rightarrow g_{3+}(t) = g_3(t) + j\,\hat{g}_3(t)$$
$$= g_2(-t) + j\,\hat{g}_2(-t)$$
$$\Rightarrow g_{3+}^*(t) = g_3(t) - j\,\hat{g}_3(t)$$
$$= g_2(-t) - j\,\hat{g}_2(-t). \qquad (1.283)$$

Clearly (1.275) is still valid with $g_{2+}(t)$ replaced by $g_{3+}(t)$ as given in (1.283), that is,

$$\int_{t=-\infty}^{\infty} \Re\left\{g_{1+}(t)\right\} \Re\left\{g_{3+}(t)\right\} dt = \frac{1}{2} \Re\left\{\int_{t=-\infty}^{\infty} g_{1+}(t) g_{3+}^{*}(t)\, dt\right\}. \quad (1.284)$$

To prove part (b) we substitute

$$g_1(t) = g_2(t) = g(t) \quad (1.285)$$

in (1.275) and note that

$$\begin{aligned} g_+(t) &= g(t) + \mathrm{j}\,\hat{g}(t) \\ &= \tilde{g}(t) \exp\left(\mathrm{j}\,2\pi f_c t\right), \end{aligned} \quad (1.286)$$

where $\tilde{g}(t)$ is the complex envelope of $g(t)$. Thus, (1.276) follows immediately.

40. Let a narrowband signal be expressed in the form:

$$g(t) = g_c(t) \cos(2\pi f_c t) - g_s(t) \sin(2\pi f_c t). \quad (1.287)$$

Using $G_+(f)$ to denote the Fourier transform of the pre-envelope of $g(t)$, express the Fourier transforms of the in-phase component $g_c(t)$ and the quadrature component $g_s(t)$ in terms of $G_+(f)$.

• *Solution*: From the basic definition of the pre-envelope

$$\begin{aligned} g_c(t) + \mathrm{j}\, g_s(t) &= g_+(t) \exp\left(-\mathrm{j}\,2\pi f_c t\right) = g_1(t) \quad \text{(say)} \\ \Rightarrow g_c(t) &= \frac{1}{2}\left[g_1(t) + g_1^*(t)\right] \\ g_s(t) &= \frac{1}{2\mathrm{j}}\left[g_1(t) - g_1^*(t)\right]. \end{aligned} \quad (1.288)$$

Taking the Fourier transform of both sides in the above equation, we get

$$\begin{aligned} G_c(f) &= \frac{1}{2}\left[G_1(f) + G_1^*(-f)\right] \\ G_s(f) &= \frac{1}{2\mathrm{j}}\left[G_1(f) - G_1^*(-f)\right]. \end{aligned} \quad (1.289)$$

Note that $G_1(f)$ is equal to

$$\begin{aligned} G_1(f) &= G_+(f + f_c) \\ \Rightarrow G_1^*(-f) &= G_+^*(-f + f_c). \end{aligned} \quad (1.290)$$

Substituting for $G_1(f)$ in (1.289), we get

$$G_c(f) = \frac{1}{2} \left[G_+(f + f_c) + G_+^*(-f + f_c) \right]$$

$$G_s(f) = \frac{1}{2j} \left[G_+(f + f_c) - G_+^*(-f + f_c) \right]. \qquad (1.291)$$

41. (Simon Haykin 1983) The duration of a signal provides a measure for describing the signal as a function of time. The bandwidth of a signal provides a measure for describing its frequency content. There is no unique set of definitions for the duration and bandwidth. However, regardless of the definition we find that their product is always a constant. The choice of a particular set of definitions merely changes the value of this constant. This problem is intended to explore these issues.

(a) The root mean square (rms) bandwidth of a lowpass energy signal $g(t)$ is defined by

$$W_{\text{rms}} = \left[\frac{\int_{f=-\infty}^{\infty} f^2 |G(f)|^2 \, df}{\int_{f=-\infty}^{\infty} |G(f)|^2 \, df} \right]^{0.5}. \qquad (1.292)$$

The corresponding rms duration of the signal is defined by

$$T_{\text{rms}} = \left[\frac{\int_{t=-\infty}^{\infty} t^2 |g(t)|^2 \, dt}{\int_{t=-\infty}^{\infty} |g(t)|^2 \, dt} \right]^{0.5}. \qquad (1.293)$$

Show that (Heisenberg–Gabor uncertainty principle)

$$T_{\text{rms}} W_{\text{rms}} \geq 1/(4\pi). \qquad (1.294)$$

(b) Consider an energy signal $g(t)$ for which $|G(f)|$ is defined for all frequencies from $-\infty$ to ∞ and symmetric about $f = 0$ with its maximum value at $f = 0$. The equivalent rectangular bandwidth is defined by

$$W_{\text{eq}} = \frac{\int_{f=-\infty}^{\infty} |G(f)|^2 \, df}{2|G(0)|^2}. \qquad (1.295)$$

The corresponding equivalent duration is defined by

$$T_{\text{eq}} = \frac{\left[\int_{t=-\infty}^{\infty} |g(t)| \, dt \right]^2}{\int_{t=-\infty}^{\infty} |g(t)|^2 \, dt}. \qquad (1.296)$$

Show that

$$T_{eq} W_{eq} \geq 0.5. \tag{1.297}$$

- *Solution*: To prove

$$T_{rms} W_{rms} \geq 1/(4\pi) \tag{1.298}$$

we need to use Schwarz's inequality which states that

$$\left[\int_{t=-\infty}^{\infty} \left(g_1^*(t)g_2(t) + g_1(t)g_2^*(t) \right) \, dt \right]^2 \leq 4 \int_{t=-\infty}^{\infty} |g_1(t)|^2 \, dt$$
$$\times \int_{t=-\infty}^{\infty} |g_2(t)|^2 \, dt. \tag{1.299}$$

For the given problem, we substitute

$$g_1(t) = tg(t)$$
$$g_2(t) = \frac{dg(t)}{dt}. \tag{1.300}$$

Thus, the right-hand side of Schwarz's inequality in (1.299) becomes

$$4 \int_{t=-\infty}^{\infty} t^2 |g(t)|^2 \, dt \int_{t=-\infty}^{\infty} \left| \frac{dg(t)}{dt} \right|^2 \, dt. \tag{1.301}$$

Let

$$g_3(t) = \frac{dg(t)}{dt} \rightleftharpoons j2\pi f G(f) = G_3(f). \tag{1.302}$$

Then, by Rayleigh's energy theorem

$$\int_{t=-\infty}^{\infty} |g_3(t)|^2 \, dt = \int_{f=-\infty}^{\infty} |G_3(f)|^2 \, df$$
$$\Rightarrow \int_{t=-\infty}^{\infty} \left| \frac{dg(t)}{dt} \right|^2 \, dt = 4\pi^2 \int_{f=-\infty}^{\infty} f^2 |G(f)|^2 \, df. \tag{1.303}$$

Thus, the right-hand side of Schwarz's inequality in (1.299) becomes

$$16\pi^2 \int_{t=-\infty}^{\infty} t^2 |g(t)|^2 \, dt \int_{f=-\infty}^{\infty} f^2 |G(f)|^2 \, df. \tag{1.304}$$

Let us now consider the square root of the left-hand side of Schwarz's inequality in (1.299). We have

$$
\int_{t=-\infty}^{\infty} t \left(g^*(t) \frac{dg(t)}{dt} + g(t) \left(\frac{dg(t)}{dt} \right)^* \right) dt = \int_{t=-\infty}^{\infty} t \left(g^*(t) \frac{dg(t)}{dt} \right.
$$
$$
\left. + g(t) \frac{dg^*(t)}{dt} \right) dt
$$
$$
= \int_{t=-\infty}^{\infty} t \frac{d}{dt} \left(g(t) g^*(t) \right) dt
$$
$$
= \int_{t=-\infty}^{\infty} t \frac{d}{dt} \left(|g(t)|^2 \right) dt.
$$

$$(1.305)$$

In the above equation, we have used the fact (this can also be proved using Fourier transforms) that

$$
\left(\frac{dg(t)}{dt} \right)^* = \frac{dg^*(t)}{dt}.
$$

$$(1.306)$$

Let

$$
g_4(t) = |g(t)|^2 \rightleftharpoons G_4(f)
$$
$$
\Rightarrow g_5(t) = \frac{dg_4(t)}{dt} \rightleftharpoons j 2\pi f G_4(f) = G_5(f)
$$
$$
\Rightarrow t g_5(t) \rightleftharpoons \frac{j}{2\pi} \frac{dG_5(f)}{df}
$$
$$
\Rightarrow \int_{t=-\infty}^{\infty} t g_5(t) \, dt = \frac{j}{2\pi} \frac{dG_5(f)}{df} \bigg|_{f=0}
$$
$$
\Rightarrow \int_{t=-\infty}^{\infty} t g_5(t) \, dt = \frac{j}{2\pi} \frac{d}{df} \left(j 2\pi f G_4(f) \right) \bigg|_{f=0}
$$
$$
\Rightarrow \int_{t=-\infty}^{\infty} t \frac{d}{dt} |g(t)|^2 \, dt = -G_4(0)
$$
$$
\Rightarrow \int_{t=-\infty}^{\infty} t \frac{d}{dt} |g(t)|^2 \, dt = - \int_{t=-\infty}^{\infty} |g(t)|^2 \, dt.
$$

$$(1.307)$$

Using Rayleigh's energy theorem, the left-hand side of Schwarz's inequality in (1.299) becomes

$$
\int_{t=-\infty}^{\infty} |g(t)|^2 \, dt \int_{f=-\infty}^{\infty} |G(f)|^2 \, df.
$$

$$(1.308)$$

Taking the square root of (1.304) and (1.308), we get the desired result in (1.298).

To prove

$$T_{eq} W_{eq} \geq 0.5, \tag{1.309}$$

we invoke Schwarz's inequality which states that

$$\left| \int_{t=-\infty}^{\infty} g(t)\, dt \right| \leq \int_{t=-\infty}^{\infty} |g(t)|\, dt. \tag{1.310}$$

Squaring both sides, we obtain

$$\left| \int_{t=-\infty}^{\infty} g(t)\, dt \right|^2 \leq \left[\int_{t=-\infty}^{\infty} |g(t)|\, dt \right]^2$$

$$\Rightarrow |G(0)|^2 \leq \left[\int_{t=-\infty}^{\infty} |g(t)|\, dt \right]^2. \tag{1.311}$$

Using the above inequality and the Rayleigh's energy theorem which states that

$$\int_{t=-\infty}^{\infty} |g(t)|^2\, dt = \int_{f=-\infty}^{\infty} |G(f)|^2\, df, \tag{1.312}$$

we get the desired result

$$T_{eq} W_{eq} \geq 0.5. \tag{1.313}$$

42. $x(t)$ is a bandpass signal whose Fourier transform is shown in Fig. 1.32. $h(t)$ is an LTI system having a bandpass frequency response of the form

$$h(t) = h_c(t) \cos(2\pi f_c t). \tag{1.314}$$

The Fourier transform of $h_c(t)$ is given by

Fig. 1.32 A bandpass spectrum

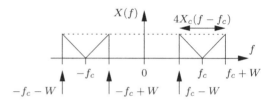

$$H_c(f) = \begin{cases} f^2 & \text{for } |f| < B \\ 0 & \text{otherwise} \end{cases}, \tag{1.315}$$

where $B > W$ and $f_c \gg B$, W. $x(t)$ is input to $h(t)$.

(a) Write down the expression of $x(t)$ in terms of $x_c(t)$.
(b) Write down the expression of $H(f)$ in terms of $H_c(f)$.
(c) Find $y(t)$ (output of $h(t)$) in terms of $x_c(t)$ without performing convolution.

• *Solution*: Note that

$$
\begin{aligned}
X(f) &= 4[X_c(f - f_c) + X_c(f + f_c)] \\
&\rightleftharpoons 8x_c(t) \cos(2\pi f_c t) \\
&= x(t).
\end{aligned} \tag{1.316}
$$

Similarly

$$H(f) = \frac{H_c(f - f_c) + H_c(f + f_c)}{2}. \tag{1.317}$$

Now

$$
\begin{aligned}
Y(f) &= X(f)H(f) \\
&= 2[X_c(f - f_c)H_c(f - f_c) + X_c(f + f_c)H_c(f + f_c)].
\end{aligned} \tag{1.318}
$$

Let

$$
\begin{aligned}
G_c(f) &= X_c(f)H_c(f) \\
&= \begin{cases} f^2 X_c(f) & \text{for } |f| < W \\ 0 & \text{otherwise} \end{cases} \\
&\rightleftharpoons \frac{-1}{4\pi^2} \frac{d^2 x_c(t)}{dt^2} \\
&= g_c(t).
\end{aligned} \tag{1.319}
$$

Then

$$
\begin{aligned}
y(t) &= 4g_c(t) \cos(2\pi f_c t) \\
&= \frac{-1}{\pi^2} \frac{d^2 x_c(t)}{dt^2} \cos(2\pi f_c t).
\end{aligned} \tag{1.320}
$$

43. Two periodic signals $g_{p1}(t)$ and $g_{p2}(t)$ are depicted in Fig. 1.33. Let

$$g_p(t) = g_{p1}(t) + g_{p2}(t). \tag{1.321}$$

Fig. 1.33 Periodic
waveforms

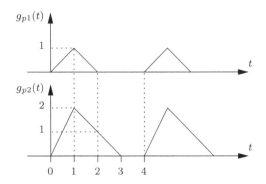

(a) Sketch $g_p(t)$.

(b) Compute the power of $dg_p(t)/dt$.

(c) Compute the Fourier transform of $d^2 g_p(t)/dt^2$. Assume that the generating
function (of $d^2 g_p(t)/dt^2$) extends over $[0, T)$ where T is the period of
$d^2 g_p(t)/dt^2$.

- *Solution*: The signal $g_p(t)$ is illustrated in Fig. 1.34.
  Let

$$g_{p3}(t) = \frac{dg_p(t)}{dt}. \qquad (1.322)$$

Then, the power of $g_{p3}(t)$ is

$$\begin{aligned}
P &= \frac{1}{T} \int_{t=0}^{T} g_{p3}^2(t)\, dt \\
&= \frac{14}{4} \\
&= 3.5. \qquad (1.323)
\end{aligned}$$

Let

$$g_{p4}(t) = \frac{d^2 g_p(t)}{dt^2} \qquad (1.324)$$

and let $g_4(t)$ denote its generating function. Then

$$\begin{aligned}
g_4(t) &= 3\delta(t) - 5\delta(t-1) + \delta(t-2) + \delta(t-3) \\
&\rightleftharpoons 3 - 5e^{-j2\pi f} + e^{-j4\pi f} + e^{-j6\pi f} \\
&= G_4(f). \qquad (1.325)
\end{aligned}$$

Fig. 1.34 Periodic
waveforms

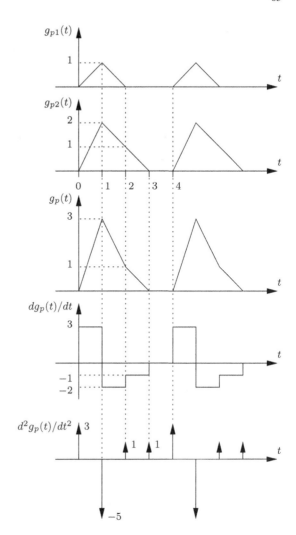

Now

$$g_{p4}(t) = \sum_{k=-\infty}^{\infty} g_4(t - kT)$$

$$\rightleftharpoons G_4(f) \sum_{k=-\infty}^{\infty} e^{-j2\pi fkT}$$

$$= \frac{1}{T} \sum_{n=-\infty}^{\infty} G_4(n/T)\delta(f - n/T), \qquad (1.326)$$

where $T = 4\,$s.

44. Using Fourier transform properties and/or any other relation, compute

$$\int_{f=-\infty}^{\infty} \left| \frac{A \operatorname{sinc}(fT)\,\mathrm{e}^{-\mathrm{j}\pi fT} - A\,\mathrm{e}^{-\mathrm{j}4\pi fT}}{\mathrm{j}2\pi f} \right|^2 df. \qquad (1.327)$$

Clearly state which Fourier transform property and/or relation is being used. The constant of any integration may be assumed to be zero.

- *Solution*: We know from Parseval's relation that

$$\int_{f=-\infty}^{\infty} |G(f)|^2\,df = \int_{t=-\infty}^{\infty} |g(t)|^2\,dt, \qquad (1.328)$$

where

$$g(t) \rightleftharpoons G(f) \qquad (1.329)$$

is a Fourier transform pair. Let

$$G(f) = \frac{A \operatorname{sinc}(fT)\,\mathrm{e}^{-\mathrm{j}\pi fT} - A\,\mathrm{e}^{-\mathrm{j}4\pi fT}}{\mathrm{j}2\pi f}. \qquad (1.330)$$

Let

$$\begin{aligned}
G_1(f) &= \mathrm{j}2\pi f G(f) \\
&= A \operatorname{sinc}(fT)\,\mathrm{e}^{-\mathrm{j}\pi fT} - A\,\mathrm{e}^{-\mathrm{j}4\pi fT} \\
&\rightleftharpoons g_1(t) \\
&= dg(t)/dt \\
&= \frac{A}{T} \operatorname{rect}\left(\frac{t - T/2}{T} \right) - A\delta(t - 2T). \qquad (1.331)
\end{aligned}$$

Both $g_1(t)$ and $g(t)$ are plotted in Fig. 1.35. Clearly

$$\begin{aligned}
\int_{t=-\infty}^{\infty} |g(t)|^2\,dt &= \int_{t=0}^{2T} |g(t)|^2\,dt \\
&= \frac{4A^2 T}{3}, \qquad (1.332)
\end{aligned}$$

which is the required answer.

45. Compute the area under $2\mathrm{e}^{-2\pi t^2} \star \mathrm{e}^{-3\pi t^2}$, where "$\star$" denotes convolution. Clearly state which Fourier transform property and/or relation is being used and show all the steps.

Fig. 1.35 Plot of $g_1(t)$ and $g(t)$

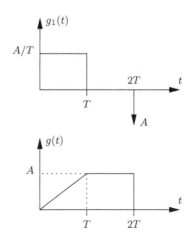

- *Solution*: We know that

$$e^{-\pi t^2} \rightleftharpoons e^{-\pi f^2}.$$ (1.333)

We also know that if

$$g(t) \rightleftharpoons G(f),$$ (1.334)

then, from the time scaling property of the Fourier transform,

$$g(at) \rightleftharpoons \frac{1}{|a|} G(f/a),$$ (1.335)

where $a \neq 0$. Therefore

$$e^{-2\pi t^2} \rightleftharpoons \frac{1}{\sqrt{2}} e^{-\pi f^2/2}$$

$$e^{-3\pi t^2} \rightleftharpoons \frac{1}{\sqrt{3}} e^{-\pi f^2/3}.$$ (1.336)

Moreover

$$2e^{-2\pi t^2} \star e^{-3\pi t^2} \rightleftharpoons \frac{2}{\sqrt{6}} e^{-\pi f^2/2} e^{-\pi f^2/3}.$$ (1.337)

Using

$$\int_{t=-\infty}^{\infty} g(t)\, dt = G(0) \tag{1.338}$$

the area under $2e^{-2\pi t^2} \star e^{-3\pi t^2}$ is $2/\sqrt{6}$.

46. State Rayleigh's energy theorem. No derivation is required.

   • *Solution*: Let $G(f)$ denote the Fourier transform of $g(t)$. Then Rayleigh's energy theorem states that

   $$\int_{t=-\infty}^{\infty} |g(t)|^2\, dt = \int_{f=-\infty}^{\infty} |G(f)|^2\, df. \tag{1.339}$$

47. Let $G(f)$ denote the Fourier transform of $g(t)$. It is given that $G(0)$ is finite and non-zero. Does $g(t)$ contain a dc component? Justify your answer.

   • *Solution*: If $g(t)$ has a dc component, then $G(f)$ contains a Dirac-delta function at $f = 0$. However, it is given that $G(0)$ is finite. Therefore, $g(t)$ does not contain any dc component.

48. Consider the following: if CONDITION then STATEMENT. The CONDITION is given to be necessary.
   Which of the following statement(s) are correct (more than one statement may be correct).

   (a) CONDITION is true implies STATEMENT is always true.
   (b) CONDITION is true implies STATEMENT is always false.
   (c) CONDITION is true implies STATEMENT may be true or false.
   (d) CONDITION is false implies STATEMENT is always true.
   (e) CONDITION is false implies STATEMENT is always false.
   (f) CONDITION is false implies STATEMENT may be true or false.

   • *Solution*: (c) and (e).
   For example, the necessary condition for the sum of an infinite series to exist is that the nth term must tend to zero as $n \to \infty$. Here:

   (a) CONDITION is: n^{th} term goes to zero as $n \to \infty$.
   (b) STATEMENT is: The sum of an infinite series exists.

   The sum given by

   $$S = \sum_{n=1}^{\infty} \frac{1}{n} \tag{1.340}$$

   does not exist even though the nth term goes to zero as $n \to \infty$.

49. Consider the following: if CONDITION then STATEMENT. The CONDITION is given to be sufficient.
Which of the following statement(s) are correct (more than one statement may be correct):

(a) CONDITION is true implies STATEMENT is always true.
(b) CONDITION is true implies STATEMENT is always false.
(c) CONDITION is true implies STATEMENT may be true or false.
(d) CONDITION is false implies STATEMENT is always true.
(e) CONDITION is false implies STATEMENT is always false.
(f) CONDITION is false implies STATEMENT may be true or false.

- *Solution*: (a) and (f).
  For example, Dirichlet's conditions for the existence of the Fourier transform are sufficient. In other words, if Dirichlet's conditions are satisfied, the Fourier transform is guaranteed to exist. However, if Dirichlet's conditions are not satisfied, the Fourier transform may or may not exist. Here:

  (a) CONDITION is: Dirichlet's conditions.
  (b) STATEMENT is: The Fourier transform exists.

50. State Parseval's power theorem for periodic signals in terms of its complex Fourier series coefficients.

- *Solution*: Let $g_p(t)$ be a periodic signal with period T_0. Then $g_p(t)$ can be represented in the form of a complex Fourier series as follows:

$$g_p(t) = \sum_{n=-\infty}^{\infty} c_n e^{j2\pi nt/T_0}. \tag{1.341}$$

  Then Parseval's power theorem states that

$$\frac{1}{T_0} \int_{t=0}^{T_0} |g_p(t)|^2 \, dt = \sum_{n=-\infty}^{\infty} |c_n|^2. \tag{1.342}$$

51. State and derive the Poisson sum formula. Hence show that

$$\sum_{n=-\infty}^{\infty} e^{j2\pi fnT_0} = \frac{1}{T_0} \sum_{n=-\infty}^{\infty} \delta\left(f - \frac{n}{T_0}\right). \tag{1.343}$$

- *Solution*: Let $g_p(t)$ be a periodic signal with period T_0. Then $g_p(t)$ can be expressed in the form of a complex Fourier series as follows:

$$g_p(t) = \sum_{n=-\infty}^{\infty} c_n e^{j2\pi nt/T_0}, \tag{1.344}$$

where

$$c_n = \frac{1}{T_0} \int_{t=0}^{T_0} g_p(t) e^{-j2\pi nt/T_0} \, dt. \tag{1.345}$$

Let $g(t)$ be the generating function of $g_p(t)$, that is,

$$g(t) = \begin{cases} g_p(t) & \text{for } 0 \le t < T_0 \\ 0 & \text{otherwise.} \end{cases} \tag{1.346}$$

Note that

$$g_p(t) = \sum_{n=-\infty}^{\infty} g(t - nT_0). \tag{1.347}$$

The Fourier transform of $g(t)$ is

$$\begin{aligned} G(f) &= \int_{t=-\infty}^{\infty} g(t) e^{-j2\pi ft} \, dt \\ &= \int_{t=0}^{T_0} g_p(t) e^{-j2\pi ft} \, dt. \end{aligned} \tag{1.348}$$

From (1.345) and (1.348), we have

$$c_n T_0 = G(n/T_0). \tag{1.349}$$

Therefore, from (1.344), (1.347), and (1.349) we have

$$\sum_{n=-\infty}^{\infty} g(t - nT_0) = \frac{1}{T_0} \sum_{n=-\infty}^{\infty} G\left(\frac{n}{T_0}\right) e^{j2\pi nt/T_0}, \tag{1.350}$$

which proves the Poisson sum formula. Taking the Fourier transform of both sides of (1.350) we obtain

$$G(f) \sum_{n=-\infty}^{\infty} e^{-j2\pi fnT_0} = \frac{1}{T_0} \sum_{n=-\infty}^{\infty} G\left(\frac{n}{T_0}\right) \delta(f - n/T_0). \tag{1.351}$$

For the particular case where $G(f) = 1$, we obtain the required result

$$\sum_{n=-\infty}^{\infty} e^{-j2\pi fnT_0} = \frac{1}{T_0} \sum_{n=-\infty}^{\infty} \delta(f - n/T_0). \tag{1.352}$$

52. Clearly define the unit step function. Derive the Fourier transform of the unit step function.

- *Solution*: The unit step function is defined as

$$u(t) = \int_{\tau=-\infty}^{t} \delta(\tau)\,d\tau$$
$$= \begin{cases} 0 & \text{for } \tau < 0 \\ 1/2 & \text{for } \tau = 0 \\ 1 & \text{for } \tau > 0 \end{cases}, \qquad (1.353)$$

where $\delta(\cdot)$ is the Dirac-delta function. Now consider

$$g(t) = \begin{cases} \exp(-at) & \text{for } t > 0 \\ 0 & \text{for } t = 0, \\ -\exp(at) & \text{for } t < 0 \end{cases} \qquad (1.354)$$

where $a > 0$ is a real-valued constant. Note that

$$\lim_{a \to 0} g(t) = \text{sgn}\,(t), \qquad (1.355)$$

where

$$\text{sgn}\,(t) = \begin{cases} 1 & \text{for } t > 0 \\ 0 & \text{for } t = 0 \\ -1 & \text{for } t < 0. \end{cases} \qquad (1.356)$$

Now, the Fourier transform of $g(t)$ is

$$G(f) = \frac{-\text{j}\,4\pi f}{a^2 + (2\pi f)^2}. \qquad (1.357)$$

Taking the limit $a \to 0$ in (1.357), we get from (1.355)

$$\text{sgn}\,(t) \rightleftharpoons \frac{1}{\text{j}\pi f}. \qquad (1.358)$$

We know that

$$2u(t) - 1 = \text{sgn}\,(t). \qquad (1.359)$$

Taking the Fourier transform of both sides of (1.359) we get

$$2U(f) - \delta(f) = \frac{1}{j\pi f}$$

$$\Rightarrow U(f) = \frac{1}{j2\pi f} + \frac{\delta(f)}{2}, \qquad (1.360)$$

which is the Fourier transform of the unit step function.
Note that if

$$h(t) = \begin{cases} \exp(-at) & \text{for } t > 0 \\ 1/2 & \text{for } t = 0, \\ 0 & \text{for } t < 0 \end{cases} \qquad (1.361)$$

where $a > 0$ is a real-valued constant, then

$$\lim_{a \to 0} h(t) = u(t). \qquad (1.362)$$

Now, the Fourier transform of $h(t)$ in (1.361) is

$$H(f) = \frac{1}{a + j2\pi f}. \qquad (1.363)$$

Taking the limit $a \to 0$ in (1.363), we obtain from (1.362), the Fourier transform of the unit step as

$$u(t) \rightleftharpoons \frac{1}{j2\pi f}, \qquad (1.364)$$

which is different from (1.360). However, we note from (1.364) that the right-hand side is purely imaginary, which implies that the unit step is an odd function, which is incorrect. Therefore, the Fourier transform of the unit step is given by (1.360).

53. Compute

$$\int_{t=-\infty}^{\infty} \text{sinc}^3(3t)\, dt, \qquad (1.365)$$

where

$$\text{sinc}(t) = \frac{\sin(\pi t)}{\pi t}. \qquad (1.366)$$

• *Solution*: We know that

$$A\, \text{rect}(t/T) \rightleftharpoons AT\, \text{sinc}(fT). \qquad (1.367)$$

Using duality we get

$$A \operatorname{rect}(f/B) \rightleftharpoons AB \operatorname{sinc}(tB). \tag{1.368}$$

Substituting $B = 3$ in (1.368) we get

$$\frac{1}{3} \operatorname{rect}(f/3) \rightleftharpoons \operatorname{sinc}(3t). \tag{1.369}$$

Using the property that multiplication in the time domain corresponds to convolution in the frequency domain, we get

$$G(f) = \frac{1}{27} (\operatorname{rect}(f/3) \star \operatorname{rect}(f/3) \star \operatorname{rect}(f/3)) \rightleftharpoons \operatorname{sinc}^3(3t) = g(t), \tag{1.370}$$

where "$\star$" denotes convolution. Using the property

$$\int_{t=-\infty}^{\infty} g(t)\, dt = G(0), \tag{1.371}$$

we get

$$\begin{aligned}
\int_{t=-\infty}^{\infty} \operatorname{sinc}^3(3t)\, dt &= \frac{1}{27} \operatorname{rect}(f/3) \star 3\,[(1 - |f|/3) \operatorname{rect}(f/6)]_{f=0} \\
&= \frac{1}{9} \int_{\alpha=-3/2}^{3/2} (1 - |\alpha|/3)\, d\alpha \\
&= \frac{2}{9} \int_{\alpha=0}^{3/2} (1 - \alpha/3)\, d\alpha \\
&= \frac{1}{4}. \tag{1.372}
\end{aligned}$$

54. A signal $g(t)$ has Fourier transform given by

$$G(F) = \sin(2\pi F t_0). \tag{1.373}$$

Compute $\hat{g}(t)$.

- *Solution*: Clearly
$$g(t) = \frac{\delta(t + t_0) - \delta(t - t_0)}{2j}, \tag{1.374}$$

where $\delta(\cdot)$ is the Dirac-delta function. The impulse response of the Hilbert transformer is

$$h(t) = \frac{1}{\pi t}. \tag{1.375}$$

Therefore

$$\hat{g}(t) = g(t) \star h(t)$$

$$= \frac{1}{2j} \left(h(t + t_0) - h(t - t_0) \right)$$

$$= \frac{1}{2\pi j} \left(\frac{1}{t + t_0} - \frac{1}{t - t_0} \right)$$

$$= \frac{1}{\pi j} \left(\frac{-t_0}{t^2 - t_0^2} \right). \tag{1.376}$$

55. Compute the Hilbert transform of

$$g(t) = \frac{24}{9 + (2\pi t)^2}. \tag{1.377}$$

- *Solution*: We start from the Fourier transform pair:

$$a e^{-b|t|} \rightleftharpoons \frac{2ab}{b^2 + (2\pi f)^2}. \tag{1.378}$$

Using duality, we get

$$G(f) = a e^{-b|f|} \rightleftharpoons \frac{2ab}{b^2 + (2\pi t)^2} = g(t). \tag{1.379}$$

Here $a = 4$ and $b = 3$. In this problem, it is best to compute the Hilbert transform from the frequency domain. We have

$$\hat{G}(f) = -j \operatorname{sgn}(f) G(f). \tag{1.380}$$

Now $\hat{g}(t)$ is the inverse Fourier transform of $\hat{G}(f)$ as given by

$$\hat{g}(t) = \int_{f=-\infty}^{\infty} \hat{G}(f) e^{j 2\pi f t} \, df$$

$$= -4j \int_{f=0}^{\infty} e^{-3f} e^{j 2\pi f t} \, df$$

$$+ 4j \int_{f=-\infty}^{0} e^{3f} e^{j 2\pi f t} \, df$$

$$= -4j \left[\frac{e^{-f(3-j 2\pi t)}}{-(3 - j 2\pi t)} \right]_{f=0}^{\infty}$$

$$+ 4j \left[\frac{e^{f(3+j 2\pi t)}}{(3 + j 2\pi t)} \right]_{f=-\infty}^{0}$$

Fig. 1.36 A signal

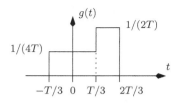

$$= 4j \left[\frac{-1}{(3 - j\,2\pi t)} \right]$$

$$+ 4j \left[\frac{1}{(3 + j\,2\pi t)} \right]$$

$$= \frac{16\pi t}{9 + (2\pi t)^2}. \tag{1.381}$$

56. Consider the signal $g(t)$ in Fig. 1.36. Which of the following statement(s) are correct?

(a)

$$g(t) \star g(t)|_{t=-T/3} = \frac{5}{48T}. \tag{1.382}$$

(b) The Fourier transform of $g(t) \star g(t)$ is real-valued and even.
(c) The $g(t) \star g(-t)$ is real-valued and even.
(d) $g(t) \star \delta(3t) = g(t)/3$.
  Here "$\star$" denotes convolution.

• *Solution*: Consider Fig. 1.37. Clearly

$$g(t) \star g(t)|_{t=-T/3} = \int_{\tau=-T/3}^{2T/3} g(\tau)g(-T/3 - \tau)\,d\tau$$

$$= \frac{5}{48T}. \tag{1.383}$$

Note that $g(t)$ is neither even nor odd. Hence, its Fourier transform $G(f)$ is complex valued. Therefore

$$g(t) \star g(t) \rightleftharpoons G^2(f). \tag{1.384}$$

Clearly, $G^2(f)$ is also complex valued. Next, since $g(t)$ is real-valued, $g(t) \star g(-t)$ is the autocorrelation of $g(t)$, and is hence real-valued and even. Finally

Fig. 1.37 A signal

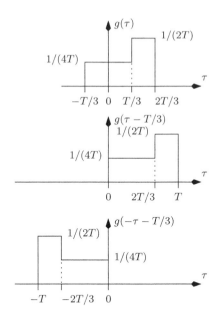

$$g(t) \star \delta(3t) = \int_{\tau=-\infty}^{\infty} \delta(3\tau)g(t-\tau)\,d\tau$$
$$= \int_{\alpha=-\infty}^{\infty} \frac{1}{3}\delta(\alpha)g(t-\alpha/3)\,d\alpha$$
$$= \frac{g(t)}{3}. \tag{1.385}$$

References

Simon Haykin. *Communication Systems*. Wiley Eastern, second edition, 1983.
Rodger E. Ziemer and William H. Tranter. *Principles of Communications*. John Wiley, fifth edition, 2002.

Chapter 2
Random Variables and Random Processes

1. Show that the magnitude of the correlation coefficient $|\rho|$ is always less than or equal to unity.

   - *Solution*: For any real number a, we have

   $$E\left[(a(X - m_X) - (Y - m_Y))^2\right] \geq 0$$
   $$\Rightarrow a^2\sigma_X^2 - 2aE\left[(X - m_X)(Y - m_Y)\right] + \sigma_Y^2 \geq 0. \qquad (2.1)$$

   Since the above quadratic equation in a is nonnegative, its discriminant is nonpositive. Hence

   $$4E^2\left[(X - m_X)(Y - m_Y)\right] - 4\sigma_X^2\sigma_Y^2 \leq 0$$
   $$\Rightarrow \frac{|E\left[(X - m_X)(Y - m_Y)\right]|}{\sigma_X\sigma_Y} \leq 1. \qquad (2.2)$$

   Hence proved.

2. (Papoulis 1991) Using characteristic functions, show that

   $$E[X_1X_2X_3X_4] = C_{12}C_{34} + C_{13}C_{24} + C_{14}C_{23}, \qquad (2.3)$$

   where $E[X_iX_j] = C_{ij}$ and the random variables X_i are jointly normal (Gaussian) with zero mean.

   - *Solution*: The joint characteristic function of X_1, X_2, X_3, and X_4 is given by

   $$E[e^{j(v_1X_1 + \cdots + v_4X_4)}]. \qquad (2.4)$$

   Expanding the exponential in the form of a power series and considering only the fourth power, we have

© The Editor(s) (if applicable) and The Author(s), under exclusive license
to Springer Nature Switzerland AG 2021
K. Vasudevan, *Analog Communications*,
https://doi.org/10.1007/978-3-030-50337-6_2

$$E\left[e^{j\,(v_1 X_1 + \cdots + v_4 X_4)}\right]$$

$$= \cdots + \frac{1}{4!}E\left[(v_1 X_1 + \cdots + v_4 X_4)^4\right] + \cdots$$

$$= \cdots + \frac{24}{4!}E[X_1 X_2 X_3 X_4]v_1 v_2 v_3 v_4 + \cdots. \tag{2.5}$$

Now, let

$$W = v_1 X_1 + \cdots + v_4 X_4. \tag{2.6}$$

Then

$$E[W] = 0$$
$$E[W^2] = \sum_{i,\,j} v_i v_j C_{ij} = \sigma_w^2. \tag{2.7}$$

Now, the characteristic function of W is

$$E\left[e^{j\,W}\right] = e^{-\sigma_w^2/2}$$
$$= e^{-\frac{1}{2}\sum_{i,\,j} v_i v_j C_{ij}}. \tag{2.8}$$

Expanding the above exponential once again we have

$$E\left[e^{j\,W}\right] = \cdots + \frac{1}{2!}\left(\frac{1}{2}\sum_{i,\,j} v_i v_j C_{ij}\right)^2 + \cdots$$

$$= \cdots + \frac{8}{8}\left(C_{12}C_{34} + C_{13}C_{24} + C_{14}C_{23}\right)v_1 v_2 v_3 v_4 + \cdots \tag{2.9}$$

Equating the coefficients in (2.5) and (2.9) we get the result.

3. (Haykin 1983) A Gaussian distributed random variable X of zero mean and variance σ_X^2 is transformed by a half-wave rectifier with input-output relation given below

$$Y = \begin{cases} X \text{ if } X \ge 0 \\ 0 \text{ if } X < 0. \end{cases} \tag{2.10}$$

Compute the pdf of Y.

- *Solution*: We start with the basic equation when a random variable X is transformed to a random variable Y through the relation

$$Y = g(X). \tag{2.11}$$

We note that the mapping between Y and X is one-to-one for $X \geq 0$ and an inverse mapping exists for $X \geq 0$

$$X = g^{-1}(Y). \tag{2.12}$$

Thus the following relation holds

$$f_Y(y) = \left. \frac{f_X(x)}{|dy/dx|} \right|_{X=g^{-1}(Y)}. \tag{2.13}$$

For the given problem

$$\frac{dy}{dx} = \begin{cases} 1 & \text{for } X > 0 \\ 0 & \text{for } X < 0. \end{cases} \tag{2.14}$$

Since $X < 0$ maps to $Y = 0$, we must have

$$P(X < 0) = P(Y = 0). \tag{2.15}$$

Thus the pdf of Y must have a Dirac delta function at $y = 0$. Let

$$f_Y(y) = \begin{cases} k\delta(y) + \frac{1}{\sqrt{2\pi\sigma_X^2}} \exp\left(-\frac{y^2}{2\sigma_X^2}\right) & \text{for } Y \geq 0 \\ 0 & \text{for } Y < 0, \end{cases} \tag{2.16}$$

where k is a constant such that the pdf of Y integrates to unity. It is easy to see that

$$\int_{y=-\infty}^{\infty} f_Y(y)\, dy = k + 1/2 = 1$$
$$\Rightarrow k = 1/2, \tag{2.17}$$

where we have used the fact that

$$\int_{y=-\infty}^{\infty} \delta(y)\, dy = 1. \tag{2.18}$$

4. Let X be a uniformly distributed random variable between 0 and 2π. Compute the pdf of $Y = \sin(X)$.

   • *Solution*: Clearly, the mapping between X and Y is not one-to-one. In fact, two distinct values of X give the same value of Y (excepting at the extreme points of $y = -1$ and $y = 1$). This is illustrated in Fig. 2.1. Let us consider two points x_1 and x_2 as shown in the figure. The probability that Y lies in the range y to $y + dy$ is equal to the probability that X lies in the range $[x_1, x_1 + dx_1]$ and $[x_2, x_2 + dx_2]$. Note that dx_2 is actually negative and

Fig. 2.1 The transformation
$Y = \sin(X)$

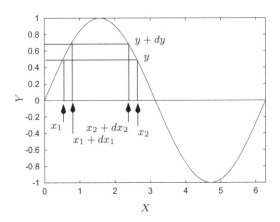

$$\frac{dy}{dx_1} \triangleq \frac{dy}{dx}\bigg|_{x=x_1} \tag{2.19}$$

Thus we have (since dx_2 is negative)

$$f_Y(y)\,dy = f_X(x_1)\,dx_1 - f_X(x_2)\,dx_2$$
$$\Rightarrow f_Y(y) = \frac{1/2\pi}{|dy/dx|}\bigg|_{x_1=\sin^{-1}(y)} + \frac{1/2\pi}{|dy/dx|}\bigg|_{x_2=\pi-\sin^{-1}(y)}$$
$$\Rightarrow f_Y(y) = \frac{1}{\pi\cos(x)}\bigg|_{x=\sin^{-1}(y)}$$
$$\Rightarrow f_Y(y) = \frac{1}{\pi\sqrt{1-y^2}} \qquad \text{for } -1 \le y \le 1. \tag{2.20}$$

Note that even though $f_Y(\pm 1) = \infty$, the probability that $Y = \pm 1$ is zero, since

$$P(Y = 1) = P(X = \pi/2)$$
$$= 0$$
$$P(Y = -1) = P(X = 3\pi/2)$$
$$= 0. \tag{2.21}$$

5. (Haykin 1983) Consider a random process $X(t)$ defined by

$$X(t) = \sin(2\pi F t) \tag{2.22}$$

in which the frequency F is a random variable with the probability density function

$$f_F(f) = \begin{cases} 1/W & \text{for } 0 \le f \le W \\ 0 & \text{elsewhere.} \end{cases} \tag{2.23}$$

Is $X(t)$ WSS?

- *Solution*: Let us first compute the mean value of $X(t)$.

$$E[X(t)] = \frac{1}{W} \int_{f=0}^{W} \sin(2\pi F t) \, dF$$

$$= \frac{-1}{2W\pi t} (\cos(2\pi W t) - 1) \tag{2.24}$$

which is a function of time. Hence $X(t)$ is not WSS.

6. (Haykin 1983) A random process $X(t)$ is defined by

$$X(t) = A \cos(2\pi f_c t), \tag{2.25}$$

where A is a Gaussian distributed random variable with zero mean and variance σ_A^2. This random process is applied to an ideal integrator producing an output $Y(t)$ defined by

$$Y(t) = \int_{\tau=0}^{t} X(\tau) \, d\tau. \tag{2.26}$$

(a) Determine the probability density function of the output $Y(t)$ at a particular time t_k.
(b) Determine whether $Y(t)$ is WSS.
(c) Determine whether $Y(t)$ is ergodic in the mean and in the autocorrelation.

- *Solution*: The random variable $Y(t_k)$ is given by

$$Y(t_k) = \frac{A}{2\pi f_c} \sin(2\pi f_c t_k). \tag{2.27}$$

Since all the terms in the above equation excepting A are constants for the time instant t_k, $Y(t_k)$ is also a Gaussian distributed random variable with mean and variance given by

$$E[Y(t_k)] = \frac{E[A]}{2\pi f_c} \sin(2\pi f_c t_k) = 0$$

$$E[Y^2(t_k)] = \frac{E[A^2]}{4\pi^2 f_c^2} \sin^2(2\pi f_c t_k)$$

$$= \frac{\sigma_A^2}{4\pi^2 f_c^2} \sin^2(2\pi f_c t_k). \tag{2.28}$$

Since the variance is a function of time, $Y(t_k)$ is not WSS. Hence it is not ergodic in the autocorrelation. However, it can be shown that the time-averaged mean is zero. Hence $Y(t)$ is ergodic in the mean.

7. (Haykin 1983) Let X and Y be statistically independent Gaussian random variables with zero mean and unit variance. Define the Gaussian process

$$Z(t) = X \cos(2\pi t) + Y \sin(2\pi t). \tag{2.29}$$

(a) Determine the joint pdf of the random variables $Z(t_1)$ and $Z(t_2)$.
(b) Is $Z(t)$ WSS?

- *Solution*: The random process $Z(t)$ has mean and autocorrelation given by

$$
\begin{aligned}
E[Z(t)] &= E[X]\cos(2\pi t) + E[Y]\sin(2\pi t) = 0 \\
E[Z(t_1)Z(t_2)] &= E[(X\cos(2\pi t_1) + Y\sin(2\pi t_2)) \\
&\quad (X\cos(2\pi t_1) + Y\sin(2\pi t_2))] \\
&= \cos(2\pi t_1)\cos(2\pi t_2) + \sin(2\pi t_1)\sin(2\pi t_2) \\
&= \cos(2\pi(t_1 - t_2)) \\
\Rightarrow E[Z^2(t)] &= 1. \tag{2.30}
\end{aligned}
$$

In the above relations we have used the fact that

$$
\begin{aligned}
E[X] &= E[Y] = 0 \\
E[X^2] &= E[Y^2] = 1 \\
E[XY] &= E[X]E[Y] = 0. \tag{2.31}
\end{aligned}
$$

Since the mean is independent of time and the autocorrelation is dependent only on the time lag, $Z(t)$ is WSS.
Since $Z(t)$ is a linear combination of two Gaussian random variables, it is also Gaussian. Therefore $Z(t_1)$ and $Z(t_2)$ are jointly Gaussian.
The joint pdf of two real-valued Gaussian random variables y_1 and y_2 is given by

$$
\begin{aligned}
&f_{Y_1, Y_2}(y_1, y_2) \\
&= \frac{1}{2\pi\sigma_1\sigma_2\sqrt{1 - \rho^2}} \\
&\quad \times \exp\left[-\frac{\sigma_2^2(y_1 - m_1)^2 - 2\sigma_1\sigma_2\rho(y_1 - m_1)(y_2 - m_2) + \sigma_1^2(y_2 - m_2)^2}{2\sigma_1^2\sigma_2^2(1 - \rho^2)}\right],
\end{aligned}
$$
$$\tag{2.32}$$

where

$$E[y_1] = m_1$$
$$E[y_2] = m_2$$
$$E[(y_1 - m_1)^2] = \sigma_1^2$$
$$E[(y_2 - m_2)^2] = \sigma_2^2$$
$$E[(y_1 - m_1)(y_2 - m_2)]/(\sigma_1\sigma_2) = \rho. \tag{2.33}$$

Thus, for the given problem

$$p_{Z(t_1), Z(t_2)}(z(t_1), z(t_2))$$

$$= \frac{1}{2\pi|\sin(2\pi(t_1 - t_2))|}$$

$$\times \exp\left[-\frac{z^2(t_1) - 2\cos(2\pi(t_1 - t_2))z(t_1)z(t_2) + z^2(t_2)}{2\sin^2(2\pi(t_1 - t_2))}\right]. \tag{2.34}$$

8. Using ensemble averaging, find the mean and the autocorrelation of the random process given by

$$X(t) = \sum_{k=-\infty}^{\infty} S_k p(t - kT - \alpha), \tag{2.35}$$

where S_k denotes a discrete random variable taking values $\pm A$ with equal probability, $p(\cdot)$ denotes a real-valued waveform, $1/T$ denotes the bit-rate and α denotes a random timing phase uniformly distributed in $[0, T)$. Assume that S_k is independent of α and S_k is independent of S_j for $k \neq j$.

• Solution: Since S_k and α are independent

$$E[X(t)] = \sum_{k=-\infty}^{\infty} E[S_k]E[p(t - kT - \alpha)]$$

$$, = 0 \tag{2.36}$$

where for the given binary phase shift keying (BPSK) constellation

$$E[S_k] = (1/2)A + (1/2)(-A)$$
$$= 0. \tag{2.37}$$

The autocorrelation of $X(t)$ is

$$R_X(\tau) = E[X(t)X(t - \tau)]$$

$$= E\left[\sum_{k=-\infty}^{\infty} S_k p(t - kT - \alpha) \sum_{j=-\infty}^{\infty} S_j p(t - \tau - jT - \alpha)\right]$$

$$= \sum_{k=-\infty}^{\infty} \sum_{j=-\infty}^{\infty} E[S_k S_j] E[p(t - kT - \alpha) p(t - \tau - jT - \alpha)]$$

$$= \sum_{k=-\infty}^{\infty} \sum_{j=-\infty}^{\infty} A^2 \delta_K(k - j)$$

$$\times \frac{1}{T} \int_{\alpha=0}^{T} p(t - kT - \alpha) p(t - \tau - jT - \alpha) \, d\alpha$$

$$= \sum_{k=-\infty}^{\infty} \frac{A^2}{T} \int_{\alpha=0}^{T} p(t - kT - \alpha) p(t - \tau - kT - \alpha) \, d\alpha, \quad (2.38)$$

where we have assumed that S_k and α arc independent and

$$\delta_K(k - j) = \begin{cases} 1 \text{ for } k = j \\ 0 \text{ for } k \neq j \end{cases} \quad (2.39)$$

is the Kronecker delta function. Let

$$x = t - kT - \alpha. \quad (2.40)$$

Substituting (2.40) in (2.38) we obtain

$$R_X(\tau) = \sum_{k=-\infty}^{\infty} \frac{A^2}{T} \int_{x=t-kT-T}^{t-kT} p(x) p(x - \tau) \, dx. \quad (2.41)$$

Combining the summation and the integral, (2.41) becomes

$$R_X(\tau) = \frac{A^2}{T} \int_{x=-\infty}^{\infty} p(x) p(x - \tau) \, dx$$

$$= \frac{A^2}{T} R_p(\tau), \quad (2.42)$$

where $R_p(\cdot)$ is the autocorrelation of $p(\cdot)$.

9. (Haykin 1983) The square wave $x(t)$ of constant amplitude A, period T_0, and delay t_d represents the sample function of a random process $X(t)$. This is illustrated in Fig. 2.2. The delay is random and is described by the pdf

Fig. 2.2 A periodic waveform with a random timing phase

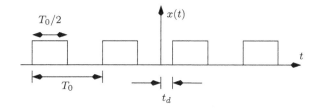

$$f_{T_d}(t_d) = \begin{cases} 1/T_0 & \text{for } 0 \le t_d < T_0 \\ 0 & \text{otherwise.} \end{cases} \tag{2.43}$$

(a) Determine the mean and autocorrelation of $X(t)$ using ensemble averaging.
(b) Determine the mean and autocorrelation of $X(t)$ using time averaging.
(c) Is $X(t)$ WSS?
(d) Is $X(t)$ ergodic in the mean and the autocorrelation?

- *Solution*: Instead of solving this particular problem we try to solve a more general problem. Let $p(t)$ denote an arbitrary pulse shape. Consider a random process $X(t)$ defined by

$$X(t) = \sum_{k=-\infty}^{\infty} p(t - kT_0 - t_d), \tag{2.44}$$

where t_d is a uniformly distributed random variable in the range $[0, T_0)$. Clearly, in the absence of t_d, $X(t)$ is no longer a random process and it simply becomes a periodic waveform. We wish to find out the mean and autocorrelation of the random process defined in (2.44). The mean of $X(t)$ is given by

$$E[X(t)] = E\left[\sum_{k=-\infty}^{\infty} p(t - kT_0 - t_d)\right]$$

$$= \frac{1}{T_0} \int_{t_d=0}^{T_0} \sum_{k=-\infty}^{\infty} p(t - kT_0 - t_d) \, dt_d. \tag{2.45}$$

Interchanging the order of integration and summation and substituting

$$t - kT_0 - t_d = x \tag{2.46}$$

we get

$$E[X(t)] = \frac{1}{T_0} \sum_{k=-\infty}^{\infty} \int_{x=t-kT_0-T_0}^{t-kT_0} p(x) \, dx. \tag{2.47}$$

Combining the summation and the integral we get

$$E[X(t)] = \frac{1}{T_0} \int_{x=-\infty}^{\infty} p(x)\, dx. \tag{2.48}$$

For the given problem:

$$E[X(t)] = \frac{A}{2}. \tag{2.49}$$

The autocorrelation can be computed as

$$E[X(t)X(t-\tau)] = E\left[\sum_{k=-\infty}^{\infty} p(t - kT_0 - t_d) \right.$$

$$\left. \sum_{j=-\infty}^{\infty} p(t - \tau - jT_0 - t_d) \right]$$

$$= \frac{1}{T_0} \sum_{k=-\infty}^{\infty} \sum_{j=-\infty}^{\infty}$$

$$\int_{t_d=0}^{T_0} p(t - kT_0 - t_d)p(t - \tau - jT_0 - t_d)\, dt_d. \tag{2.50}$$

Substituting

$$t - kT_0 - t_d = x \tag{2.51}$$

we get

$$E[X(t)X(t-\tau)] = \frac{1}{T_0} \sum_{k=-\infty}^{\infty} \sum_{j=-\infty}^{\infty}$$

$$\int_{x=t-kT_0-T_0}^{t-kT_0} p(x)p(x + kT_0 - \tau - jT_0)\, dx. \tag{2.52}$$

Let

$$kT_0 - jT_0 = mT_0. \tag{2.53}$$

Substituting for j we get

$$E[X(t)X(t-\tau)] = \frac{1}{T_0} \sum_{k=-\infty}^{\infty} \sum_{m=-\infty}^{\infty}$$

$$\int_{x=t-kT_0-T_0}^{t-kT_0} p(x)p(x+mT_0-\tau)\,dx. \quad (2.54)$$

Now we interchange the order of summation and combine the summation over k and the integral to obtain

$$E[X(t)X(t-\tau)] = \frac{1}{T_0} \sum_{m=-\infty}^{\infty} \int_{x=-\infty}^{\infty} p(x)p(x+mT_0-\tau)\,dx$$

$$= \frac{1}{T_0} \sum_{m=-\infty}^{\infty} R_p(\tau-mT_0) = R_X(\tau), \quad (2.55)$$

where $R_p(\tau)$ is the autocorrelation of $p(t)$. Thus, the autocorrelation of $X(t)$ is also periodic with a period T_0, hence $X(t)$ is a cyclostationary random process. This is illustrated in Fig. 2.3c.

Since the random process is periodic, the time-averaged mean is given by

$$< x(t) > = \frac{1}{T_0} \int_{t=t_d}^{t_d+T_0} x(t)\,dt$$

$$= \frac{A}{2} \quad (2.56)$$

independent of t_d. Comparing (2.49) and (2.56) we find that $X(t)$ is ergodic in the mean.

The time-averaged autocorrelation is given by

$$< x(t)x(t-\tau) > = \frac{1}{T_0} \int_{t=-T_0/2}^{T_0/2} x(t)x(t-\tau)\,dt$$

$$= \frac{1}{T_0} \sum_{m=-\infty}^{\infty} R_g(\tau-mT_0), \quad (2.57)$$

where $R_g(\tau)$ is the autocorrelation of the generating function of $x(t)$ (the generating function has been discussed earlier in this chapter of Haykin 2nd ed). The generating function can be conveniently taken to be

$$g(t) = \begin{cases} p(t-t_d) & \text{for } t_d \le t < t_d + T_0 \\ 0 & \text{elsewhere} \end{cases}, \quad (2.58)$$

where $p(t)$ is illustrated in Fig. 2.3a. Let $P(f)$ be the Fourier transform of $p(t)$. We have

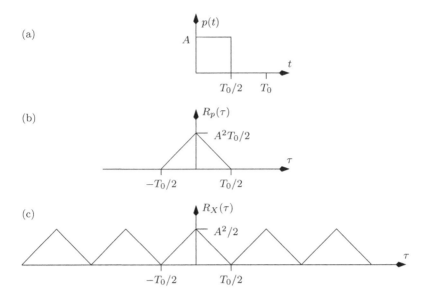

Fig. 2.3 Computing the autocorrelation of a periodic wave with random timing phase

Fig. 2.4 A periodic
waveform with a random
timing phase

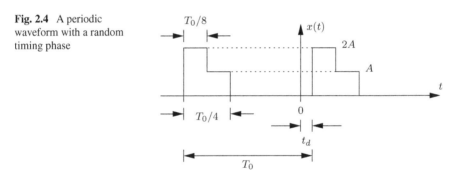

$$R_g(\tau) = g(t) \star g(-t) \rightleftharpoons |P(f)|^2 \rightleftharpoons p(t) \star p(-t) = R_p(\tau). \quad (2.59)$$

Therefore, comparing (2.55) and (2.57) we find that $X(t)$ is ergodic in the autocorrelation.

10. A signal $x(t)$ with period T_0 and delay t_d represents the sample function of a random process $X(t)$. This is illustrated in Fig. 2.4. The delay t_d is a random variable which is uniformly distributed in $[0, T_0)$.

(a) Determine the mean and autocorrelation of $X(t)$ using ensemble averaging.
(b) Determine the mean and autocorrelation of $X(t)$ using time averaging.
(c) Is $X(t)$ WSS?
(d) Is $X(t)$ ergodic in the mean and the autocorrelation?

- *Solution*: Instead of solving this particular problem we try to solve a more general problem. Let $p(t)$ denote an arbitrary pulse shape. Consider a random process $X(t)$ defined by

$$X(t) = \sum_{k=-\infty}^{\infty} p(t - kT_0 - t_d), \qquad (2.60)$$

where t_d is a uniformly distributed random variable in the range $[0, T_0)$. Clearly, in the absence of t_d, $X(t)$ is no longer a random process and it simply becomes a periodic waveform. We wish to find out the mean and autocorrelation of the random process defined in (2.60). The mean of $X(t)$ is given by

$$E[X(t)] = E\left[\sum_{k=-\infty}^{\infty} p(t - kT_0 - t_d)\right]$$

$$= \frac{1}{T_0} \int_{t_d=0}^{T_0} \sum_{k=-\infty}^{\infty} p(t - kT_0 - t_d) \, dt_d. \qquad (2.61)$$

Interchanging the order of integration and summation and substituting

$$t - kT_0 - t_d = x \qquad (2.62)$$

we get

$$E[X(t)] = \frac{1}{T_0} \sum_{k=-\infty}^{\infty} \int_{x=t-kT_0-T_0}^{t-kT_0} p(x) \, dx. \qquad (2.63)$$

Combining the summation and the integral we get

$$E[X(t)] = \frac{1}{T_0} \int_{x=-\infty}^{\infty} p(x) \, dx. \qquad (2.64)$$

For the given problem

$$E[X(t)] = \frac{3A}{8}. \qquad (2.65)$$

The autocorrelation can be computed as

$$E[X(t)X(t - \tau)] = E\left[\sum_{k=-\infty}^{\infty} p(t - kT_0 - t_d)\right.$$

$$\sum_{j=-\infty}^{\infty} p(t - \tau - jT_0 - t_d) \Bigg]$$

$$= \frac{1}{T_0} \sum_{k=-\infty}^{\infty} \sum_{j=-\infty}^{\infty}$$

$$\int_{t_d=0}^{T_0} p(t - kT_0 - t_d) p(t - \tau - jT_0 - t_d) \, dt_d \quad (2.66)$$

Substituting

$$t - kT_0 - t_d = x \tag{2.67}$$

we get

$$E[X(t)X(t-\tau)] = \frac{1}{T_0} \sum_{k=-\infty}^{\infty} \sum_{j=-\infty}^{\infty}$$

$$\int_{x=t-kT_0-T_0}^{t-kT_0} p(x) p(x + kT_0 - \tau - jT_0) \, dx. \tag{2.68}$$

Let

$$kT_0 - jT_0 = mT_0. \tag{2.69}$$

Substituting for j we get

$$E[X(t)X(t-\tau)] = \frac{1}{T_0} \sum_{k=-\infty}^{\infty} \sum_{m=-\infty}^{\infty}$$

$$\int_{x=t-kT_0-T_0}^{t-kT_0} p(x) p(x + mT_0 - \tau) \, dx. \tag{2.70}$$

Now we interchange the order of summation and combine the summation over k and the integral to obtain

$$E[X(t)X(t-\tau)] = \frac{1}{T_0} \sum_{m=-\infty}^{\infty} \int_{x=-\infty}^{\infty} p(x) p(x + mT_0 - \tau) \, dx$$

$$= \frac{1}{T_0} \sum_{m=-\infty}^{\infty} R_p(\tau - mT_0) = R_X(\tau), \tag{2.71}$$

where $R_p(\tau)$ is the autocorrelation of $p(t)$. Thus, the autocorrelation of $X(t)$ is also periodic with a period T_0, hence $X(t)$ is a cyclostationary random

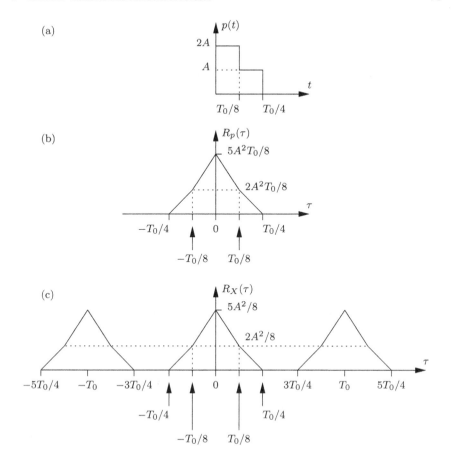

Fig. 2.5 Computing the autocorrelation of a periodic wave with random timing phase

process. This is illustrated in Fig. 2.5c.

Since the random process is periodic, the time-averaged mean is given by

$$< x(t) > = \frac{1}{T_0} \int_{t=t_d}^{t_d+T_0} x(t) \, dt$$

$$= \frac{3A}{8} \tag{2.72}$$

independent of t_d. Comparing (2.65) and (2.72) we find that $X(t)$ is ergodic in the mean.

The time-averaged autocorrelation is given by

$$< x(t)x(t - \tau) > = \frac{1}{T_0} \int_{t=-T_0/2}^{T_0/2} x(t)x(t - \tau) \, dt$$

$$= \frac{1}{T_0} \sum_{m=-\infty}^{\infty} R_g(\tau - mT_0), \qquad (2.73)$$

where $R_g(\tau)$ is the autocorrelation of the generating function of $x(t)$ (the generating function has been discussed earlier in this chapter of Haykin 2nd ed). The generating function can be conveniently taken to be

$$g(t) = \begin{cases} p(t - t_d) & \text{for } t_d \leq t < t_d + T_0 \\ 0 & \text{elsewhere} \end{cases} \qquad (2.74)$$

where $p(t)$ is illustrated in Fig. 2.5a. Let $P(f)$ be the Fourier transform of $p(t)$. We have

$$R_g(\tau) = g(t) \star g(-t) \rightleftharpoons |P(f)|^2 \rightleftharpoons p(t) \star p(-t) = R_p(\tau). \quad (2.75)$$

Therefore, comparing (2.71) and (2.73) we find that $X(t)$ is ergodic in the autocorrelation.

11. For two jointly Gaussian random variables X and Y with means m_X and m_Y, variances σ_X^2 and σ_Y^2 and coefficient of correlation ρ, compute the conditional pdf of X given Y. Hence compute the values of $E[X|Y]$ and $\text{var}(X|Y)$.

- *Solution*: The joint pdf of two real-valued Gaussian random variables X and Y is given by

$$f_{X,Y}(x, y) = \frac{1}{2\pi\sigma_X\sigma_Y\sqrt{1 - \rho^2}}$$
$$\times \exp\left[-\frac{\sigma_Y^2(x - m_X)^2 - 2\sigma_X\sigma_Y\rho(x - m_X)(y - m_Y) + \sigma_X^2(y - m_Y)^2}{2\sigma_X^2\sigma_Y^2(1 - \rho^2)}\right].$$
$$(2.76)$$

We also know that

$$f_Y(y) = \frac{1}{\sigma_Y\sqrt{2\pi}} \exp\left(-\frac{(y - m_Y)^2}{2\sigma_Y^2}\right). \qquad (2.77)$$

Therefore

$$f_{X|Y}(x|y) = \frac{f_{XY}(x, y)}{f_Y(y)}$$

$$= \frac{1}{\sigma_X\sqrt{2\pi(1-\rho^2)}}$$

$$\times \exp\left[-\frac{\sigma_Y^2 A^2 - 2\sigma_X\sigma_Y\rho AB + \rho^2\sigma_X^2 B^2}{2\sigma_X^2\sigma_Y^2(1-\rho^2)}\right], \qquad (2.78)$$

where we have made the substitution

$$A = x - m_X$$
$$B = y - m_Y. \qquad (2.79)$$

Simplifying (2.78) further we get

$$f_{X|Y}(x|y) = \frac{1}{\sigma_X\sqrt{2\pi(1-\rho^2)}}$$

$$\times \exp\left[-\frac{(\sigma_Y(x - m_X) - \rho\sigma_X(y - m_Y))^2}{2\sigma_X^2\sigma_Y^2(1-\rho^2)}\right]$$

$$= \frac{1}{\sigma_X\sqrt{2\pi(1-\rho^2)}}$$

$$\times \exp\left[-\frac{(x - (m_X + \rho(\sigma_X/\sigma_Y)(y - m_Y)))^2}{2\sigma_X^2(1-\rho^2)}\right]. \qquad (2.80)$$

By inspection we conclude from the above equation that $X|Y$ is also a Gaussian random variable with mean and variance given by

$$E[X|Y] = m_X + \rho\frac{\sigma_X}{\sigma_Y}(y - m_Y)$$

$$\text{var}(X|Y) = \sigma_X^2(1 - \rho^2). \qquad (2.81)$$

12. If

$$f_X(x) = \frac{1}{\sigma\sqrt{2\pi}}\exp\left(-\frac{(x-m)^2}{2\sigma^2}\right) \qquad (2.82)$$

compute (a) $E\left[(X-m)^{2n}\right]$ and (b) $E\left[(X-m)^{2n-1}\right]$.

• *Solution*: Clearly

$$E\left[(X-m)^{2n-1}\right] = 0 \qquad (2.83)$$

since the integrand is an odd function. To compute $E[(X-m)^{2n}]$ we use the method of integration by parts. We have

$$E\left[(X-m)^{2n}\right] = \frac{1}{\sigma\sqrt{2\pi}} \int_{x=-\infty}^{\infty} (x-m)^{2n-1}(x-m)e^{-(x-m)^2/(2\sigma^2)}\,dx.$$

$$(2.84)$$

The first function is taken as $(x-m)^{2n-1}$. The integral of the second function is

$$\int_x (x-m)\exp\left(-\frac{(x-m)^2}{2\sigma^2}\right)dx = -\sigma^2\exp\left(-\frac{(x-m)^2}{2\sigma^2}\right). \quad (2.85)$$

Therefore (2.84) becomes

$$E\left[(X-m)^{2n}\right]$$
$$= \frac{1}{\sigma\sqrt{2\pi}}\left[-\sigma^2(x-m)^{2n-1}\exp\left(-\frac{(x-m)^2}{2\sigma^2}\right)\Big|_{x=-\infty}^{\infty}\right.$$
$$\left. - \int_{x=-\infty}^{\infty}(2n-1)(x-m)^{2n-2}(-\sigma^2)\exp\left(-\frac{(x-m)^2}{2\sigma^2}\right)dx\right]$$
$$= (2n-1)\sigma^2 E\left[(x-m)^{2n-2}\right]. \qquad (2.86)$$

Using the above recursion we get

$$E\left[(X-m)^{2n}\right] = 1\cdot 3\cdot 5\cdots(2n-1)\sigma^{2n}. \qquad (2.87)$$

13. X is a uniformly distributed random variable between zero and one. Find out the transformation $Y = g(X)$ such that Y is a Rayleigh distributed random variable. You can assume that the mapping between X and Y is one-to-one and Y monotonically increases with X. There should not be any unknown constants in your answer. The Rayleigh pdf is given by

$$f_Y(y) = \frac{y}{\sigma^2}\exp\left(-\frac{y^2}{2\sigma^2}\right) \qquad \text{for } y \geq 0. \qquad (2.88)$$

Note that $f_Y(y)$ is zero for $y < 0$.

- *Solution*: Note that

$$f_X(x) = \begin{cases} 1 \text{ for } 0 \leq x \leq 1 \\ 0 \text{ elsewhere} \end{cases}. \qquad (2.89)$$

Since the mapping is assumed to be one-to-one and monotonically increasing $(dy/dx$ is always positive), we have

$$f_Y(y)\,dy = f_X(x)\,dx$$
$$\Rightarrow \int_{y=0}^{Y} f_Y(y)\,dy = \int_0^X f_X(x)\,dx. \qquad (2.90)$$

Substituting for $f_Y(y)$ and $f_X(x)$ we get

$$1 - e^{-Y^2/(2\sigma^2)} = X$$
$$\Rightarrow Y^2 = -2\sigma^2 \ln(1 - X). \tag{2.91}$$

14. Given that Y is a Gaussian distributed random variable with zero mean and variance σ^2, use the Chernoff bound to compute the probability that $Y \geq \delta$ for $\delta > 0$.

   - *Solution*: We know that

$$P[Y \geq \delta] \leq E\left[e^{\mu(Y-\delta)}\right], \tag{2.92}$$

   where μ needs to be found out such that the RHS is minimized. Hence we set

$$\frac{d}{d\mu} E\left[e^{\mu(Y-\delta)}\right] = 0. \tag{2.93}$$

   The optimum value of $\mu = \mu_0$ satisfies

$$E\left[Ye^{\mu_0 Y}\right] = \delta E\left[e^{\mu_0 Y}\right]$$
$$\Rightarrow \frac{1}{\sigma\sqrt{2\pi}} \int_{y=-\infty}^{\infty} ye^{\mu_0 y}e^{-y^2/2\sigma^2}\, dy = \frac{\delta}{\sigma\sqrt{2\pi}} \int_{y=-\infty}^{\infty} e^{\mu_0 y}e^{-y^2/2\sigma^2}\, dy$$
$$\Rightarrow \mu_0 = \frac{\delta}{\sigma^2}. \tag{2.94}$$

   Therefore

$$P[Y \geq \delta] \leq e^{-\mu_0\delta} E\left[e^{\mu_0 Y}\right]$$
$$\Rightarrow P[Y \geq \delta] \leq e^{-\delta^2/\sigma^2} e^{\delta^2/(2\sigma^2)}$$
$$\Rightarrow P[Y \geq \delta] \leq e^{-\delta^2/(2\sigma^2)}. \tag{2.95}$$

15. If two Gaussian distributed random variables, X and Y are uncorrelated, are they statistically independent? Justify your statement.

   - *Solution*: Since X and Y are uncorrelated

$$E[(X - m_x)(Y - m_Y)] = 0$$
$$\Rightarrow \rho = E[(X - m_X)(Y - m_Y)]/\sigma_X\sigma_Y = 0. \tag{2.96}$$

   Therefore the joint pdf is given by

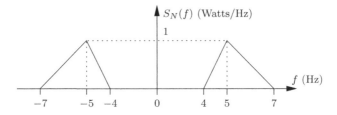

Fig. 2.6 Power spectral density of a narrowband noise process

$$f_{XY}(x, y) = \frac{1}{\sigma_X \sqrt{2\pi}} e^{-(x-m_X)^2/(2\sigma_X^2)} \frac{1}{\sigma_Y \sqrt{2\pi}} e^{-(y-m_Y)^2/(2\sigma_Y^2)}$$
$$= f_X(x) f_Y(y). \tag{2.97}$$

Hence X and Y are also statistically independent.

16. (Haykin 1983) The power spectral density of a narrowband noise process is shown in Fig. 2.6. The carrier frequency is 5 Hz.

   (a) Plot the power spectral density of $N_c(t)$ and $N_s(t)$.
   (b) Plot the cross-spectral density $S_{N_c N_s}(f)$.

   • *Solution*: We know that

$$S_{N_c}(f) = S_{N_s}(f) = \begin{cases} S_N(f - f_c) + S_N(f + f_c) & \text{for } |f| < f_c \\ 0 & \text{otherwise.} \end{cases} \tag{2.98}$$

   The psd of the in-phase and quadrature components is plotted in Fig. 2.7. We also know that the cross-spectral densities are given by

$$S_{N_c N_s}(f) = \begin{cases} j[S_N(f + f_c) - S_N(f - f_c)] & \text{for } |f| < f_c \\ 0 & \text{otherwise.} \end{cases}$$
$$= -S_{N_s N_c}(f). \tag{2.99}$$

   The cross-spectral density $S_{N_c N_s}(f)$ is plotted in Fig. 2.8.

17. Let a random variable X have a uniform pdf over $-1 \leq x \leq 3$. Compute the pdf of $Z = \sqrt{|X|}$. The transformation is plotted in Fig. 2.9.

   • *Solution*: Using z and x as dummy variables corresponding to Z and X, respectively, we note that

$$z^2 = \begin{cases} x & \text{for } x > 0 \\ -x & \text{for } x < 0. \end{cases} \tag{2.100}$$

   Differentiating both sides we get

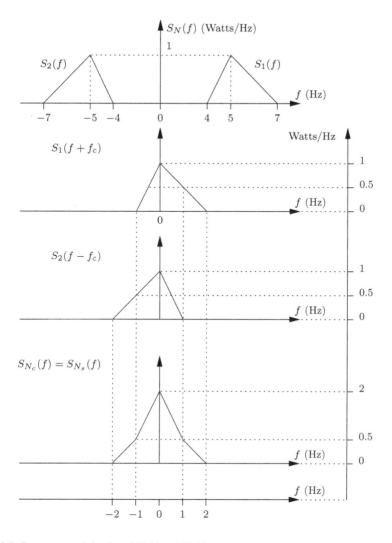

Fig. 2.7 Power spectral density of $N_c(t)$ and $N_s(t)$

$$2z\,dz = \begin{cases} dx & \text{for } x > 0 \\ -dx & \text{for } x < 0. \end{cases} \tag{2.101}$$

This implies that

$$\frac{dz}{dx} = \begin{cases} 1/(2z) & \text{for } x > 0 \\ -1/(2z) & \text{for } x < 0 \end{cases}$$

$$\Rightarrow \left| \frac{dz}{dx} \right| = \frac{1}{2z}. \tag{2.102}$$

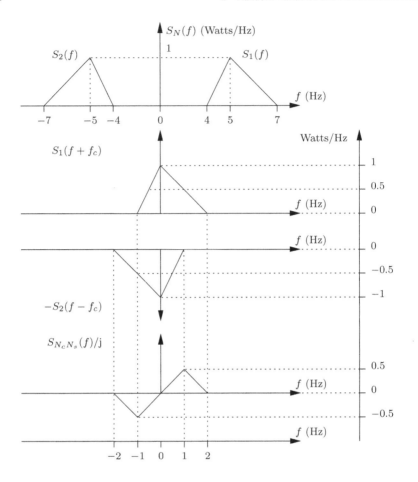

Fig. 2.8 Cross-spectral densities of $N_c(t)$ and $N_s(t)$

Now, in the range $-1 \leq x \leq 1$ corresponding to $0 \leq z \leq 1$, the mapping from X to Z is surjective (many-to-one) as indicated in Fig. 2.9. Hence

$$f_Z(z) = \left.\frac{f_X(x)}{|dz/dx|}\right|_{x=-z^2} + \left.\frac{f_X(x)}{|dz/dx|}\right|_{x=z^2} \qquad \text{for } 0 \leq z \leq 1$$

$$\Rightarrow f_Z(z) = 2\frac{1/4}{1/(2z)}$$

$$= z \qquad \text{for } 0 \leq z \leq 1. \tag{2.103}$$

In the range $1 \leq x \leq 3$ corresponding to $1 \leq z \leq \sqrt{3}$ the mapping from X to Z is one-to-one (injective). Hence

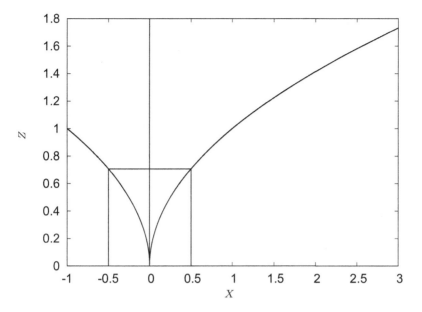

Fig. 2.9 The transformation $Z = \sqrt{|X|}$

$$f_Z(z) = \frac{f_X(x)}{|dz/dx|}\Big|_{x=z^2}$$

$$\Rightarrow f_Z(z) = \frac{1/4}{1/(2z)}$$

$$= \frac{z}{2} \quad \text{for } 1 \le z \le \sqrt{3}. \tag{2.104}$$

To summarize

$$f_Z(z) = \begin{cases} z & \text{for } 0 \le z \le 1 \\ \frac{z}{2} & \text{for } 1 \le z \le \sqrt{3}. \end{cases} \tag{2.105}$$

18. Given that

$$F_X(x) = \frac{\pi + 2\tan^{-1}(x)}{2\pi} \quad \text{for } -\infty < x < \infty \tag{2.106}$$

find the pdf of the random variable Z given by

$$Z = \begin{cases} X^2 & \text{for } X \ge 0 \\ -1 & \text{for } X < 0 \end{cases} \tag{2.107}$$

- *Solution*: Note that

$$F_X(-\infty) = 0$$
$$F_X(\infty) = 1. \tag{2.108}$$

The pdf of X is given by

$$f_X(x) = \frac{d F_X(x)}{dx}$$
$$= \frac{1}{\pi(1 + x^2)}. \tag{2.109}$$

Note that

$$\frac{dz}{dx} = \begin{cases} 2x & \text{for } x \geq 0 \\ 0 & \text{for } x < 0 \end{cases}$$
$$\Rightarrow \frac{dz}{dx} = \begin{cases} 2\sqrt{z} & \text{for } x \geq 0 \\ 0 & \text{for } x < 0. \end{cases} \tag{2.110}$$

We also note that

$$P(Z = -1) = P(X < 0) = F_X(0) = 1/2. \tag{2.111}$$

Therefore $f_Z(z)$ must have a delta function at $Z = -1$. Further, the mapping from X to Z is one-to-one (injective) in the range $0 \leq x < \infty$. Hence, the pdf of Z is given by

$$f_Z(z) = \begin{cases} k\delta(z + 1) & \text{for } z = -1 \\ f_X(x)/(dz/dx)|_{x=\sqrt{z}} & \text{for } 0 \leq z < \infty, \end{cases} \tag{2.112}$$

where k is a constant such that the pdf of Z integrates to unity. Simplifying the above expression we get

$$f_Z(z) = \begin{cases} k\delta(z + 1) & \text{for } z = -1 \\ 1/(2\pi(1 + z)\sqrt{z}) & \text{for } 0 \leq z < \infty. \end{cases} \tag{2.113}$$

Since

$$\int_{z=0}^{\infty} f_Z(z)\, dz = P(Z > 0) = P(X > 0) = 1 - F_X(0) = 1/2 \tag{2.114}$$

we must have $k = 1/2$.

19. (Papoulis 1991) Let X be a nonnegative continuous random variable and let a be any positive constant. Prove the Markov inequality given by

Fig. 2.10 Illustration of the
Markov's inequality

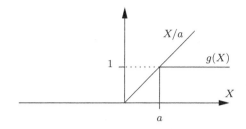

Fig. 2.11 The
transformation $Y = g(X)$

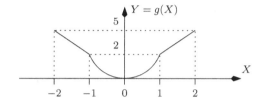

$$P(X \geq a) \leq m_X/a, \qquad (2.115)$$

where $m_X = E[X]$.

- *Solution*: Consider a function $g(X)$ given by

$$g(X) = \begin{cases} 1 \text{ for } X > a \\ 0 \text{ for } X < a \end{cases} \qquad (2.116)$$

as shown in Fig. 2.10. Clearly

$$g(X) \leq X/a \qquad \text{for } X, \, a \geq 0$$
$$\Rightarrow \int_{x=0}^{\infty} g(x) f_X(x) \, dx \leq \int_{x=0}^{\infty} (x/a) f_X(x) \, dx$$
$$\Rightarrow \int_{x=a}^{\infty} f_X(x) \, dx \leq m_X/a$$
$$\Rightarrow P(X > a) \leq m_X/a \qquad \text{(proved)}. \qquad (2.117)$$

20. Let a random variable X have a uniform pdf over $-2 \leq X \leq 2$. Compute the
    pdf of $Y = g(X)$, where

$$g(X) = \begin{cases} 2X^2 & \text{for } |X| \leq 1 \\ 3|X| - 1 & \text{for } 1 \leq |X| \leq 2. \end{cases} \qquad (2.118)$$

- *Solution*: We first note that the mapping $g(x)$ is many-to-one. This is illustrated
  in Fig. 2.11. Let y and x denote particular values taken by the RVs Y and X,
  respectively. Therefore we have

$$[f_X(x) + f_X(-x)] \, |dx| = f_Y(y) \, |dy|. \tag{2.119}$$

Note that

$$f_X(x) = \begin{cases} 1/4 & \text{for } -2 < x < 2 \\ 0 & \text{elsewhere.} \end{cases} \tag{2.120}$$

Substituting (2.120) in (2.119) we have:

$$\frac{1}{2} \left| \frac{dx}{dy} \right| = f_Y(y). \tag{2.121}$$

Next we observe that

$$\left| \frac{dx}{dy} \right| = \begin{cases} 1/(2\sqrt{2y}) & \text{for } 0 < y < 2 \\ 1/3 & \text{for } 2 < y < 5. \end{cases} \tag{2.122}$$

Substituting (2.122) in (2.121) we get

$$f_Y(y) = \begin{cases} 1/(4\sqrt{2y}) & \text{for } 0 < y < 2 \\ 1/6 & \text{for } 2 < y < 5. \end{cases} \tag{2.123}$$

21. A Gaussian distributed random variable X having zero mean and variance σ_X^2 is transformed by a square-law device defined by $Y = X^2$.

   (a) Compute the pdf of Y.
   (b) Compute $E[Y]$.

   • *Solution*: We first note that the mapping is many-to-one. Thus the probability that Y lies in the range $[y, \, y + dy]$ is equal to the probability that X lies in the range $[x_1, \, x_1 + dx_1]$ and $[x_2, \, x_2 + dx_2]$, as illustrated in Fig. 2.12. Observe that dx_2 is negative. Mathematically

$$f_Y(y) \, dy = f_X(x_1) \, dx_1 - f_X(x_2) \, dx_2$$
$$\Rightarrow f_Y(y) = \frac{f_X(x)}{|dy/dx|} \bigg|_{x=\sqrt{y}} + \frac{f_X(x)}{|dy/dx|} \bigg|_{x=-\sqrt{y}}, \tag{2.124}$$

where

$$\frac{dy}{dx_1} \triangleq \frac{dy}{dx} \bigg|_{x=x_1} \tag{2.125}$$

with $x_1 = \sqrt{y}$ and $x_2 = -\sqrt{y}$. Since

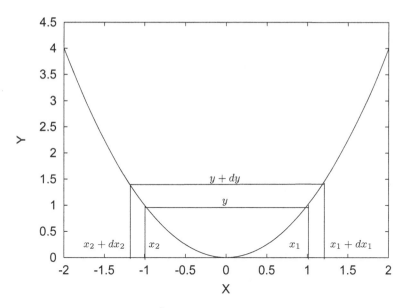

Fig. 2.12 The transformation $Y = X^2$

$$f_X(x) = \frac{1}{\sigma_X\sqrt{2\pi}} e^{-x^2/(2\sigma_X^2)}$$

$$\frac{dy}{dx} = 2x \tag{2.126}$$

we have

$$f_Y(y) = \frac{1}{\sigma_X\sqrt{2\pi y}} e^{-y/(2\sigma_X^2)} \qquad \text{for } y \geq 0. \tag{2.127}$$

Finally

$$E[Y] = E[X^2] = \sigma_X^2. \tag{2.128}$$

22. Consider two real-valued random variables A and B.

   (a) Prove that $E^2[AB] \leq E[A^2]E[B^2]$.
   (b) Now, let $X(t) = A$ and $X(t + \tau + \tau_p) - X(t + \tau) = B$. Prove that if $R_X(\tau_p)$ $= R_X(0)$ for $\tau_p \neq 0$, then $R_X(\tau)$ is periodic with period τ_p.

   • *Solution*: Note that

$$E\left[(Ax \pm B)^2\right] \geq 0$$
$$\Rightarrow x^2 E\left[A^2\right] \pm 2x E[AB] + E\left[B^2\right] \geq 0. \tag{2.129}$$

Thus we get a quadratic equation in x whose discriminant is nonpositive. Therefore

$$E^2[AB] - E\left[A^2\right] E\left[B^2\right] \leq 0 \tag{2.130}$$

which proves the first part. Substituting

$$A = X(t)$$
$$B = X(t + \tau + \tau_p) - X(t + \tau) \tag{2.131}$$

in we obtain

$$E[AB] = E[X(t)[X(t + \tau + \tau_p) - X(t + \tau)]]$$
$$= R_X(\tau + \tau_p) - R_X(\tau)$$
$$E\left[A^2\right] = R_X(0)$$
$$E\left[B^2\right] = 2[R_X(0) - R_X(\tau_p)]. \tag{2.132}$$

Substituting (2.132) in (2.130) we get

$$\left[R_X(\tau + \tau_p) - R_X(\tau)\right]^2 \leq 2[R_X(0) - R_X(\tau_p)]. \tag{2.133}$$

Since it is given that $R_X(0) = R_X(\tau_p)$, (2.133) becomes

$$\left[R_X(\tau + \tau_p) - R_X(\tau)\right]^2 \leq 0. \tag{2.134}$$

However (2.134) cannot be less than zero; it can only be equal to zero. Therefore

$$R_X(\tau + \tau_p) = R_X(\tau) \tag{2.135}$$

which implies that $R_X(\tau)$ is periodic with period τ_p.

23. Let X and Y be two independent and uniformly distributed random variables over $[-a, a]$.

   (a) Compute the pdf of $Z = X + Y$.
   (b) If $a = \pi$ and X, Y, Z denote the phase over $[-\pi, \pi]$, recompute the pdf of Z. Assume that X and Y are independent.

   • *Solution*: We note that the pdf of the sum of two independent random variables is equal to the convolution of the individual pdfs. Therefore

$$f_Z(z) = f_X(z) \star f_Y(z)$$
$$= \int_{\alpha=-\infty}^{\infty} f_X(\alpha) f_Y(z - \alpha) \, d\alpha. \tag{2.136}$$

Fig. 2.13 The pdf of
$Z = X + Y$

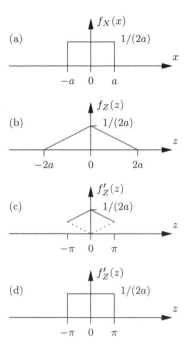

For the given problem, $f_Z(z)$ is illustrated in Fig. 2.13b.

When $a = \pi$, we note that the pdf of Z extends over $[-2\pi, 2\pi]$. Since it is given that the phase extends only over $[-\pi, \pi]$, the values of Z over $[-2\pi, -\pi]$ is identical to $[0, \pi]$. Similarly, the values of Z over $[\pi, 2\pi]$ is identical to $[-\pi, 0]$. Therefore, the modified pdf of Z over $[-\pi, \pi]$ is computed as follows:

$$f_Z'(z) = \begin{cases} f_Z(z) + f_Z(z - 2\pi) & \text{for } 0 \le z \le \pi \\ f_Z(z) + f_Z(z + 2\pi) & \text{for } -\pi \le z \le 0. \end{cases} \tag{2.137}$$

This is illustrated in Fig. 2.13c. The resulting pdf is shown in Fig. 2.13d. Thus we find that the modified pdf of Z is also uniform in $[-\pi, \pi]$.

24. Let X and Y be two independent RVs with pdfs given by

$$f_X(x) = \begin{cases} 1/4 & \text{for } -2 \le x \le 2 \\ 0 & \text{otherwise} \end{cases} \tag{2.138}$$

and

$$f_Y(y) = \begin{cases} Ae^{-3y} & \text{for } 0 \le y < \infty \\ 0 & \text{otherwise.} \end{cases} \tag{2.139}$$

(a) Find A.

(b) Find the pdf of $Z = 3X + 4Y$.

- *Solution*: We have

$$\int_{y=0}^{\infty} f_Y(y)\,dy = 1$$

$$\Rightarrow A \int_{y=0}^{\infty} e^{-3y}\,dy = 1$$

$$\Rightarrow A = 3. \tag{2.140}$$

Let $U = 3X$. Then

$$f_U(u) = \begin{cases} 1/12 & \text{for } -6 \le u \le 6 \\ 0 & \text{otherwise.} \end{cases} \tag{2.141}$$

Similarly, let $V = 4Y$. Clearly $0 \le v < \infty$. Since the mapping between V and Y is one-to-one, we have

$$f_V(v)\,dv = f_Y(y)\,dy$$

$$\Rightarrow f_V(v) = \left. \frac{f_Y(y)}{dv/dy} \right|_{y=v/4}$$

$$\Rightarrow f_V(v) = \frac{3}{4} e^{-3v/4} \qquad \text{for } 0 \le v < \infty$$

$$= \frac{3}{4} e^{-3v/4} S(v), \tag{2.142}$$

where $S(v)$ denotes the modified unit step function defined by

$$S(v) = \begin{cases} 1 & \text{for } v \ge 0 \\ 0 & \text{for } v < 0. \end{cases} \tag{2.143}$$

Since X and Y are independent, so are U and V. Moreover, $Z = U + V$, therefore $-6 \le z < \infty$. Hence

$$f_Z(z) = f_U(z) \star f_V(z)$$

$$= \int_{\alpha=-\infty}^{\infty} f_U(\alpha) f_V(z - \alpha)\,d\alpha$$

$$= \begin{cases} \frac{1}{16} e^{-3z/4} \int_{\alpha=-6}^{z} e^{3\alpha/4}\,d\alpha & \text{for } z \le 6 \\ \frac{1}{16} e^{-3z/4} \int_{\alpha=-6}^{6} e^{3\alpha/4}\,d\alpha & \text{for } z \ge 6 \end{cases}$$

$$= \begin{cases} \frac{1}{12} e^{-3z/4} \left[e^{3z/4} - e^{-9/2} \right] & \text{for } z \le 6 \\ \frac{1}{12} e^{-3z/4} \left[e^{9/2} - e^{-9/2} \right] & \text{for } z \ge 6. \end{cases} \tag{2.144}$$

Fig. 2.14 Two filters
connected in cascade

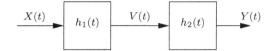

It can be verified that

$$\int_{z=-6}^{\infty} f_Z(z)\, dz = 1. \tag{2.145}$$

25. (Haykin 1983) Consider two linear filters $h_1(t)$ and $h_2(t)$ connected in cascade
    as shown in Fig. 2.14. Let $X(t)$ be a WSS process with autocorrelation $R_X(\tau)$.

    (a) Find the autocorrelation function of $Y(t)$.
    (b) Find the cross-correlation function $R_{VY}(t)$.

- *Solution*: Let

$$g(t) = h_1(t) \star h_2(t). \tag{2.146}$$

Then

$$Y(t) = X(t) \star g(t). \tag{2.147}$$

Hence

$$\begin{aligned}
R_Y(\tau) &= E\left[Y(t)Y(t-\tau)\right] \\
&= E\left[\int_{\alpha=-\infty}^{\infty} g(\alpha)X(t-\alpha)\, d\alpha \int_{\beta=-\infty}^{\infty} g(\beta)X(t-\tau-\beta)\, d\beta\right] \\
&= \int_{\alpha=-\infty}^{\infty}\int_{\beta=-\infty}^{\infty} g(\alpha)g(\beta)R_X(\tau+\beta-\alpha)\, d\alpha\, d\beta.
\end{aligned} \tag{2.148}$$

The cross-correlation is given by

$$\begin{aligned}
R_{VY}(\tau) &= E\left[V(t)Y(t-\tau)\right] \\
&= E\left[\int_{\alpha=-\infty}^{\infty} h_1(\alpha)X(t-\alpha)\, d\alpha \right. \\
&\qquad \left. \times \int_{\beta=-\infty}^{\infty} h_2(\beta)V(t-\tau-\beta)\, d\beta\right] \\
&= E\left[\int_{\alpha=-\infty}^{\infty}\int_{\beta=-\infty}^{\infty} h_1(\alpha)h_2(\beta)X(t-\alpha) \right. \\
&\qquad \left. \times \int_{\gamma=-\infty}^{\infty} h_1(\gamma)X(t-\tau-\beta-\gamma)\, d\gamma\, d\alpha\, d\beta\right]
\end{aligned}$$

$$= \int_{\alpha=-\infty}^{\infty} \int_{\beta=-\infty}^{\infty} \int_{\gamma=-\infty}^{\infty} h_1(\alpha) h_2(\beta) h_1(\gamma)$$
$$\times R_X(\tau + \beta + \gamma - \alpha) \, d\gamma \, d\alpha \, d\beta. \qquad (2.149)$$

26. (Haykin 1983) Consider a pair of WSS processes $X(t)$ and $Y(t)$. Show that their cross-correlations $R_{XY}(\tau)$ and $R_{YX}(\tau)$ have the following properties:

(a) $R_{YX}(\tau) = R_{XY}(-\tau)$.
(b) $|R_{XY}(\tau)| \leq \frac{1}{2}[R_X(0) + R_Y(0)]$.

- *Solution*: We know that

$$R_{YX}(\tau) = E\left[Y(t)X(t - \tau)\right]. \qquad (2.150)$$

Now substitute $t - \tau = \alpha$ to get

$$R_{YX}(\tau) = E\left[Y(\alpha + \tau)X(\alpha).\right]. \qquad (2.151)$$

Substituting $\tau = -\beta$ we get

$$R_{YX}(-\beta) = E\left[Y(\alpha - \beta)X(\alpha)\right] = R_{XY}(\beta). \qquad (2.152)$$

Thus the first part is proved. To prove the second part we note that

$$E\left[(X(t) \pm Y(t - \tau))^2\right] \geq 0$$
$$\Rightarrow \frac{1}{2}\left[E\left[X(t)^2\right] + E\left[Y(t - \tau)^2\right]\right] \geq \mp E\left[X(t)Y(t - \tau)\right]$$
$$\Rightarrow \frac{1}{2}\left[R_X(0) + R_Y(0)\right] \geq \mp R_{XY}(\tau)$$
$$\Rightarrow \frac{1}{2}\left[R_X(0) + R_Y(0)\right] \geq |R_{XY}(\tau)|. \qquad (2.153)$$

27. (Haykin 1983) Given that a stationary random process $X(t)$ has the autocorrelation function $R_X(\tau)$ and power spectral density $S_X(f)$ show that

(a) The autocorrelation function of $dX(t)/dt$ is equal to minus the second derivative of $R_X(\tau)$.
(b) The power spectral density of $dX(t)/dt$ is equal to $4\pi^2 f^2 S_X(f)$.
(c) If $S_X(f) = 2\,\text{rect}\,(f/W)$, compute the power of $dX(t)/dt$.

- *Solution*: Let

$$Y(t) = \frac{dX(t)}{dt}. \qquad (2.154)$$

It is clear that $Y(t)$ can be obtained by passing $X(t)$ through an ideal differentiator. We know that the Fourier transform of an ideal differentiator is

$$H(f) = j 2\pi f. \tag{2.155}$$

Hence the psd of $Y(t)$ is

$$
\begin{aligned}
S_Y(f) &= S_X(f)|H(f)|^2 \\
&= 4\pi^2 f^2 S_X(f) \\
\Rightarrow R_Y(\tau) &= \int_{f=-\infty}^{\infty} S_Y(f) \exp(j 2\pi f \tau) \, df \\
&= 4\pi^2 \int_{f=-\infty}^{\infty} f^2 S_X(f) \exp(j 2\pi f \tau) \, df. \tag{2.156}
\end{aligned}
$$

Thus, the second part of the problem is proved. To prove the first part, we note that

$$
\begin{aligned}
R_X(\tau) &= \int_{f=-\infty}^{\infty} S_X(f) \exp(j 2\pi f \tau) \, df \\
\Rightarrow \frac{d^2 R_X(\tau)}{d\tau^2} &= -4\pi^2 \int_{f=-\infty}^{\infty} f^2 S_X(f) \exp(j 2\pi f \tau) \, df. \tag{2.157}
\end{aligned}
$$

Comparing (2.156) and (2.157) we see that

$$R_Y(\tau) = -\frac{d^2 R_X(\tau)}{d\tau^2}. \tag{2.158}$$

The power of $Y(t)$ is given by

$$
\begin{aligned}
E\left[Y^2(t)\right] &= \int_{f=-\infty}^{\infty} S_Y(f) \, df \\
&= \int_{f=-W/2}^{W/2} 8\pi^2 f^2 \, df \\
&= \frac{2}{3} \pi^2 W^3. \tag{2.159}
\end{aligned}
$$

28. (Haykin 1983) The psd of a random process $X(t)$ is shown in Fig. 2.15.

Fig. 2.15 Power spectral density of a random process $X(t)$

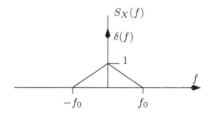

(a) Determine $R_X(\tau)$.
(b) Determine the dc power in $X(t)$.
(c) Determine the ac power in $X(t)$.
(d) What sampling-rates will give uncorrelated samples of $X(t)$? Are the samples statistically independent?

- *Solution*: The psd of $X(t)$ can be written as

$$S_X(f) = \delta(f) + \left(1 - \frac{|f|}{f_0}\right). \tag{2.160}$$

We know that

$$A \operatorname{rect}(t/T_0) \rightleftharpoons A T_0 \operatorname{sinc}(f T_0)$$
$$\Rightarrow A \operatorname{rect}(t/T_0) \star A \operatorname{rect}(-t/T_0) \rightleftharpoons A^2 T_0^2 \operatorname{sinc}^2(f T_0)$$
$$\Rightarrow A^2 T_0 \left(1 - \frac{|t|}{T_0}\right) \rightleftharpoons A^2 T_0^2 \operatorname{sinc}^2(f T_0). \tag{2.161}$$

Applying duality

$$A^2 f_0 \left(1 - \frac{|f|}{f_0}\right) \rightleftharpoons A^2 f_0^2 \operatorname{sinc}^2(t f_0). \tag{2.162}$$

Given that $A^2 f_0 = 1$. Hence

$$\left(1 - \frac{|f|}{f_0}\right) \rightleftharpoons f_0 \operatorname{sinc}^2(t f_0). \tag{2.163}$$

Thus

$$R_X(\tau) = 1 + f_0 \operatorname{sinc}^2(f_0 \tau). \tag{2.164}$$

Now, consider a real-valued random process $Z(t)$ given by

$$Z(t) = A + Y(t), \tag{2.165}$$

where A is a constant and $Y(t)$ is a zero-mean random process. Clearly

$$E[Z(t)] = A$$
$$R_Z(\tau) = E[Z(t)Z(t - \tau)] = A^2 + R_Y(\tau). \tag{2.166}$$

Thus we conclude that if a random process has a DC component equal to A, then the autocorrelation has a constant component equal to A^2. Hence from (2.164) we get

$$E[X(t)] = \pm 1 = m_X \quad \text{(say).} \tag{2.167}$$

Hence the DC power (contributed by the delta function of the psd) is unity. The AC power (contributed by the triangular part of the psd) is f_0. The covariance of $X(t)$ is

$$K_X(\tau) = \text{cov}(X(t)X(t-\tau)) = E[(X(t) - m_X)(X(t-\tau) - m_X)]$$
$$= f_0 \, \text{sinc}^2(f_0\tau). \tag{2.168}$$

It is clear that

$$K_X(\tau) = \begin{cases} f_0 \text{ for } \tau = 0 \\ 0 \text{ for } \tau = n(k/f_0), n, \ k \neq 0, \end{cases} \tag{2.169}$$

where n and k are positive integers. Thus, when $X(t)$ is sampled at a rate equal to f_0/k, the samples are uncorrelated. However the samples may not be statistically independent.

29. The random variable X has a uniform distribution over $0 \leq x \leq 2$. For the random process defined by $V(t) = 6e^{Xt}$ compute

(a) $E[V(t)]$
(b) $E[V(t)V(t-\tau)]$
(c) $E\left[V^2(t)\right]$

• *Solution*: Note that

$$E[V(t)] = 6E\left[e^{Xt}\right]$$
$$= \frac{6}{2}\int_{x=0}^{2} e^{xt}\,dx$$
$$= \frac{3}{t}\left(e^{2t} - 1\right). \tag{2.170}$$

The autocorrelation is given by

$$E[V(t)V(t-\tau)] = 36E\left[e^{Xt}e^{X(t-\tau)}\right]$$
$$= \frac{36}{2}\int_{x=0}^{2} e^{x(2t-\tau)}\,dx$$
$$= \frac{36}{2(2t-\tau)}\left(e^{2(2t-\tau)} - 1\right)$$
$$= \frac{18}{2t-\tau}\left(e^{2(2t-\tau)} - 1\right). \tag{2.171}$$

Hence

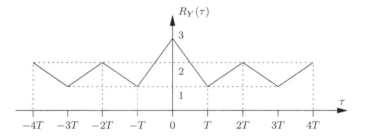

Fig. 2.16 Autocorrelation of a random process $Y(t)$

$$E\left[V^2(t)\right] = E\left[V(t)V(t-\tau)\right]_{\tau=0}$$
$$= \frac{9}{t}\left(e^{4t} - 1\right). \tag{2.172}$$

30. (Haykin 1983) A random process $Y(t)$ consists of a DC component equal to $\sqrt{3/2}$ V, a periodic component $G(t)$ having a random timing phase, and a random component $X(t)$. Both $G(t)$ and $X(t)$ have zero mean and are independent of each other. The autocorrelation of $Y(t)$ is shown in Fig. 2.16. Compute

(a) The DC power.
(b) The average power of $G(t)$. Sketch a sample function of $G(t)$.
(c) The average power of $X(t)$. Sketch a sample function of $X(t)$.

- *Solution*: The random process $Y(t)$ can be written as:

$$Y(t) = A + G(t) + X(t) \tag{2.173}$$

where $G(t)$ and $X(t)$ have zero mean and are independent of each other. The autocorrelation of $Y(t)$ can be written as

$$\begin{aligned} R_Y(\tau) &= E[Y(t)Y(t-\tau)] \\ &= E[(A + G(t) + X(t))(A + G(t-\tau) + X(t-\tau))] \\ &= A^2 + R_G(\tau) + R_X(\tau), \end{aligned} \tag{2.174}$$

where $R_G(\tau)$ and $R_X(\tau)$ denote the autocorrelation of $G(t)$ and $X(t)$, respectively, and A^2 denotes the DC power. Note that since $G(t)$ is periodic with a period $T_0 = 2T$, we have

$$\begin{aligned} R_G(\tau) &= E[G(t)G(t-\tau)] \\ &= E[G(t)G(t-\tau-kT_0)] \\ &= R_G(\tau + kT_0). \end{aligned} \tag{2.175}$$

Therefore $R_G(\tau)$ is also periodic with a period of kT_0. Moreover, since A is given to be equal to $\sqrt{3/2}$, we have $A^2 = 3/2$. Therefore the DC power is $3/2$.

By inspecting Fig. 2.16 we conclude that $G(t)$ and $X(t)$ have the autocorrelation as indicated in Fig. 2.17a, b, respectively. Thus, the power in the periodic component is $R_G(0) = 0.5$ and the power in the random component is $R_X(0) = 1$.

Recall that a periodic signal with period T_0 and random timing phase t_d can be expressed as a random process:

$$G(t) = \sum_{k=-\infty}^{\infty} p(t - kT_0 - t_d), \qquad (2.176)$$

where t_d is a uniformly distributed random variable in the range $[0, T_0)$ and $p(\cdot)$ is a real-valued waveform. A sample function of $G(t)$ and the corresponding $p(t)$ is shown in Fig. 2.17c. Recall that the autocorrelation of $G(t)$ is

$$R_G(\tau) = \frac{1}{T_0} \sum_{m=-\infty}^{\infty} R_p(\tau - mT_0), \qquad (2.177)$$

where $R_p(\cdot)$ denotes the autocorrelation of $p(t)$ in (2.176).

Similarly, recall that a random binary wave is given by

$$X(t) = \sum_{k=-\infty}^{\infty} S_k p(t - kT_0 - \alpha), \qquad (2.178)$$

where S_k denotes a discrete random variable taking values $\pm A$ with equal probability, $p(\cdot)$ denotes a real-valued waveform, $1/T_0$ denotes the bit-rate and α denotes a random timing phase uniformly distributed in $[0, T_0)$. It is assumed that S_k is independent of α and S_k is independent of S_j for $k \neq j$. A sample function of $X(t)$ along with the corresponding $p(t)$ is shown in Fig. 2.17d. Recall that the autocorrelation of $X(t)$ is

$$R_X(\tau) = \frac{A^2}{T_0} R_p(\tau), \qquad (2.179)$$

where $R_p(\cdot)$ is the autocorrelation of $p(\cdot)$ in (2.178). In this problem, $S_k = \pm 1$.

31. The moving average of a random signal $X(t)$ is defined as

$$Y(t) = \frac{1}{T} \int_{\lambda=t-T/2}^{t+T/2} X(\lambda) \, d\lambda. \qquad (2.180)$$

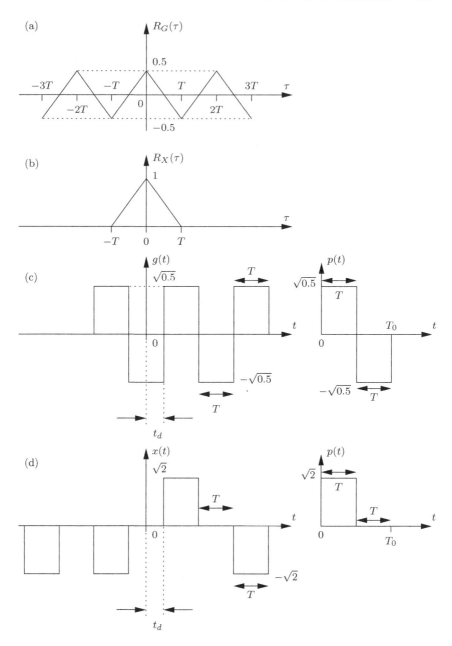

Fig. 2.17 **a** Autocorrelation of $G(t)$. **b** Autocorrelation of $X(t)$. **c** Sample function of $G(t)$ and the corresponding $p(t)$. **d** Sample function of $X(t)$ and the corresponding $p(t)$

(a) The output $Y(t)$ can be written as

$$Y(t) = X(t) \star h(t). \tag{2.181}$$

Identify $h(t)$.

(b) Obtain a general expression for the autocorrelation of $Y(t)$ in terms of the autocorrelation of $X(t)$ and the autocorrelation of $h(t)$.

(c) Using the above relation, compute $R_Y(\tau)$.

• *Solution*: The random process $Y(t)$ can be written as

$$Y(t) = X(t) \star h(t), \tag{2.182}$$

where

$$h(t) = \frac{1}{T} \text{rect}\left(\frac{t}{T}\right). \tag{2.183}$$

We know that

$$R_Y(\tau) = \int_{\alpha=-\infty}^{\infty} \int_{\beta=-\infty}^{\infty} h(\alpha) h(\beta) R_X(\tau - \alpha + \beta) \, d\alpha \, d\beta. \tag{2.184}$$

Let

$$\alpha - \beta = \lambda. \tag{2.185}$$

Substituting for β we get

$$R_Y(\tau) = \int_{\alpha=-\infty}^{\infty} \int_{\lambda=-\infty}^{\infty} h(\alpha) h(\alpha - \lambda) R_X(\tau - \lambda) \, d\alpha \, d\lambda. \tag{2.186}$$

Interchanging the order of the integrals we get

$$R_Y(\tau) = \int_{\lambda=-\infty}^{\infty} R_X(\tau - \lambda) \, d\lambda. \int_{\alpha=-\infty}^{\infty} h(\alpha) h(\alpha - \lambda) \, d\alpha$$

$$= \int_{\lambda=-\infty}^{\infty} R_X(\tau - \lambda) R_h(\lambda) \, d\lambda. \tag{2.187}$$

For the given problem

$$R_h(\tau) = \begin{cases} (1/T)\,(1 - (|\tau|/T)) & \text{for } |\tau| < T \\ 0 & \text{elsewhere.} \end{cases} \tag{2.188}$$

Therefore

$$R_Y(\tau) = \frac{1}{T} \int_{\lambda=-T}^{T} R_X(\tau - \lambda) \left(1 - \frac{|\lambda|}{T}\right) d\lambda. \qquad (2.189)$$

32. Let a random process $X(t)$ be applied at the input of a filter with impulse response $h(t)$. Let the output process be denoted by $Y(t)$.

(a) Show that

$$R_{YX}(\tau) = h(\tau) \star R_X(\tau). \qquad (2.190)$$

(b) Show that $S_{YX}(f) = H(f)S_X(f)$.

• *Solution*: Note that

$$Y(t) = \int_{\alpha=-\infty}^{\infty} h(\alpha)X(t - \alpha) \, d\alpha. \qquad (2.191)$$

Therefore

$$R_{YX}(\tau) = E[Y(t)X(t - \tau)]$$
$$= E\left[\int_{\alpha=-\infty}^{\infty} h(\alpha)X(t - \alpha)X(t - \tau) \, d\alpha\right]. \qquad (2.192)$$

Interchanging the order of the expectation and the integral, we get

$$R_{YX}(\tau) = \int_{\alpha=-\infty}^{\infty} h(\alpha)E\left[X(t - \alpha)X(t - \tau)\right] d\alpha$$
$$= \int_{\alpha=-\infty}^{\infty} h(\alpha)R_X(\tau - \alpha) \, d\alpha$$
$$= h(\tau) \star R_X(\tau). \qquad (2.193)$$

Taking the Fourier transform of both sides and noting that convolution in the time-domain is equivalent to multiplication in the frequency domain, we get

$$S_{YX}(f) = H(f)S_X(f). \qquad (2.194)$$

33. Let $X(t_1)$ and $X(t_2)$ be two random variables obtained by observing the random process $X(t)$ at times t_1 and t_2, respectively. It is given that $X(t_1)$ and $X(t_2)$ are uncorrelated. Does this imply that $X(t)$ is WSS?

• *Solution*: Since $X(t_1)$ and $X(t_2)$ are uncorrelated we have

$$E[(X(t_1) - m_X(t_1))(X(t_2) - m_X(t_2))] = 0$$
$$\Rightarrow R_X(t_1, t_2) = m_X(t_1)m_X(t_2), \qquad (2.195)$$

Fig. 2.18 PSD of $N(t)$

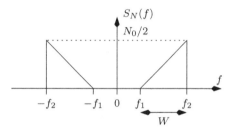

where $m_X(t_1)$ and $m_X(t_2)$ are the means at times t_1 and t_2, respectively.
From the above equation, it is clear that $X(t)$ need not be WSS.

34. The voltage at the output of a noise generator, whose statistics is known to be a
WSS Gaussian process, is measured with a DC voltmeter and a true rms (root
mean square) meter. The DC voltmeter reads 3 V. The rms meter, when it is AC
coupled (DC component is removed), reads 2 V.

(a) Find out the pdf of the voltage at any time.
(b) What would be the reading of the rms meter if it is DC coupled?

• *Solution*: The DC meter reads the average (mean) value of the voltage. Hence
the mean of the voltage is $m = 3$ V.
When the rms meter is AC coupled, it reads the standard deviation (σ). Hence
$\sigma = 2$ V.
Thus the pdf of the voltage takes the form

$$p_V(v) = \frac{1}{\sigma\sqrt{2\pi}}e^{-(v-m)^2/(2\sigma^2)}. \tag{2.196}$$

We know that

$$E[V^2] = \sigma^2 + m^2 = 13. \tag{2.197}$$

Thus, when the rms meter is DC coupled, the reading would be $\sqrt{13}$.

35. (Ziemer and Tranter 2002) This problem demonstrates the non-uniqueness of
the representation of narrowband noise. Here we show that the in-phase and
quadrature components depend on the choice of the carrier frequency f_c.
A narrowband noise process of the form

$$N(t) = N_c(t)\cos(2\pi f_c t) - N_s(t)\sin(2\pi f_c t) \tag{2.198}$$

has a psd as indicated in Fig. 2.18, where f_1, $f_2 \gg W$. Sketch the psd of $N_c(t)$
and $N_s(t)$ for the following cases:

(a) $f_c = f_1$

(b) $f_c = f_2$
(c) $f_c = (f_1 + f_2)/2$

• *Solution*: It is given that

$$f_2 - f_1 = W, \qquad (2.199)$$

where f_1, $f_2 \gg W$. We know that

$$S_{N_c}(f) = \begin{cases} S_N(f - f_c) + S_N(f + f_c) & \text{for } |f| < f_c \\ 0 & \text{otherwise.} \end{cases}$$
$$= S_{N_s}(f). \qquad (2.200)$$

The psd of $N_c(t)$ and $N_s(t)$ for $f_c = f_1$ is indicated in Fig. 2.19. The psd of $N_c(t)$ and $N_s(t)$ for $f_c = f_2$ is shown in Fig. 2.20. The psd of $N_c(t)$ and $N_s(t)$ for $f_c = (f_1 + f_2)/2$ is shown in Fig. 2.21.

36. Consider the system used for generating narrowband noise, as shown in Fig. 2.22. The Fourier transform of the bandpass filter is also shown. Assume that the narrowband noise representation is of the form

$$N(t) = N_c(t)\cos(2\pi f_c t) - N_s(t)\sin(2\pi f_c t). \qquad (2.201)$$

The input $W(t)$ is a zero mean, WSS, white Gaussian noise process with a psd equal to 2×10^{-8} W/Hz.

(a) Sketch and label the psd of $N_c(t)$ when $f_c = 22$ MHz. Show the steps, starting from $S_N(f)$.
(b) Write down the pdf of the random variable X obtained by observing $N_c(t)$ at $t = 2$ s.
(c) Sketch and label the cross-spectral density $S_{N_c N_s}(f)$ when $f_c = 22$ MHz.
(d) Derive the expression for the correlation coefficient $(\rho(\tau))$ between $N_c(t)$ and $N_s(t - \tau)$ for $f_c = 22$ MHz.
(e) For what carrier frequency f_c, is $\rho(\tau) = 0$ for all values of τ?

Assume the following:

$$R_{N_c N_s}(\tau) \overset{\Delta}{=} E[N_c(t)N_s(t - \tau)]$$
$$R_{N_c}(\tau) \overset{\Delta}{=} E[N_c(t)N_c(t - \tau)]. \qquad (2.202)$$

• *Solution*: The psd of $N(t)$ is given by

$$S_N(f) = S_W(f)|H(f)|^2$$
$$= \frac{N_0}{2}|H(f)|^2. \qquad (2.203)$$

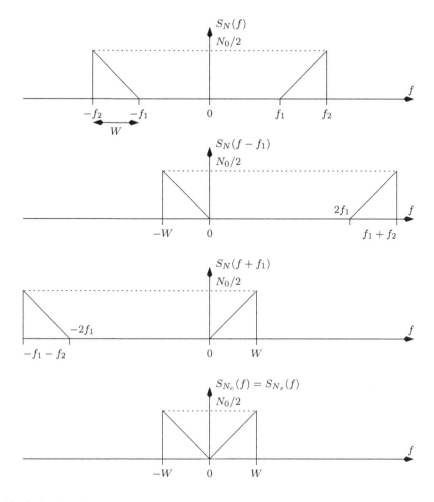

Fig. 2.19 PSD of $N_c(t)$ and $N_s(t)$ for $f_c = f_1$

The psd of $N_c(t)$ is shown in Fig. 2.23.

Since $W(t)$ is zero mean, WSS Gaussian random process and $H(f)$ is an LTI system, $N(t)$ is also zero mean, WSS Gaussian random process. Thus $N_c(t)$ and $N_s(t)$ are also zero mean, WSS Gaussian random processes. Therefore $X = N_c(2)$ is a Gaussian RV with pdf

$$f_X(x) = \frac{1}{\sigma\sqrt{2\pi}} e^{-(x-m)^2/(2\sigma^2)}. \tag{2.204}$$

Here

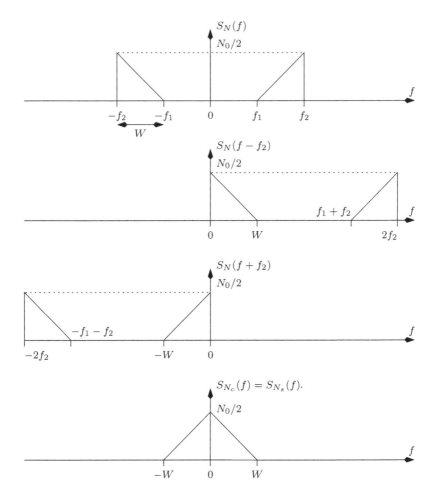

Fig. 2.20 PSD of $N_c(t)$ and $N_s(t)$ for $f_c = f_2$

$$m = 0$$
$$\sigma^2 = \int_{f=-\infty}^{\infty} S_{N_c}(f)\, df = 0.68. \tag{2.205}$$

The cross-spectral density $S_{N_c N_s}(f)$ is plotted in Fig. 2.24.
The cross-spectral density is of the form

$$S_{N_c N_s}(f) = j\,[A_1 \operatorname{rect}((f - f_1)/B) + A_2 \operatorname{rect}((f - f_2)/B)$$
$$- A_1 \operatorname{rect}((f + f_1)/B) - A_2 \operatorname{rect}((f + f_2)/B)] \tag{2.206}$$

where

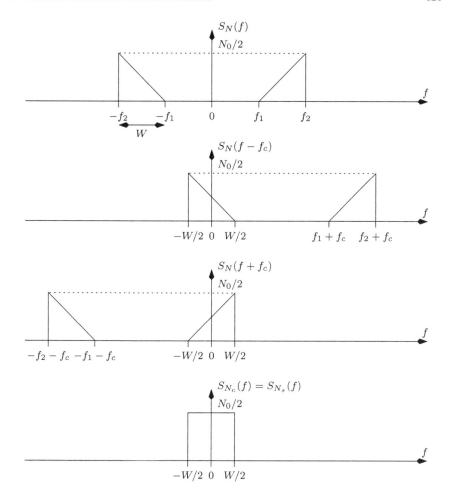

Fig. 2.21 PSD of $N_c(t)$ and $N_s(t)$ for $f_c = (f_1 + f_2)/2$

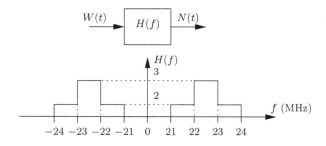

Fig. 2.22 System used for generating narrowband noise

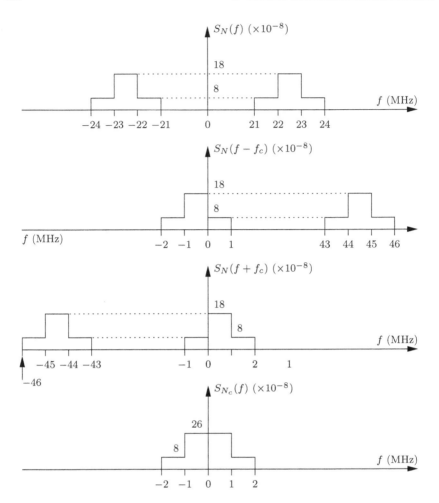

Fig. 2.23 PSD of $N_c(t)$

$$f_1 = 0.5 \times 10^6$$
$$f_2 = 1.5 \times 10^6$$
$$B = 10^6$$
$$A_1 = 10 \times 10^{-8}$$
$$A_2 = 8 \times 10^{-8}. \tag{2.207}$$

The cross-correlation $R_{N_c N_s}(\tau)$ can be obtained by taking the inverse Fourier transform of $S_{N_c N_s}(f)$ and is given by

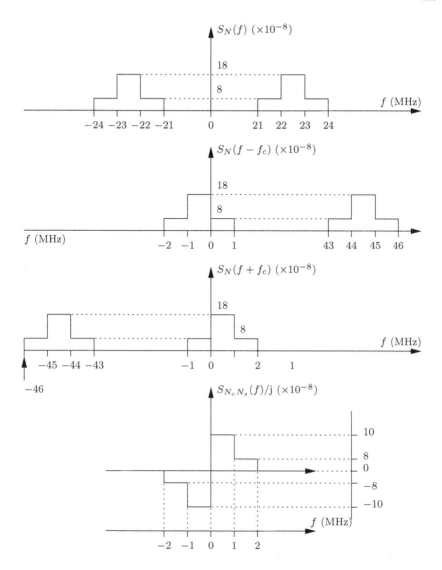

Fig. 2.24 The cross-spectral density $S_{N_c N_s}(f)$

$$
\begin{aligned}
R_{N_c N_s}(\tau) &= jA_1 B \operatorname{sinc}(Bt) \left[e^{j2\pi f_1 t} - e^{-j2\pi f_1 t} \right] \\
&\quad + jA_2 B \operatorname{sinc}(Bt) \left[e^{j2\pi f_2 t} - e^{-j2\pi f_2 t} \right] \\
&= -\operatorname{sinc}(Bt) \left[0.2 \sin(2\pi f_1 t) + 0.16 \sin(2\pi f_2 t) \right]. \quad (2.208)
\end{aligned}
$$

Since $N_c(t)$ and $N_s(t - \tau)$ are both $\mathcal{N}(0, \sigma^2)$ we have

Fig. 2.25 $R(\tau)$ versus τ

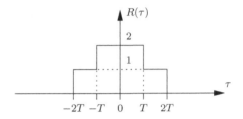

$$\rho(\tau) = R_{N_c N_s}(\tau)/\sigma^2, \tag{2.209}$$

where $\sigma^2 = 0.68$.

$R_{N_c N_s}(\tau) = 0$ for all τ when $f_c = 22.5\,\text{MHz}$. Hence $\rho(\tau) = 0$ for all τ when $f_c = 22.5\,\text{MHz}$.

37. Consider the function $R(\tau)$ as illustrated in Fig. 2.25. Is it a valid autocorrelation function? Justify your answer.

   • *Solution*: Note that

$$R(\tau) = \text{rect}\left(\frac{t}{2T}\right) + \text{rect}\left(\frac{t}{4T}\right)$$
$$\rightleftharpoons 2T\,\text{sinc}\,(2fT) + 4T\,\text{sinc}\,(4fT). \tag{2.210}$$

Let $2\pi fT = \theta$. Hence

$$R(\tau) \quad \rightleftharpoons \quad 2T\frac{\sin(\theta)}{\theta} + 4T\frac{\sin(2\theta)}{2\theta}$$
$$\rightleftharpoons \quad 2T\frac{\sin(\theta)}{\theta}(1 + 2\cos(\theta))$$
$$\triangleq S(\theta). \tag{2.211}$$

Clearly, $S(\theta)$ is negative for $3\pi/2 \le |\theta| < 2\pi$, as illustrated in Fig. 2.26 ($\sin(\theta)$ is negative and $\cos(\theta)$ is positive). Therefore $S(\theta)$ cannot be the power spectral density, and $R(\tau)$ is not a valid autocorrelation.

38. Consider the function

$$R(\tau) = \text{rect}\,(\tau/T). \tag{2.212}$$

Is it a valid autocorrelation function? Justify your answer.

   • *Solution*: The Fourier transform of $R(\tau)$ is

$$\text{rect}\,(\tau/T) \rightleftharpoons T\,\text{sinc}\,(fT) \tag{2.213}$$

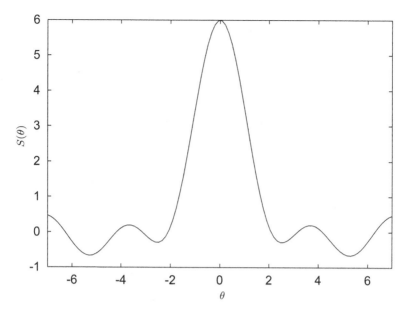

Fig. 2.26 Fourier transform of $R(\tau)$ given in Fig. 2.25

Fig. 2.27 Power spectral
density of $n_1(t)$

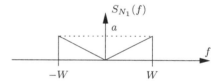

which can take negative values. Since the power spectral density cannot be
negative, $R(\tau)$ is not a valid autocorrelation, even though it is an even function.

39. (Haykin 1983) A pair of noise processes are related by

$$N_2(t) = N_1(t)\cos(2\pi f_c t + \theta) - N_1(t)\sin(2\pi f_c t + \theta), \qquad (2.214)$$

where f_c is a constant and θ is a uniformly distributed random variable between
0 and 2π. The noise process $N_1(t)$ is stationary and its psd is shown in Fig. 2.27.
Find the psd of $N_2(t)$.

- *Solution*: The autocorrelation of $N_2(t)$ is given by

$$\begin{aligned}
&E[N_2(t)N_2(t-\tau)] \\
&= E\left[(N_1(t)\cos(2\pi f_c t + \theta) - N_1(t)\sin(2\pi f_c t + \theta))\right. \\
&\quad \left.(N_1(t-\tau)\cos(2\pi f_c(t-\tau) + \theta) - N_1(t-\tau)\sin(2\pi f_c(t-\tau) + \theta))\right]
\end{aligned}$$

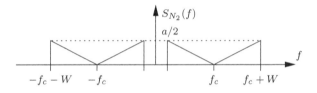

Fig. 2.28 Power spectral density of $n_2(t)$

$$
= \frac{R_{N_1}(\tau)\cos(2\pi f_c \tau)}{2} + \frac{R_{N_1}(\tau)\cos(2\pi f_c \tau)}{2}
$$
$$
+ \frac{R_{N_1}(\tau)\sin(2\pi f_c \tau)}{2} - \frac{R_{N_1}(\tau)\sin(2\pi f_c \tau)}{2}.
$$
$$
= R_{N_1}(\tau)\cos(2\pi f_c \tau). \tag{2.215}
$$

Hence

$$
S_{N_2}(f) = \frac{S_{N_1}(f - f_c) + S_{N_1}(f + f_c)}{2}. \tag{2.216}
$$

The psd for $N_2(t)$ is illustrated in Fig. 2.28.

40. The output of an oscillator is described by

$$
X(t) = A\cos(2\pi f t + \theta), \tag{2.217}
$$

where A is a constant, f and θ are independent random variables. The pdf of f is denoted by $f_F(f)$ for $-\infty < f < \infty$ and the pdf of θ is uniformly distributed between 0 and 2π. It is given that $f_F(f)$ is an even function of f.

(a) Find the psd of $X(t)$ in terms of $f_F(f)$.
(b) What happens to the psd of $X(t)$ when f is a constant equal to f_c.

• *Solution*: The autocorrelation of $X(t)$ is given by

$$
E[X(t)X(t-\tau)] = A^2 E\left[\cos(2\pi f t + \Theta)\cos(2\pi f(t-\tau) + \Theta)\right]
$$
$$
= \frac{A^2}{2} E\left[\cos(2\pi f \tau) + \cos(4\pi f t - 2\pi f \tau + 2\Theta)\right]
$$
$$
= \frac{A^2}{2} \int_{f=-\infty}^{\infty} \cos(2\pi f \tau) f_F(f)\, df
$$
$$
+ \frac{1}{2\pi} \int_{f=-\infty}^{\infty} f_F(f)\, df
$$
$$
\int_{\theta=0}^{2\pi} \cos(4\pi f t - 2\pi f \tau + 2\Theta)\, d\theta
$$

$$= \frac{A^2}{2} \int_{f=-\infty}^{\infty} \cos(2\pi f \tau) f_F(f) \, df$$

$$= R_X(\tau). \tag{2.218}$$

Since $R_X(\tau)$ is real-valued, $S_X(f)$ is an even function. Hence

$$R_X(\tau) = \int_{f=-\infty}^{\infty} S_X(f) \exp\left(\mathrm{j}\, 2\pi f \tau\right) \, df$$

$$= \int_{f=-\infty}^{\infty} S_X(f) \cos\left(2\pi f \tau\right) \, df. \tag{2.219}$$

Comparing (2.218) and (2.219) we get

$$S_X(f) = \frac{A^2}{2} f_F(f). \tag{2.220}$$

When f is a constant, say equal to f_c, then

$$R_X(\tau) = \frac{A^2}{2} \cos(2\pi f_c \tau)$$

$$\Rightarrow S_X(f) = \frac{A^2}{4} \left[\delta(f - f_c) + \delta(f + f_c)\right]. \tag{2.221}$$

41. (Haykin 1983) A real-valued stationary Gaussian process $X(t)$ with mean m_X, variance σ_X^2 and power spectral density $S_X(f)$ is passed through two real-valued linear time invariant filters $h_1(t)$ and $h_2(t)$, yielding the output processes $Y(t)$ and $Z(t)$, respectively.

(a) Determine the joint pdf of $Y(t)$ and $Z(t - \tau)$.
(b) State the conditions, in terms of the frequency response of $h_1(t)$ and $h_2(t)$, that are sufficient to ensure that $Y(t)$ and $Z(t - \tau)$ are statistically independent.

• *Solution*: The mean values of $Y(t)$ and $Z(t - \tau)$ are given by

$$E[Y(t)] = E\left[\int_{\alpha=-\infty}^{\infty} h_1(\alpha) X(t - \alpha) \, d\alpha\right]$$

$$= m_X \int_{\alpha=-\infty}^{\infty} h_1(\alpha) \, d\alpha$$

$$= m_X H_1(0)$$

$$= m_Y$$

$$E[Z(t - \tau)] = m_X \int_{\alpha=-\infty}^{\infty} h_2(\alpha) \, d\alpha$$

$$= m_X H_2(0)$$

$$= m_Z. \tag{2.222}$$

The variance of $Y(t)$ and $Z(t - \tau)$ are given by:

$$E[(Y(t) - m_Y)^2] = E\left[Y^2(t)\right] - m_Y^2$$

$$= \int_{f=-\infty}^{\infty} S_X(f)|H_1(f)|^2 \, df - m_Y^2$$

$$= \sigma_Y^2$$

$$E[(Z(t - \tau) - m_Z)^2] = E\left[Z^2(t)\right] - m_Z^2$$

$$= \int_{f=-\infty}^{\infty} S_X(f)|H_2(f)|^2 \, df - m_Z^2$$

$$= \sigma_Z^2. \tag{2.223}$$

The cross covariance between $Y(t)$ and $Z(t - \tau)$ is given by

$$E\left[(Y(t) - m_Y)(Z(t - \tau) - m_Z)\right]$$

$$= E[Y(t)Z(t - \tau)] - m_Y m_Z$$

$$= E\left[\int_{\alpha=-\infty}^{\infty} h_1(\alpha)X(t - \alpha) \, d\alpha \int_{\beta=-\infty}^{\infty} h_2(\beta)X(t - \tau - \beta) \, d\beta\right]$$

$$- m_Y m_Z$$

$$= \int_{\alpha=-\infty}^{\infty} \int_{\beta=-\infty}^{\infty} h_1(\alpha)h_2(\beta)R_X(\tau + \beta - \alpha) \, d\alpha \, d\beta - m_Y m_Z$$

$$= K_{YZ}(\tau). \tag{2.224}$$

The correlation coefficient between $Y(t)$ and $Z(t)$ is given by

$$\rho = \frac{K_{YZ}(\tau)}{\sigma_Y \sigma_Z}. \tag{2.225}$$

Now, the joint pdf of two Gaussian random variables y_1 and y_2 is given by

$$p_{Y_1, Y_2}(y_1, y_2)$$

$$= \frac{1}{2\pi\sigma_1\sigma_2\sqrt{1 - \rho^2}}$$

$$\times \exp\left[-\frac{\sigma_2^2(y_1 - m_1)^2 - 2\sigma_1\sigma_2\rho(y_1 - m_1)(y_2 - m_2) + \sigma_1^2(y_2 - m_2)^2}{2\sigma_1^2\sigma_2^2(1 - \rho^2)}\right]. \tag{2.226}$$

The joint pdf of $Y(t)$ and $Z(t - \tau)$ can be similarly found out by substituting the appropriate values from (2.222)–(2.225). The random variables $Y(t)$ and $Z(t - \tau)$ are uncorrelated if their covariance is zero. This is possible when

(a) $H_1(0) = 0$ or $H_2(0) = 0$ AND

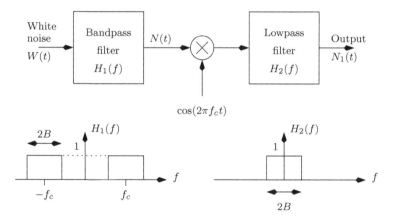

Fig. 2.29 System block diagram

(b) The product $H_1(f)H_2(f) = 0$ for all f. This can be easily shown by computing the Fourier transform of $K_{YZ}(\tau)$, which is equal to (assuming $m_Y m_Z = 0$)

$$K_{YZ}(\tau) \rightleftharpoons H_1(f)H_2^*(f)S_X(f). \tag{2.227}$$

Thus if $H_1(f)H_2(f) = 0$, then $K_{YZ}(\tau) = 0$, implying that $Y(t)$ and $Z(t - \tau)$ are uncorrelated, and being Gaussian, they are statistically independent.

42. (Haykin 1983) Consider a white Gaussian noise process of zero mean and psd $N_0/2$ which is applied to the input of the system shown in Fig. 2.29.

(a) Find the psd of the random process at the output of the system.
(b) Find the mean and variance of this output.

- *Solution*: We know that narrow-band noise can be represented by

$$n(t) = n_c(t)\cos(2\pi f_c t) - n_s(t)\sin(2\pi f_c t). \tag{2.228}$$

We also know that the psd of $n_c(t)$ and $n_s(t)$ are given by

$$S_{N_c}(f) = S_{N_s}(f) = \begin{cases} S_N(f - f_c) + S_N(f + f_c) & \text{for } -B \le f \le B \\ 0 & \text{elsewhere,} \end{cases} \tag{2.229}$$

where it is assumed that $S_N(f)$ occupies the frequency band $f_c - B \le |f| \le f_c + B$. For the given problem $S_N(f)$ is shown in Fig. 2.30a. Thus

$$R_{N_c}(\tau) \rightleftharpoons S_{N_c}(f) = \begin{cases} N_0 & \text{for } -B \le f \le B \\ 0 & \text{elsewhere.} \end{cases} \tag{2.230}$$

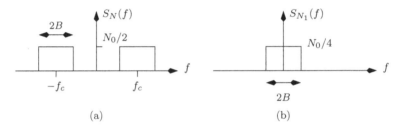

Fig. 2.30 a Noise psd at the output of $H_1(f)$. b Noise psd at the output of $H_2(f)$

The output of $H_2(f)$ is clearly

$$n(t)\cos(2\pi f_c t) \xrightarrow{H_2(f)} n_c(t)/2 \tag{2.231}$$

whose autocorrelation is

$$E\left[\frac{n_c(t)}{2}\frac{n_c(t-\tau)}{2}\right] = \frac{R_{N_c}(\tau)}{4} \rightleftharpoons \frac{S_{N_c}(f)}{4}. \tag{2.232}$$

The psd of noise at the output of $H_2(f)$ is illustrated in Fig. 2.30b. The output noise is zero-mean Gaussian with variance

$$\frac{N_0}{4} \times 2B = \frac{N_0 B}{2}. \tag{2.233}$$

43. (Haykin 1983) Consider a narrowband noise process $N(t)$ with its Hilbert transform denoted by $\hat{N}(t)$. Show that the cross-correlation functions of $N(t)$ and $\hat{N}(t)$ are given by

$$R_{N\hat{N}}(\tau) = -\hat{R}_N(\tau)$$
$$R_{\hat{N}N}(\tau) = \hat{R}_N(\tau), \tag{2.234}$$

where $\hat{R}_N(\tau)$ is the Hilbert transform of $R_N(\tau)$.

• *Solution*: From the basic definition of the autocorrelation function we have

$$R_{N\hat{N}}(\tau) = E\left[N(t)\hat{N}(t-\tau)\right]$$
$$= E\left[N(t)\frac{1}{\pi}\int_{\alpha=-\infty}^{\infty}\frac{N(\alpha)}{t-\tau-\alpha}\,d\alpha\right]$$
$$= \frac{1}{\pi}\int_{\alpha=-\infty}^{\infty}\frac{E[N(t)N(\alpha)]}{t-\tau-\alpha}\,d\alpha$$
$$= \frac{1}{\pi}\int_{\alpha=-\infty}^{\infty}\frac{R_N(t-\alpha)}{t-\tau-\alpha}\,d\alpha$$

$$= \frac{1}{\pi} \int_{\beta=-\infty}^{\infty} \frac{R_N(\beta)}{\beta - \tau} \, d\beta$$

$$= -\hat{R}_N(\tau). \tag{2.235}$$

The second part can be similarly proved.

44. (Haykin 1983) A random process $X(t)$ characterized by the autocorrelation function

$$R_X(\tau) = e^{-2\nu|\tau|}, \tag{2.236}$$

where ν is a constant, is applied to a first-order RC-lowpass filter. Determine the psd and the autocorrelation of the random process at the filter output.

- *Solution*: The psd of $X(t)$ is the Fourier transform of the autocorrelation function and is given by

$$S_X(f) = \int_{\tau=-\infty}^{0} e^{2\nu\tau} e^{-j2\pi f\tau} \, d\tau + \int_{\tau=0}^{\infty} e^{-2\nu\tau} e^{-j2\pi f\tau} \, d\tau$$

$$= \frac{1}{2(\nu - j\pi f)} \left[e^{2\tau(\nu - j\pi f)} \right]_{\tau=-\infty}^{0}$$

$$- \frac{1}{2(\nu + j\pi f)} \left[e^{-2\tau(\nu + j\pi f)} \right]_{\tau=0}^{\infty}$$

$$= \frac{\nu}{\nu^2 + \pi^2 f^2}$$

$$= \frac{1/\nu}{1 + \pi^2 f^2/\nu^2}. \tag{2.237}$$

Since the transfer function of the LPF is

$$H(f) = \frac{1}{1 + j2\pi f RC}. \tag{2.238}$$

Therefore the psd of the output random process is

$$S_Y(f) = S_X(f)|H(f)|^2$$

$$= \frac{\nu}{\nu^2 + \pi^2 f^2} \times \frac{1}{1 + (2\pi f RC)^2}$$

$$= \frac{A}{\nu^2 + \pi^2 f^2} + \frac{B}{1 + (2\pi f RC)^2}. \tag{2.239}$$

Solving for A and B we get

$$A = \frac{\nu}{1 - 4R^2C^2\nu^2}$$

$$B = \frac{-4R^2C^2\nu}{1 - 4R^2C^2\nu^2}. \tag{2.240}$$

The autocorrelation of the output random process is the inverse Fourier transform of (2.239), and is given by

$$R_Y(\tau) = \frac{A}{\nu}e^{-2\nu|\tau|} + \frac{B}{2RC}e^{-|\tau|/(RC)}. \tag{2.241}$$

45. Let X and Y be independent random variables with

$$f_X(x) = \begin{cases} \alpha e^{-\alpha x} & x > 0 \\ 0 & \text{otherwise} \end{cases}$$

$$f_Y(y) = \begin{cases} \beta e^{-\beta y} & y > 0 \\ 0 & \text{otherwise,} \end{cases} \tag{2.242}$$

where α and β are positive constants. Find the pdf of $Z = X + Y$ when $\alpha = \beta$ and $\alpha \neq \beta$.

- *Solution*: We know that the pdf of Z is the convolution of the pdfs of X and Y and is given by (for $\alpha \neq \beta$)

$$\begin{aligned} f_Z(z) &= \int_{a=-\infty}^{\infty} f_X(a)f_Y(z-a)\,da \\ &= \int_{a=0}^{z} f_X(a)f_Y(z-a)\,da \\ &= \int_{a=0}^{z} \alpha\beta e^{-\alpha a}e^{-\beta(z-a)}\,da \\ &= \begin{cases} \frac{\alpha\beta}{\alpha-\beta}\left[e^{-\beta z} - e^{-\alpha z}\right] & \text{for } z > 0 \\ 0 & \text{for } z < 0. \end{cases} \end{aligned} \tag{2.243}$$

When $\alpha = \beta$ we have

$$\begin{aligned} f_Z(z) &= \alpha^2 e^{-\alpha z}\int_{a=0}^{z} da \\ &= \begin{cases} z\alpha^2 e^{-\alpha z} & \text{for } z > 0 \\ 0 & \text{otherwise.} \end{cases} \end{aligned} \tag{2.244}$$

46. A WSS random process $X(t)$ with psd $N_0/2$ is passed through a first-order RC highpass filter. Determine the psd and autocorrelation of the filter output.

- *Solution*: The transfer function of the first-order RC highpass filter is

$$H(f) = \frac{j2\pi f RC}{1 + j2\pi f RC}. \tag{2.245}$$

Let $Y(t)$ denote the random process at the filter output. Therefore the psd of the filter output is

$$
\begin{aligned}
S_Y(f) &= \frac{N_0}{2} \times \frac{(2\pi f RC)^2}{1 + (2\pi f RC)^2} \\
&= \frac{N_0}{2} \left[1 - \frac{1}{1 + (2\pi f RC)^2} \right].
\end{aligned} \tag{2.246}
$$

The autocorrelation of $Y(t)$ is the inverse Fourier transform of $S_Y(f)$. Consider the Fourier transform pair

$$\frac{1}{2}e^{-|t|} \rightleftharpoons \frac{1}{1 + (2\pi f)^2}. \tag{2.247}$$

Using the time scaling property

$$
\begin{aligned}
g(t) &\rightleftharpoons G(f) \\
\Rightarrow g(at) &\rightleftharpoons \frac{1}{|a|} G(f/a)
\end{aligned} \tag{2.248}
$$

with $a = 1/(RC)$, (2.247) becomes

$$\frac{1}{2}e^{-|t|/(RC)} \rightleftharpoons \frac{RC}{1 + (2\pi f RC)^2}. \tag{2.249}$$

Thus from (2.246) and (2.249) we have

$$R_Y(\tau) = \frac{N_0}{2}\delta(\tau) - \frac{N_0}{4RC}e^{-|t|/(RC)}. \tag{2.250}$$

47. Let X be a random variable having pdf

$$f_X(x) = ae^{-b|x|} \quad \text{for } a, b > 0, -\infty < x < \infty. \tag{2.251}$$

(a) Find the relation between a and b.
(b) Let $Y = 2X^2 - c$, where $c > 0$. Compute $P(Y < 0)$.

- *Solution*: We have

$$2 \int_{x=0}^{\infty} ae^{-bx} = 1$$
$$\Rightarrow 2a = b. \tag{2.252}$$

Now

$$P(Y < 0) = P(2X^2 - c < 0)$$
$$= P(X^2 < c/2)$$
$$= P(|X| < \sqrt{c/2})$$
$$= 2 \int_{x=0}^{\sqrt{c/2}} a e^{-bx} \, dx$$
$$= \frac{2a}{b} \left[1 - e^{-b\sqrt{c/2}} \right]$$
$$= \left[1 - e^{-b\sqrt{c/2}} \right]. \tag{2.253}$$

48. A narrowband noise process $N(t)$ has zero mean and autocorrelation function $R_N(\tau)$. Its psd $S_N(f)$ is centered about $\pm f_c$. Its quadrature components $N_c(t)$ and $N_s(t)$ are defined by

$$N_c(t) = N(t) \cos(2\pi f_c t) + \hat{N}(t) \sin(2\pi f_c t)$$
$$N_s(t) = \hat{N}(t) \cos(2\pi f_c t) - N(t) \sin(2\pi f_c t). \tag{2.254}$$

Using (2.234) show that

$$R_{N_c}(\tau) = R_{N_s}(\tau) = R_N(\tau) \cos(2\pi f_c \tau) + \hat{R}_N(\tau) \sin(2\pi f_c \tau)$$
$$R_{N_c N_s}(\tau) = -R_{N_s N_c}(\tau) = R_N(\tau) \sin(2\pi f_c \tau) - \hat{R}_N(\tau) \cos(2\pi f_c \tau). \tag{2.255}$$

- *Solution*: From the basic definition of the autocorrelation

$$R_{N_c}(\tau) = E[N_c(t) N_c(t - \tau)]$$
$$= E\left[\left(N(t) \cos(2\pi f_c t) + \hat{N}(t) \sin(2\pi f_c t) \right) \right.$$
$$\left. \left(N(t - \tau) \cos(2\pi f_c (t - \tau)) + \hat{N}(t - \tau) \sin(2\pi f_c (t - \tau)) \right) \right]$$
$$= \frac{R_N(\tau)}{2} [\cos(2\pi f_c \tau) + \cos(4\pi f_c t - 2\pi f_c \tau)]$$
$$+ \frac{R_{N\hat{N}}(\tau)}{2} [\sin(4\pi f_c t - 2\pi f_c \tau) - \sin(2\pi f_c \tau)]$$
$$+ \frac{R_{\hat{N}N}(\tau)}{2} [\sin(4\pi f_c t - 2\pi f_c \tau) + \sin(2\pi f_c \tau)]$$
$$+ \frac{R_{\hat{N}}(\tau)}{2} [\cos(2\pi f_c \tau) - \cos(4\pi f_c t - 2\pi f_c \tau)]. \tag{2.256}$$

Now

$$S_{\hat{N}}(f) = S_N(f)\,|-j\,\mathrm{sgn}\,(f)|^2$$
$$= S_N(f)$$
$$\Rightarrow R_{\hat{N}}(\tau) = R_N(\tau). \tag{2.257}$$

In the above equation, we have assumed that $S_N(0) = 0$. Moreover from (2.234)

$$R_{\hat{N}N}(\tau) = -R_{N\hat{N}}(\tau) = \hat{R}_N(\tau). \tag{2.258}$$

Substituting (2.257) and (2.258) in (2.256) we get

$$R_{N_c}(\tau) = R_N(\tau)\cos(2\pi f_c\tau) + \hat{R}_N(\tau)\sin(2\pi f_c\tau). \tag{2.259}$$

The expression for $R_{N_s}(\tau)$ can be similarly proved.
Similarly we have

$$R_{N_cN_s}(\tau)$$
$$= E\left[N_c(t)N_s(t-\tau)\right]$$
$$= E\left[\left(N(t)\cos(2\pi f_c t) + \hat{N}(t)\sin(2\pi f_c t)\right)\right.$$
$$\left.\left(\hat{N}(t-\tau)\cos(2\pi f_c(t-\tau)) - N(t-\tau)\sin(2\pi f_c(t-\tau))\right)\right]$$
$$= \frac{R_{N\hat{N}}(\tau)}{2}[\cos(2\pi f_c\tau) + \cos(4\pi f_c t - 2\pi f_c\tau)]$$
$$- \frac{R_N(\tau)}{2}[\sin(4\pi f_c t - 2\pi f_c\tau) - \sin(2\pi f_c\tau)]$$
$$+ \frac{R_{\hat{N}}(\tau)}{2}[\sin(4\pi f_c t - 2\pi f_c\tau) + \sin(2\pi f_c\tau)]$$
$$- \frac{R_{\hat{N}N}(\tau)}{2}[\cos(2\pi f_c\tau) - \cos(4\pi f_c t - 2\pi f_c\tau)]. \tag{2.260}$$

Once again substituting (2.257) and (2.258) in (2.260) we get

$$R_{N_cN_s}(\tau) = R_N(\tau)\sin(2\pi f_c\tau) - \hat{R}_N(\tau)\cos(2\pi f_c\tau). \tag{2.261}$$

The expression for $R_{N_sN_c}(\tau)$ can be similarly proved.

49. Let $X(t)$ be a stationary process with zero mean, autocorrelation function $R_X(\tau)$ and psd $S_X(f)$. Find a linear filter with impulse response $h(t)$ such that the filter output is $X(t)$ when the input is white noise with psd $N_0/2$. What is the corresponding condition on the transfer function $H(f)$?

- *Solution*: Let $Y(t)$ denote the input process. It is given that

$$R_Y(\tau) = \frac{N_0}{2}\delta(\tau). \tag{2.262}$$

The autocorrelation of the output is given by

$$
\begin{aligned}
R_X(\tau) &= E\left[\int_{\alpha=-\infty}^{\infty} h(\alpha)Y(t-\alpha)\,d\alpha \int_{\beta=-\infty}^{\infty} h(\beta)Y(t-\tau-\beta)\,d\beta\right] \\
&= \int_{\alpha=-\infty}^{\infty}\int_{\beta=-\infty}^{\infty} h(\alpha)h(\beta)R_Y(\tau+\beta-\alpha)\,d\alpha\,d\beta \\
&= \frac{N_0}{2}\int_{\alpha=-\infty}^{\infty}\int_{\beta=-\infty}^{\infty} h(\alpha)h(\beta)\delta(\tau+\beta-\alpha)\,d\alpha\,d\beta \\
&= \frac{N_0}{2}\int_{\alpha=-\infty}^{\infty} h(\alpha)h(\alpha-\tau)\,d\alpha \\
&= \frac{N_0}{2}\left(h(t)\star h(-t)\right). \tag{2.263}
\end{aligned}
$$

The transfer function must satisfy the following relationship:

$$S_X(f) = \frac{N_0}{2}|H(f)|^2. \tag{2.264}$$

50. White Gaussian noise $W(t)$ of zero mean and psd $N_0/2$ is applied to a first-order RC-lowpass filter, producing noise $N(t)$. Determine the pdf of the random variable obtained by observing $N(t)$ at time t_k.

- *Solution*: Clearly, the first-order RC lowpass filter is stable and linear, hence $N(t)$ is also a Gaussian random process. Hence $N(t_k)$ is a Gaussian distributed random variable with pdf given by

$$p_{N(t_k)}(x) = \frac{1}{\sigma\sqrt{2\pi}}e^{-(x-m)^2/(2\sigma^2)}, \tag{2.265}$$

where m is the mean and σ^2 denotes the variance.

Moreover, since $W(t)$ is WSS, $N(t)$ is also WSS. It only remains to compute the mean and variance of $N(t_k)$. The mean value of $N(t_k)$ is computed as follows:

$$
\begin{aligned}
N(t_k) &= \int_{\tau=0}^{\infty} h(\tau)W(t_k-\tau)\,d\tau \\
\Rightarrow m = E\left[N(t_k)\right] &= \int_{\tau=0}^{\infty} h(\tau)E[W(t_k-\tau)]\,d\tau \\
&= 0. \tag{2.266}
\end{aligned}
$$

The variance can be computed from the psd as

$$\sigma^2 = E\left[N^2(t_k)\right] = \frac{N_0}{2} \int_{f=-\infty}^{\infty} |H(f)|^2 \, df. \qquad (2.267)$$

For the given problem $H(f)$ is given by

$$H(f) = \frac{1/(j\,2\pi f C)}{R + 1/(j\,2\pi f C)}$$

$$= \frac{1}{1 + j\,2\pi f RC}$$

$$\Rightarrow |H(f)|^2 = \frac{1}{1 + (2\pi f RC)^2}. \qquad (2.268)$$

Hence

$$\sigma^2 = E\left[N^2(t_k)\right] = \frac{N_0}{2} \frac{1}{2\pi RC} \left[\tan^{-1}(2\pi f RC)\right]_{f=-\infty}^{\infty} = \frac{N_0}{4RC} \quad (2.269)$$

51. Let X and Y be two independent random variables. X is uniformly distributed between $[-1, 1]$ and Y is uniformly distributed between $[-2, 2]$. Find the pdf of $Z = \max(X, Y)$.

- *Solution*: We begin by noting that Z lies in the range $[-1, 2]$. Next, we compute the cumulative distribution function of Z and proceed to compute the pdf by differentiating the distribution function.
  To compute the cumulative distribution function, we note that

$$Z = \begin{cases} X \text{ if } X > Y \\ Y \text{ if } Y > X. \end{cases} \qquad (2.270)$$

The pdfs of X and Y are given by

$$f_X(x) = \begin{cases} 1/2 \text{ for } -1 \le x \le 1 \\ 0 \quad \text{otherwise} \end{cases}$$

$$f_Y(y) = \begin{cases} 1/4 \text{ for } -2 \le y \le 2 \\ 0 \quad \text{otherwise.} \end{cases} \qquad (2.271)$$

We also note that since the pdfs of X and Y are different in the range $[-1, 1]$ and $[1, 2]$, we expect the distribution function of Z to be different in these two regions.
Using the fact that X and Y are independent and the events $X > Y$ and $Y > X$ are mutually exclusive we have for $-1 \le z \le 1$

$$P(Z \le z) = P(X \le z \text{ AND } X > y)$$
$$+ P(Y \le z \text{ AND } Y > x)$$
$$= \frac{1}{8} \int_{x=-1}^{z} \int_{y=-2}^{x} dy \, dx$$
$$+ \frac{1}{8} \int_{y=-1}^{z} \int_{x=-1}^{y} dy \, dx$$
$$= \frac{1}{8} \left[z^2 + 3z + 2 \right]. \tag{2.272}$$

When $1 \le z \le 2$, then $Z = Y$. Hence the cumulative distribution takes the form

$$P(Z \le z) = P(Z < 1) + \frac{1}{8} \int_{y=1}^{z} \int_{x=-1}^{1} dy \, dx$$
$$= \frac{1}{8} \left[z^2 + 3z + 2 \right]_{z=1}$$
$$+ \frac{1}{8} \int_{y=1}^{z} \int_{x=-1}^{1} dy \, dx$$
$$= \frac{1}{8} \left[2z + 4 \right]. \tag{2.273}$$

Thus

$$f_Z(z) = \begin{cases} (1/8) \left[2z + 3 \right] & \text{for } -1 \le z \le 1 \\ 1/4 & \text{for } 1 \le z \le 2. \end{cases} \tag{2.274}$$

52. Consider a random variable given by $Z = aX + bY$. Here X and Y are statistically independent Gaussian random variables with mean m_X and m_Y, respectively. The variance of X and Y is σ^2.

(a) Is Z a Gaussian distributed random variable?
(b) Compute the mean and the variance of Z.

• *Solution*: Z is Gaussian distributed because it is a linear combination of two Gaussian random variables. The mean value of Z is

$$m_Z = E[Z] = am_X + bm_Y. \tag{2.275}$$

The variance of Z is

$$E\left[(Z - m_Z)^2\right] = E\left[Z^2\right] - m_Z^2$$
$$= a^2 E\left[X^2\right] + b^2 E\left[Y^2\right] + 2ab E[X]E[Y]$$
$$- (am_X + bm_Y)^2$$

$$
\begin{aligned}
&= a^2(\sigma^2 + m_X^2) + b^2(\sigma^2 + m_Y^2) + 2abm_X m_Y \\
&\quad - (am_X + bm_Y)^2 \\
&= \sigma^2(a^2 + b^2).
\end{aligned}
\tag{2.276}
$$

53. Let X and Y be two independent random variables. X is uniformly distributed between $[-1, 1]$ and Y is uniformly distributed between $[-2, 2]$. Find the pdf of $Z = \min(X, Y)$.

- *Solution*: We begin by noting that Z lies in the range $[-2, 1]$. Next, we compute the probability distribution function of Z and proceed to compute the pdf by differentiating the cumulative distribution function with respect to z. To compute the cumulative distribution function, we note that

$$
Z = \begin{cases} X \text{ if } X < Y \\ Y \text{ if } Y < X. \end{cases}
\tag{2.277}
$$

The pdfs of X and Y are given by

$$
f_X(x) = \begin{cases} 1/2 \text{ for } -1 \le x \le 1 \\ 0 \quad \text{otherwise} \end{cases}
$$

$$
f_Y(y) = \begin{cases} 1/4 \text{ for } -2 \le y \le 2 \\ 0 \quad \text{otherwise.} \end{cases}
\tag{2.278}
$$

Moreover, since the pdfs of X and Y are different in the region $[-2, -1]$ and $[-1, 1]$, we also expect the cumulative distribution function of Z to be different in these two regions.

Let us first consider the region $[-2, -1]$. Using the fact that X and Y are independent and the events $X < Y$ and $Y < X$ are mutually exclusive we have for $-2 \le z \le -1$

$$
\begin{aligned}
P(Z \le z) = {}& P(X \le z \text{ AND } X < y) \\
& + P(Y \le z \text{ AND } Y < x).
\end{aligned}
\tag{2.279}
$$

However we note that for $-2 \le z \le -1$

$$
P(X \le z \text{ AND } X < y) = 0
\tag{2.280}
$$

since X is always greater than -1. Hence

$$
\begin{aligned}
P(Z \le z) &= P(Y \le z \text{ AND } Y < x) \\
&= \frac{1}{8} \int_{y=-2}^{z} \int_{x=-1}^{1} dy \, dx \\
&= \frac{1}{8} [2z + 4].
\end{aligned}
\tag{2.281}
$$

Similarly we have for $-1 \leq z \leq 1$.

$$
\begin{aligned}
P(Z \leq z) &= P(X \leq z \text{ AND } X < y) \\
&\quad + P(Y \leq z \text{ AND } Y < x) \\
&= P(Z < -1) + \frac{1}{8} \int_{x=-1}^{z} \int_{y=x}^{2} dy\, dx \\
&\quad + \frac{1}{8} \int_{y=-1}^{z} \int_{x=y}^{1} dy\, dx \\
&= \frac{1}{8} \int_{y=-2}^{-1} \int_{x=-1}^{1} dy\, dx \\
&\quad + \frac{1}{8} \int_{x=-1}^{z} \int_{y=x}^{2} dy\, dx \\
&\quad + \frac{1}{8} \int_{y=-1}^{z} \int_{x=y}^{1} dy\, dx \\
&= \frac{1}{8} \left[-z^2 + 3z + 6 \right].
\end{aligned}
\tag{2.282}
$$

Thus

$$
f_Z(z) = \begin{cases} 1/4 & \text{for } -2 \leq z \leq -1 \\ (1/8)\,[-2z + 3] & \text{for } -1 \leq z \leq 1 \end{cases}.
\tag{2.283}
$$

54. (Haykin 1983) Let $X(t)$ be a stationary Gaussian process with zero mean and variance σ^2 and autocorrelation $R_X(\tau)$. This process is applied to an ideal half-wave rectifier. The random process at the rectifier output is denoted by $Y(t)$.

(a) Compute the mean value of $Y(t)$.
(b) Compute the autocorrelation of $Y(t)$. Make use of the result

$$
\int_{u=0}^{\infty} \int_{v=0}^{\infty} uv \exp\left(-u^2 - v^2 - 2uv\cos(\theta) \right) du\, dv = \frac{1 - \theta \cot(\theta)}{4\sin^2(\theta)}
\tag{2.284}
$$

• *Solution*: We begin by noting that $Y(t)$ is some function of $X(t)$, which we denote by $g(X)$. For the given problem

$$
Y(t) = g(X) = \begin{cases} X(t) & \text{if } X(t) \geq 0 \\ 0 & \text{if } X(t) < 0. \end{cases}
\tag{2.285}
$$

We also know that since $X(t)$ is given to be stationary, $f_X(x)$ is independent of time, hence

$$E\left[g(X)\right] = \int_{x=-\infty}^{\infty} g(x) f_X(x) \, dx$$

$$= \frac{1}{\sigma\sqrt{2\pi}} \int_{x=0}^{\infty} x \exp\left(-\frac{x^2}{2\sigma^2}\right) dx$$

$$= \frac{\sigma}{\sqrt{2\pi}} \int_{z=0}^{\infty} \exp(-z) \, dz$$

$$= \frac{\sigma}{\sqrt{2\pi}}. \tag{2.286}$$

The autocorrelation of $Y(t)$ is given by

$$E\left[Y(t)Y(t-\tau)\right]. \tag{2.287}$$

For convenience of representation, let

$$X(t) = \alpha$$

$$X(t-\tau) = \beta. \tag{2.288}$$

To compute the autocorrelation we make use of the result

$$E\left[g(\alpha, \beta)\right] = \int_{\alpha=-\infty}^{\infty} \int_{\beta=-\infty}^{\infty} g(\alpha, \beta) f_{\alpha, \beta}(\alpha, \beta) \, d\alpha \, d\beta. \tag{2.289}$$

The joint pdf of two real-valued Gaussian random variables Y_1 and Y_2 is given by

$$f_{Y_1, Y_2}(y_1, y_2)$$

$$= \frac{1}{2\pi\sigma_1\sigma_2\sqrt{1-\rho^2}}$$

$$\times \exp\left[-\frac{\sigma_2^2(y_1 - m_1)^2 - 2\sigma_1\sigma_2\rho(y_1 - m_1)(y_2 - m_2) + \sigma_1^2(y_2 - m_2)^2}{2\sigma_1^2\sigma_2^2(1 - \rho^2)}\right]. \tag{2.290}$$

For the given problem, the joint pdf of α and β is given by

$$f_{\alpha, \beta}(\alpha, \beta) = \frac{1}{2\pi\sigma^2\sqrt{1-\rho^2}} \exp\left[-\frac{\alpha^2 - 2\rho\alpha\beta + \beta^2}{2\sigma^2(1 - \rho^2)}\right], \tag{2.291}$$

where we have made use of the fact that

$$E[\alpha] = E[\beta] = 0$$
$$E[\alpha^2] = E[\beta^2] = \sigma^2$$
$$\rho = \rho(\tau) = \frac{E[\alpha\beta]}{\sigma^2} = \frac{R_X(\tau)}{\sigma^2}. \tag{2.292}$$

Thus the autocorrelation of $Y(t)$ can be written as

$$R_Y(\tau) = \frac{1}{2\pi\sigma^2\sqrt{1-\rho^2}} \int_{\alpha=0}^{\infty} \int_{\beta=0}^{\infty} \alpha\beta \exp\left[-\frac{\alpha^2 - 2\rho\alpha\beta + \beta^2}{2\sigma^2(1-\rho^2)}\right] d\alpha\, d\beta. \tag{2.293}$$

Let

$$A = \frac{1}{\sqrt{2\sigma^2(1-\rho^2)}}$$
$$B = \frac{1}{\sqrt{2\sigma^2(1-\rho^2)}}$$
$$u = A\alpha$$
$$v = -B\beta. \tag{2.294}$$

Thus

$$R_Y(\tau) = \frac{2\sigma^2(1-\rho^2)^{3/2}}{\pi} \int_{u=0}^{\infty} \int_{v=0}^{\infty} uv \exp\left[-(u^2 + 2uv\rho + v^2)\right] du\, dv \tag{2.295}$$

Using (2.284) we get

$$R_Y(\tau) = \frac{\sigma^2}{2\pi}\left[\sqrt{1-\rho^2} - \rho\cos^{-1}(\rho)\right]. \tag{2.296}$$

55. (Haykin 1983) Let $X(t)$ be a real-valued zero-mean stationary Gaussian process with autocorrelation function $R_X(\tau)$. This process is applied to a square-law device. Denote the output process as $Y(t)$. Compute the mean and autocovariance of $Y(t)$.

- *Solution*: It is given that

$$Y(t) = X^2(t). \tag{2.297}$$

Since $X(t)$ is given to be stationary, we have

$$m_Y = E\left[X^2(t)\right] = R_X(0). \tag{2.298}$$

The autocorrelation of $Y(t)$ is given by

$$R_Y(\tau) = E[Y(t)Y(t-\tau)] = E\left[X^2(t)X^2(t-\tau)\right]. \qquad (2.299)$$

Let

$$X(t) = X_1$$
$$X(t-\tau) = X_2. \qquad (2.300)$$

We know that

$$E[X_1 X_2 X_3 X_4] = C_{12}C_{34} + C_{13}C_{24} + C_{14}C_{23}, \qquad (2.301)$$

where $E[X_i X_j] = C_{ij}$ and the random variables X_i are jointly normal (Gaussian) with zero mean. For the given problem we note that $X_3 = X_1$ and $X_4 = X_2$. Thus

$$R_Y(\tau) = 2R_X^2(\tau) + R_X^2(0). \qquad (2.302)$$

The autocovariance of $Y(t)$ is given by

$$\begin{aligned} K_Y(\tau) &= E[(Y(t) - m_Y)(Y(t-\tau) - m_Y)] \\ &= R_Y(\tau) - m_Y^2 \\ &= 2R_X^2(\tau). \end{aligned} \qquad (2.303)$$

56. A random variable X is uniformly distributed in $[-1, 4]$ and another random variable Y is uniformly distributed in $[7, 8]$. X and Y are statistically independent. Find the pdf of $Z = |X| - Y$.

   • *Solution*: The pdf of X and Y are shown in Fig. 2.31a, b, respectively. Let

$$U = |X|$$
$$V = -Y. \qquad (2.304)$$

Then the pdf of U and V are illustrated in Fig. 2.31c, d. Thus

$$Z = U + V, \qquad (2.305)$$

where U and V are independent. Clearly Z lies in the range $[-8, -3]$. Moreover, the pdf of Z is given by the convolution of the pdfs of U and V and is equal to

$$f_Z(z) = \int_{\alpha=-\infty}^{\infty} f_U(\alpha) f_V(z-\alpha) \, d\alpha. \qquad (2.306)$$

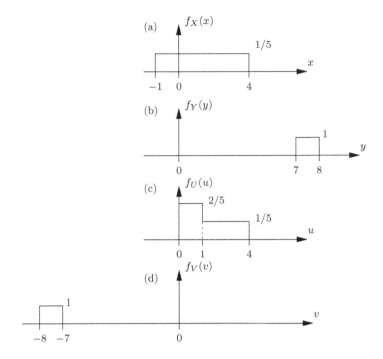

Fig. 2.31 PDF of various random variables

The various steps in the convolution of U and V are illustrated in Fig. 2.32. We have

$$f_Z(-8) = 0$$
$$f_Z(-7) = 2/5$$
$$f_Z(-6) = 1/5$$
$$f_Z(-4) = 1/5$$
$$f_Z(-3) = 0. \tag{2.307}$$

The pdf of Z is illustrated in Fig. 2.33.

57. A random variable X is uniformly distributed in [2, 3] and [−3, −2]. Y is a discrete random variable which takes values −1 and 7 with probabilities 3/5 and 2/5, respectively. X and Y are statistically independent. Find the pdf of $Z = X^2 + |Y|$.

   • *Solution*: Note that

$$f_X(x) = 1/2 \quad \text{for } 2 \le |x| \le 3 \tag{2.308}$$

Fig. 2.32 Various steps in
the convolution of U and V

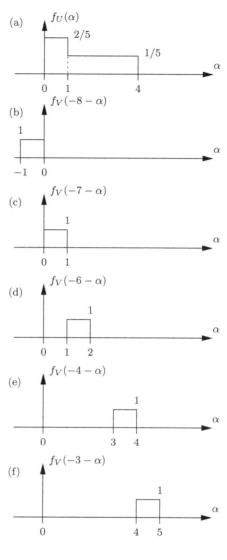

Fig. 2.33 PDF of Z

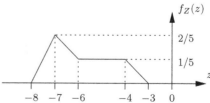

Fig. 2.34 PDF of $U = X^2$

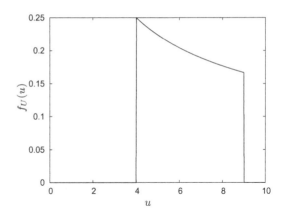

and

$$f_Y(y) = \frac{3}{5}\delta(y+1) + \frac{2}{5}\delta(y-7). \tag{2.309}$$

Now, the pdf of $V = |Y|$ is

$$f_V(v) = \frac{3}{5}\delta(v-1) + \frac{2}{5}\delta(v-7). \tag{2.310}$$

Let $U = X^2$. The pdf of U is

$$f_U(u) = f_X(x)/|du/dx|_{x=\sqrt{u}} + f_X(x)/|du/dx|_{x=-\sqrt{u}}$$
$$= \frac{1}{2\sqrt{u}} \qquad \text{for } 4 \le u \le 9 \tag{2.311}$$

which is illustrated in Fig. 2.34. Therefore

$$Z = U + V. \tag{2.312}$$

Since X and Y are independent, U and V are also independent. Moreover, the pdf of Z is given by the convolution of the pdfs of U and V and is equal to

$$f_Z(z) = \int_{\alpha=-\infty}^{\infty} f_U(\alpha) f_V(z-\alpha) \, d\alpha$$
$$= \frac{3}{5} f_U(z-1) + \frac{2}{5} f_U(z-7). \tag{2.313}$$

The pdf of Z is shown in Fig. 2.35.

Fig. 2.35 PDF of
$U = X^2 + |Y|$

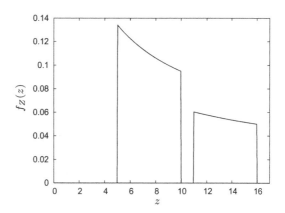

58. Let $X(t)$ be a zero-mean Gaussian random process with psd $N_0/2$. $X(t)$ is passed
    through a filter with impulse response $h(t) = \exp(-\pi t^2)$, followed by an ideal
    differentiator. Let the output process be denoted by $Y(t)$. Determine the pdf of
    $Y(1)$.

    • *Solution*: At the outset, we emphasize that $Y(t)$ can be considered to be a
      random process as well as a random variable. Recall that a random variable is
      obtained by observing a random process at a particular instant of time.
      Note that both $h(t)$ and the ideal differentiator are LTI filters. Hence $Y(t)$ is
      also a Gaussian random process. Since $X(t)$ is WSS, so is $Y(t)$. Therefore the
      random variable $Y(t)$ has a Gaussian pdf, independent of t. It only remains
      to compute the mean and the variance of $Y(t)$. Since $X(t)$ is zero mean, so is
      $Y(t)$. Therefore

    $$E[Y(t)] = 0 \qquad (2.314)$$

    independent of t. The variance of $Y(t)$ is computed as follows.
    Let $H(f)$ denote the Fourier transform of $h(t)$. Clearly

    $$H(f) = \exp(-\pi f^2). \qquad (2.315)$$

    The overall frequency response of $h(t)$ in cascade with the ideal differentiator
    is

    $$G(f) = j 2\pi f H(f). \qquad (2.316)$$

    Therefore the psd of the random process $Y(t)$ is

    $$S_Y(f) = (N_0/2) 4\pi^2 f^2 \exp(-2\pi f^2). \qquad (2.317)$$

The variance of the random variable $Y(t)$ (this is also the power of the random process $Y(t)$) is

$$\sigma_Y^2 = E\left[|Y(t)|^2\right]$$
$$= \int_{f=-\infty}^{\infty} S_Y(f)\,df$$
$$= \frac{\sqrt{2}}{4} N_0\pi. \tag{2.318}$$

Hence the pdf of the random variable $Y(t)$ is

$$p_Y(y) = \frac{1}{\sigma_Y\sqrt{2\pi}} \exp(-y^2/(2\sigma_Y^2)) \tag{2.319}$$

independent of t, and is hence also the pdf of the random variable $Y(1)$.

59. Let X_1 and X_2 be two independent random variables. X_1 is uniformly distributed between $[-1,\ 1]$ and X_2 has a triangular pdf between $[-3,\ 3]$ with a peak at zero. Find the pdf of $Z = \max(X_1,\ X_2)$.

- *Solution*: Let us consider a more general problem given by

$$Z = \max(X_1,\ \ldots,\ X_n), \tag{2.320}$$

where $X_1,\ \ldots,\ X_n$ are independent random variables. Now

$$P(Z < z) = P((X_1 < z)\,\text{AND}\,\ldots\,\text{AND}\,(X_n < z))$$
$$= P(X_1 < z)\,\ldots\,P(X_n < z). \tag{2.321}$$

Therefore

$$f_Z(z) = \frac{d}{dz}P(Z < z)$$
$$= \frac{d}{dz}\left[P(X_1 < z)\,\ldots\,P(X_n < z)\right]. \tag{2.322}$$

In the given problem

$$P(Z < z) = P(X_1 < z)P(X_2 < z). \tag{2.323}$$

Moreover, it can be seen that Z lies in the range $[-1,\ 3]$. In order to compute (2.323), we need to partion the range of Z into three intervals, given by $[-1,\ 0]$, $[0,\ 1]$ and $[1,\ 3]$. Now

$$P(X_1 < z) = \begin{cases} (z+1)/2 & \text{for } -1 \le z \le 0 \\ (z+1)/2 & \text{for } 0 \le z \le 1 \\ 1 & \text{for } 1 \le z \le 3 \end{cases}, \qquad (2.324)$$

where we have used the fact that

$$\begin{aligned} P(X_1 < z) &= \int_{x_1=-1}^{z} f_{X_1}(x_1) \, dx_1 \\ &= \frac{1}{2} \int_{x_1=-1}^{z} dx_1 \\ &= \frac{z+1}{2}. \end{aligned} \qquad (2.325)$$

Similarly, noting that $f_{X_2}(0) = 1/3$, we obtain

$$P(X_2 < z) = \begin{cases} z^2/18 + z/3 + 1/2 & \text{for } -1 \le z \le 0 \\ -z^2/18 + z/3 + 1/2 & \text{for } 0 \le z \le 1 \\ -z^2/18 + z/3 + 1/2 & \text{for } 1 \le z \le 3. \end{cases} \qquad (2.326)$$

Substituting (2.324) and (2.326) in (2.323) and using (2.322) we get

$$f_Z(z) = \begin{cases} z^2/12 + 7z/18 + 5/12 & \text{for } -1 \le z \le 0 \\ -z^2/12 + 5z/18 + 5/12 & \text{for } 0 \le z \le 1 \\ -z/9 + 1/3 & \text{for } 1 \le z \le 3. \end{cases} \qquad (2.327)$$

60. Let X_1 and X_2 be two independent random variables. X_1 is uniformly distributed between $[-1, 1]$ and X_2 has a triangular pdf between $[-3, 3]$ with a peak at zero. Find the pdf of $Z = \min(X_1, X_2)$.

- *Solution*: Let us consider a more general problem given by

$$Z = \min(X_1, \ldots, X_n), \qquad (2.328)$$

where $X_1, \ldots, X_n$ are independent random variables. Now

$$\begin{aligned} P(Z > z) &= P((X_1 > z) \text{ AND } \ldots \text{ AND } (X_n > z)) \\ &= P(X_1 > z) \ldots P(X_n > z). \end{aligned} \qquad (2.329)$$

Therefore

$$\begin{aligned} f_Z(z) &= \frac{d}{dz} P(Z < z) \\ &= \frac{d}{dz} [1 - P(Z > z)] \end{aligned}$$

$$= \frac{d}{dz} [1 - P(X_1 > z) \ldots P(X_n > z)]$$

$$= \frac{d}{dz} [1 - (1 - P(X_1 < z)) \ldots (1 - P(X_n < z))]. \quad (2.330)$$

In the given problem

$$P(Z > z) = P(X_1 > z) P(X_2 > z). \quad (2.331)$$

Moreover, it can be seen that Z lies in the range $[-3, 1]$. In order to compute (2.331), we need to partion the range of Z into three intervals, given by $[-3, -1], [-1, 0]$ and $[0, 1]$. Now

$$P(X_1 > z) = 1 - P(X_1 < z)$$

$$= \begin{cases} 1 & \text{for } -3 \le z \le -1 \\ 1 - (z+1)/2 & \text{for } -1 \le z \le 0 \\ 1 - (z+1)/2 & \text{for } 0 \le z \le 1 \end{cases}$$

$$= \begin{cases} 1 & \text{for } -3 \le z \le -1 \\ (1-z)/2 & \text{for } -1 \le z \le 0 \\ (1-z)/2 & \text{for } 0 \le z \le 1 \end{cases} \quad (2.332)$$

where we have used the fact that

$$P(X_1 < z) = \int_{x_1=-1}^{z} f_{X_1}(x_1) \, dx_1$$

$$= \frac{1}{2} \int_{x_1=-1}^{z} dx_1$$

$$= \frac{z+1}{2}. \quad (2.333)$$

Similarly, noting that $f_{X_2}(0) = 1/3$, we obtain

$$P(X_2 > z) = 1 - P(X_2 < z)$$

$$= \begin{cases} 1 - z^2/18 - z/3 - 1/2 & \text{for } -3 \le z \le -1 \\ 1 - z^2/18 - z/3 - 1/2 & \text{for } -1 \le z \le 0 \\ 1 + z^2/18 - z/3 - 1/2 & \text{for } 0 \le z \le 1 \end{cases}$$

$$= \begin{cases} -z^2/18 - z/3 + 1/2 & \text{for } -3 \le z \le -1 \\ -z^2/18 - z/3 + 1/2 & \text{for } -1 \le z \le 0 \\ z^2/18 - z/3 + 1/2 & \text{for } 0 \le z \le 1. \end{cases} \quad (2.334)$$

Substituting (2.332) and (2.334) in (2.331) and using (2.330) we get

$$f_Z(z) = \begin{cases} z/9 + 1/3 & \text{for } -3 \leq z \leq -1 \\ -z^2/12 - 5z/18 + 5/12 & \text{for } -1 \leq z \leq 0 \\ z^2/12 - 7z/18 + 5/12 & \text{for } 0 \leq z \leq 1. \end{cases} \qquad (2.335)$$

References

Simon Haykin. *Communication Systems*. Wiley Eastern, second edition, 1983.

A. Papoulis. *Probability, Random Variables and Stochastic Processes*. McGraw-Hill, third edition, 1991.

Rodger E. Ziemer and William H. Tranter. *Principles of Communications*. John Wiley, fifth edition, 2002.

Chapter 3
Amplitude Modulation

1. (Haykin 1983) Suppose that a non-linear device is available for which the output current i_0 and the input voltage v_i are related by:

$$i_0(t) = a_1 v_i(t) + a_3 v_i^3(t), \tag{3.1}$$

where a_1 and a_3 are constants. Explain how this device may be used to provide (a) a product modulator (b) an amplitude modulator.

- *Solution*: Let

$$v_i(t) = A_c \cos(\pi f_c t) + m(t), \tag{3.2}$$

where $m(t)$ occupies the frequency band $[-W, +W]$. Then

$$
\begin{aligned}
i_0(t) = {} & a_1 \left(A_c \cos(\pi f_c t) + m(t) \right) + \frac{a_3 A_c^3}{4} \left(3 \cos(\pi f_c t) + \cos(3\pi f_c t) \right) \\
& + \frac{3}{2} a_3 A_c^2 \left(1 + \cos(2\pi f_c t) m(t) \right) + 3 a_3 A_c \cos(\pi f_c t) m^2(t) \\
& + a_3 m^3(t) \\
= {} & \left(a_1 + \frac{3}{2} a_3 A_c^2 \right) m(t) + a_3 m^3(t) + \left(a_1 A_c + \frac{3}{4} a_3 A_c^3 \right) \cos(\pi f_c t) \\
& + 3 a_3 A_c \cos(\pi f_c t) m^2(t) + \frac{3}{2} a_3 A_c^2 m(t) \cos(2\pi f_c t) \\
& + \frac{a_3 A_c^3}{4} \cos(3\pi f_c t). \tag{3.3}
\end{aligned}
$$

Using the fact that multiplication in the time domain is equivalent to convolution in the frequency domain, we know that $m^2(t)$ occupies the frequency

© The Editor(s) (if applicable) and The Author(s), under exclusive license to Springer Nature Switzerland AG 2021
K. Vasudevan, *Analog Communications*,
https://doi.org/10.1007/978-3-030-50337-6_3

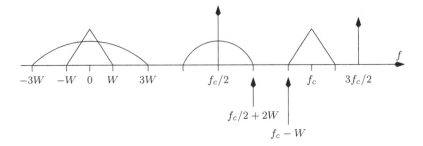

Fig. 3.1 Fourier transform of $i_0(t)$

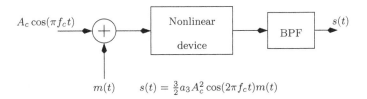

Fig. 3.2 Method of obtaining the DSBSC signal

band $[-2W, +2W]$ and $m^3(t)$ occupies $[-3W, 3W]$. The Fourier transform of $i_0(t)$ is illustrated in Fig. 3.1. To extract the DSBSC component at f_c using a bandpass filter, the following conditions must be satisfied:

$$f_c/2 + 2W < f_c - W$$
$$f_c + W < 3f_c/2. \tag{3.4}$$

The two inequalities in the above equation can be combined to get

$$f_c > 6W. \tag{3.5}$$

The procedure for obtaining the DSBSC signal is illustrated in Fig. 3.2.

The procedure for obtaining the AM signal using two identical nonlinear devices and bandpass filters is illustrated in Fig. 3.3. The amplitude sensitivity is controlled by A_0.

Note that with this method, the amplitude sensitivity and the carrier power can be independently controlled. Moreover, the amplitude sensitivity is independent of a_3, which is not under user control (a_3 is device dependent).

2. (Haykin 1983) In this problem we consider the switching modulator shown in Fig. 3.4. Assume that the carrier wave $c(t)$ applied to the diode is large in amplitude compared to $|m(t)|$, so that the diode acts like an ideal switch, that is

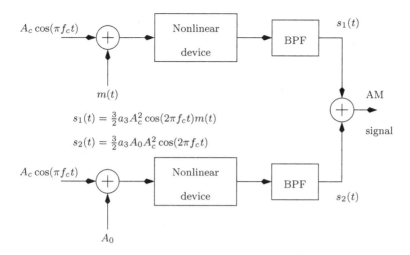

Fig. 3.3 Method of obtaining the AM signal

Fig. 3.4 Block diagram of a
switching modulator

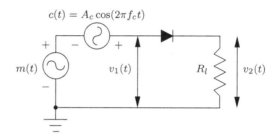

$$v_2(t) = \begin{cases} v_1(t) & c(t) > 0 \\ 0 & c(t) < 0. \end{cases} \tag{3.6}$$

Hence we may write

$$v_2(t) = (A_c \cos(2\pi f_c t) + m(t))\, g_p(t), \tag{3.7}$$

where

$$g_p(t) = \frac{1}{2} + \frac{2}{\pi} \sum_{n=1}^{\infty} \frac{(-1)^{n-1}}{2n-1} \cos(2\pi f_c(2n-1)t). \tag{3.8}$$

Find out the AM signal centered at f_c contained in $v_2(t)$.

- *Solution*: We have

$$v_2(t) = \frac{A_c}{2} \cos(2\pi f_c t)$$

$$+ \frac{A_c}{\pi} \sum_{n=1}^{\infty} \frac{(-1)^{n-1}}{2n-1} \left(\cos(4\pi(n-1)f_c t) + \cos(4\pi n f_c t) \right)$$

$$+ \frac{m(t)}{2} + \frac{2}{\pi} \sum_{n=1}^{\infty} \frac{(-1)^{n-1}}{2n-1} \cos(2\pi(2n-1)f_c t)m(t). \qquad (3.9)$$

Clearly, the desired AM signal is

$$\frac{A_c}{2} \cos(2\pi f_c t) + \frac{2}{\pi} \cos(2\pi f_c t)m(t). \qquad (3.10)$$

3. (Haykin 1983) Consider the AM signal

$$s(t) = [1 + 2\cos(2\pi f_m t)]\cos(2\pi f_c t). \qquad (3.11)$$

The AM signal $s(t)$ is applied to an ideal envelope detector, producing output $v(t)$. Determine the real (not complex) Fourier series representation of $v(t)$.

- *Solution*: The output of the ideal envelope detector is given by

$$v(t) = |1 + 2\cos(2\pi f_m t)|, \qquad (3.12)$$

which is periodic with a period of $1/f_m$. The envelope detector output, $v(x)$ is plotted in Fig. 3.5 where

$$x = 2\pi f_m t. \qquad (3.13)$$

Fig. 3.5 Output of the ideal envelope detector

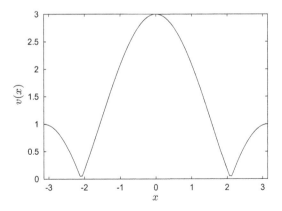

Now

$$1 + 2\cos(x) < 0 \quad \text{for } \pi - \pi/3 < x < \pi + \pi/3. \tag{3.14}$$

This implies that

$$v(x) = \begin{cases} 1 + 2\cos(x) & \text{for } 0 < |x| < \pi - \pi/3 \\ -1 - 2\cos(x) & \text{for } \pi - \pi/3 < |x| < \pi \end{cases} \tag{3.15}$$

Since $v(x)$ is an even function, it has a Fourier series representation given by:

$$v(x) = a_0 + 2 \sum_{n=1}^{\infty} a_n \cos(nx), \tag{3.16}$$

where

$$a_0 = \frac{2}{2\pi} \int_{x=0}^{\pi-\pi/3} (1 + 2\cos(x)) \, dx - \frac{2}{2\pi} \int_{\pi-\pi/3}^{\pi} (1 + 2\cos(x)) \, dx$$

$$= \frac{1}{3} + \frac{2\sqrt{3}}{\pi}. \tag{3.17}$$

Similarly

$$a_n = \frac{2}{2\pi} \int_{x=0}^{\pi-\pi/3} (1 + 2\cos(x)) \cos(nx) \, dx$$

$$\quad - \frac{2}{2\pi} \int_{\pi-\pi/3}^{\pi} (1 + 2\cos(x)) \cos(nx) \, dx$$

$$= \frac{2}{\pi} \left[\frac{\sin(2n\pi/3)}{n} + \frac{\sin(2(n-1)\pi/3)}{n-1} + \frac{\sin(2(n+1)\pi/3)}{n+1} \right]. \tag{3.18}$$

4. Consider the AM signal

$$s(t) = [1 + 3\sin(2\pi f_m t + \alpha)] \sin(2\pi f_c t + \theta). \tag{3.19}$$

The AM signal $s(t)$ is applied to an ideal envelope detector, producing output $v(t)$.

(a) Find $v(t)$.
(b) Find α, $0 \le \alpha < \pi$, such that the real Fourier series representation of $v(t)$ contains only dc and cosine terms.
(c) Determine the real Fourier series representation of $v(t)$ for α obtained in (b).

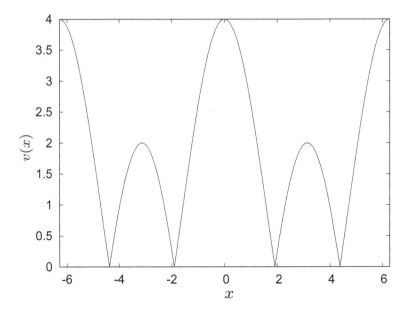

Fig. 3.6 Output of the ideal envelope detector for $\alpha = \pi/2$

- *Solution*: The output of the ideal envelope detector is given by

$$
\begin{aligned}
v(t) &= |1 + 3\sin(2\pi f_m t + \alpha)| \sqrt{\sin^2(\theta) + \cos^2(\theta)} \\
&= |1 + 3\sin(2\pi f_m t + \alpha)|
\end{aligned}
\tag{3.20}
$$

which is periodic with a period of $1/f_m$. The envelope detector output, $v(x)$ is plotted in Fig. 3.6 where

$$
x = 2\pi f_m t.
\tag{3.21}
$$

Note that $v(x)$ is periodic with a period 2π. If the real Fourier series representation of $v(x)$ is to contain only dc and cosine terms, we require:

$$
\alpha = \pi/2.
\tag{3.22}
$$

Therefore

$$
\begin{aligned}
v(x) &= |1 + 3\cos(x)| \\
&= a_0 + 2\sum_{n=1}^{\infty} a_n \cos(nx).
\end{aligned}
\tag{3.23}
$$

Let

$$v(x_1) = |1 + 3\cos(x_1)|$$
$$= 0$$
$$\Rightarrow x_1 = \pi - \cos^{-1}(1/3)$$
$$= 1.911\,\text{rad}. \tag{3.24}$$

Now

$$a_0 = \frac{2}{2\pi} \int_{x=0}^{x_1} (1 + 3\cos(x))\, dx - \frac{2}{2\pi} \int_{x_1}^{\pi} (1 + 3\cos(x))\, dx$$
$$= \frac{4x_1 - 2\pi + 12\sin(x_1)}{2\pi}$$
$$= 2.02 \tag{3.25}$$

and

$$a_n = \frac{2}{2\pi} \int_{x=0}^{x_1} (1 + 3\cos(x))\cos(nx)\, dx$$
$$- \frac{2}{2\pi} \int_{x_1}^{\pi} (1 + 3\cos(x))\cos(nx)\, dx$$
$$= \frac{2}{\pi} \frac{\sin(nx_1)}{n} + \frac{3}{\pi} \frac{\sin((n-1)x_1)}{n-1} + \frac{3}{\pi} \frac{\sin((n+1)x_1)}{n+1} \quad \text{for } n > 1 \tag{3.26}$$

and

$$a_1 = \frac{2}{2\pi} \int_{x=0}^{x_1} (1 + 3\cos(x))\cos(x)\, dx$$
$$- \frac{2}{2\pi} \int_{x_1}^{\pi} (1 + 3\cos(x))\cos(x)\, dx$$
$$= \frac{2}{\pi} \sin(x_1) + \frac{3}{\pi} x_1 + \frac{3}{\pi} \frac{\sin(2x_1)}{2} - \frac{3}{2}. \tag{3.27}$$

5. (Haykin 1983) The AM signal

$$s(t) = A_c [1 + k_a m(t)] \cos(2\pi f_c t) \tag{3.28}$$

is applied to the system in Fig. 3.7. Assuming that $|k_a m(t)| < 1$ for all t and $m(t)$ is bandlimited to $[-W,\ W]$, and that the carrier frequency $f_c > 2W$, which show that $m(t)$ can be obtained from the square rooter output, $v_3(t)$.

Fig. 3.7 Nonlinear demodulation of AM signals

Fig. 3.8 Spectrum of the message signal

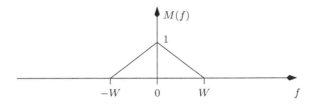

- *Solution*: We have

$$v_1(t) = \frac{A_c^2}{2} \left[1 + \cos(4\pi f_c t)\right]\left[1 + k_a m(t)\right]^2. \tag{3.29}$$

We also have (assuming an ideal LPF)

$$v_2(t) = \frac{A_c^2}{2} \left[1 + k_a m(t)\right]^2. \tag{3.30}$$

Therefore

$$v_3(t) = \frac{A_c}{\sqrt{2}} \left[1 + k_a m(t)\right]. \tag{3.31}$$

6. (Haykin 1983) Consider a message signal $m(t)$ with spectrum shown in Fig. 3.8, with $W = 1\,\text{kHz}$. This message is DSB-SC modulated using a carrier of the form $A_c \cos(2\pi f_c t)$, producing the signal $s(t)$. The modulated signal is next applied to a coherent detector with carrier $A_0 \cos(2\pi f_c t)$. Determine the spectrum of the detector output when the carrier is (a) $f_c = 1.25\,\text{kHz}$ (b) $f_c = 0.75\,\text{kHz}$. Assume that the LPF in the demodulator is ideal with unity gain. What is the lowest carrier frequency for which there is no aliasing (no overlap in the frequency spectrum) in the modulated signal $s(t)$?

- *Solution*: We have

$$s(t) = A_c \cos(2\pi f_c t) m(t)$$

$$\Rightarrow S(f) = \frac{A_c}{2} \left[M(f - f_c) + M(f + f_c)\right]. \tag{3.32}$$

The block diagram of the coherent demodulator is shown in Fig. 3.9. We have

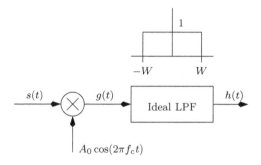

Fig. 3.9 A coherent demodulator for DSB-SC signals. An ideal LPF is assumed

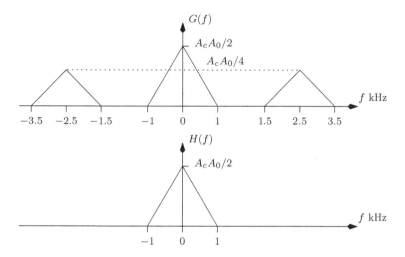

Fig. 3.10 Spectrum at the outputs of the multiplier and the LPF for $f_c = 1.25$

$$g(t) = A_c A_0 m(t) \cos^2(2\pi f_c t)$$

$$= \frac{A_c A_0 m(t)}{2} [1 + \cos(4\pi f_c t)]$$

$$\Rightarrow G(f) = \frac{A_c A_0 M(f)}{2} + \frac{A_c A_0}{4} [M(f - 2f_c) + M(f + 2f_c)].$$

$$(3.33)$$

The spectrum of $H(f)$ for $f_c = 1.25$ and $f_c = 0.75$ kHz are shown in Figs. 3.10 and 3.11, respectively.

For no aliasing the following condition must be satisfied:

$$2f_c - W > W$$

$$\Rightarrow f_c > W.$$

$$(3.34)$$

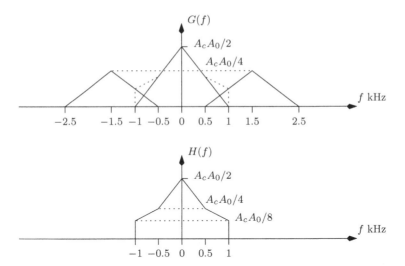

Fig. 3.11 Spectrum at the outputs of the multiplier and the LPF for $f_c = 0.75$

7. (Ziemer and Tranter 2002) An amplitude modulator has output

$$s(t) = A\cos(\pi 400t) + B\cos(\pi 360t) + B\cos(\pi 440t). \qquad (3.35)$$

The carrier power is 100 W and the power efficiency (ratio of sideband power to total power) is 40%. Compute A, B and the modulation factor μ.

- *Solution*: The general expression for a tone modulated AM signal is:

$$\begin{aligned} s(t) &= A_c[1 + \mu\cos(2\pi f_m t)]\cos(2\pi f_c t) \\ &= A_c\cos(2\pi f_c t) + \frac{\mu A_c}{2}\cos(2\pi(f_c - f_m)t) \\ &\quad + \frac{\mu A_c}{2}\cos(2\pi(f_c + f_m)t). \end{aligned} \qquad (3.36)$$

In the given problem

$$\begin{aligned} A &= A_c \\ B &= \frac{\mu A}{2} \\ f_c &= 200\,\text{Hz} \\ f_c - f_m &= 180\,\text{Hz} \\ f_c + f_m &= 220\,\text{Hz}. \end{aligned} \qquad (3.37)$$

The carrier power is given by:

$$\frac{A^2}{2} = 100W$$

$$\Rightarrow A = \sqrt{200}. \tag{3.38}$$

Since the given AM wave is tone modulated, the power efficiency is given by:

$$\frac{\mu^2}{2 + \mu^2} = \frac{2}{5}$$

$$\Rightarrow \mu = 1.155. \tag{3.39}$$

The constant B is given by:

$$B = \frac{\mu A}{2} = 8.165. \tag{3.40}$$

8. Consider a modulating wave $m(t)$ such that $(1 + k_a m(t)) > 0$ for all t. Assume that the spectrum of $m(t)$ is zero for $|f| > W$. Let

$$s(t) = A_c[1 + k_a m(t)] \cos(2\pi f_c t), \tag{3.41}$$

where $f_c > W$. The modulated wave $s(t)$ is applied to a series combination of an ideal full-wave rectifier and an ideal lowpass filter. The transfer function of the ideal lowpass filter is:

$$H(f) = \begin{cases} 1 \text{ for } |f| < W \\ 0 \text{ otherwise.} \end{cases} \tag{3.42}$$

Compute the time-domain output $v(t)$ of the lowpass filter.

- *Solution*: The output of the full-wave rectifier is given by:

$$|s(t)| = A_c[1 + k_a m(t)]|\cos(2\pi f_c t)|. \tag{3.43}$$

Let

$$2\pi f_c t = x. \tag{3.44}$$

Then $|\cos(x)|$ has a Fourier series representation:

$$|\cos(x)| = a_0 + 2\sum_{n=1}^{\infty} (a_n \cos(nx) + b_n \sin(nx)). \tag{3.45}$$

Note that $b_n = 0$ since $|\cos(x)|$ is an even function. Moreover $|\cos(x)|$ is periodic with period π. The coefficients a_n are given by:

$$a_0 = \frac{1}{\pi} \int_{x=-\pi/2}^{\pi/2} \cos(x)\,dx$$

$$= \frac{2}{\pi}$$

$$a_n = \frac{1}{\pi} \int_{x=-\pi/2}^{\pi/2} \cos(x)\cos(nx)\,dx$$

$$= \begin{cases} 0 & \text{for } n = 2m+1 \\ K_n & \text{for } n = 2m \end{cases} \tag{3.46}$$

where K_n is given by:

$$K_n = \frac{1}{\pi(n-1)}\sin((n-1)\pi/2) + \frac{1}{\pi(n+1)}\sin((n+1)\pi/2). \tag{3.47}$$

Thus the output of the LPF is clearly:

$$v(t) = A_c[1 + k_a m(t)]a_0 = \frac{2A_c}{\pi}[1 + k_a m(t)]. \tag{3.48}$$

9. (Haykin 1983) This problem is related to the 2two-stage approach for generating SSB (frequency discrimination method) signals. A voice signal $m(t)$ occupying the frequency band 0.3–3.4 kHz is to be SSB modulated with only the upper sideband transmitted. Assume the availability of bandpass filters which provide an attenuation of 50 dB in a transition band that is one percent of the center frequency of the bandpass filter, as illustrated in Fig. 3.12a. Assume that the first stage eliminates the lower sideband and the product modulator in the second stage uses a carrier frequency of 11.6 MHz. The message spectrum is shown in Fig. 3.12b and the spectrum of the final SSB signal must be as shown in Fig. 3.12c.

   Find the range of the carrier frequencies that can be used by the product modulator in the first stage, so that the unwanted sideband is attenuated by no less than 50 dB.

   • *Solution*: Firstly, we note that if only a single stage were was employed, the required Q-factor of the BPF would be very high, since

   $$\text{Q-factor} \approx \frac{\text{centre frequency}}{\text{transition bandwidth}}. \tag{3.49}$$

   Secondly, we would like to align the center frequency of the bandpass filter in the center of the message band to be transmitted. The transmitted message band could be the upper or the lower sideband. Let us denote the carrier frequency of the first stage by f_1 and that of the second stage by f_2. Then, the

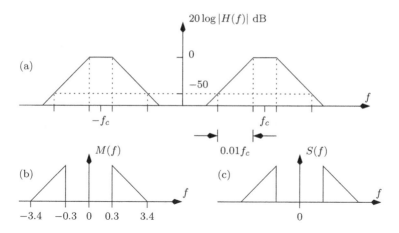

Fig. 3.12 a Magnitude response of the bandpass filters. **b** Message spectrum. **c** Spectrum of the final SSB signal

center frequency of the BPF at the second stage is given by (see Fig. 3.13):

$$f_{c_2} = \frac{1}{2}[f_2 + f_1 + f_a + f_2 + f_1 + f_b]. \tag{3.50}$$

In the above equation:

$$f_2 = 11600\,\text{kHz}$$
$$f_a = 0.3\,\text{kHz}$$
$$f_b = 3.4\,\text{kHz}. \tag{3.51}$$

The transition bandwidth of the second BPF is $0.01 f_{c_2}$. The actual transition bandwidth at the output of the first SSB modulator is $2(f_1 + f_a)$. We require that

$$2(f_1 + f_a) \geq 0.01 f_{c_2}$$
$$\Rightarrow f_1(1 - 0.005) \geq 58 + 0.01 \times 0.925 - 0.3$$
$$f_1 \geq 57.999\,\text{kHz}. \tag{3.52}$$

The center frequency of the first BPF is given by:

$$f_{c_1} = \frac{1}{2}[f_1 + f_a + f_1 + f_b]$$
$$= f_1 + 1.85. \tag{3.53}$$

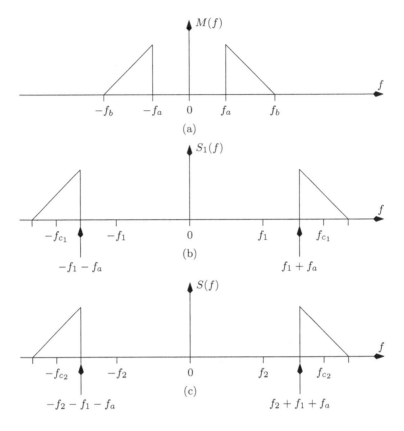

Fig. 3.13 **a** Message spectrum. **b** SSB signal at the output of the first stage. **c** SSB signal at the output of the second stage

The transition bandwidth of the first BPF is $0.01 f_{c_1}$ and the actual transition bandwidth of the input message is $2 f_a$. We require that:

$$2 f_a \geq 0.01 f_{c_1}$$
$$\Rightarrow f_1 \leq 58.15 \, \text{kHz}. \tag{3.54}$$

Thus the range of f_1 is given by $57.999 \leq f_1 \leq 58.15 \, \text{kHz}$.

10. (Haykin 1983) This problem is related to the 2two-stage approach for generating SSB signals. A voice signal $m(t)$ occupying the frequency band 0.3–3 kHz is to be SSB modulated with only the lower sideband transmitted. Assume the availability of bandpass filters which provide an attenuation of 60 dB in a transition band that is two percent of the center frequency of the bandpass filter, as illustrated in Fig. 3.14a. Assume that the first stage eliminates the upper sideband and the product modulator in the second stage uses a carrier frequency of 1 MHz.

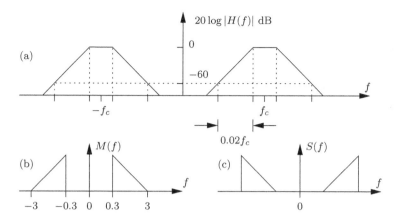

Fig. 3.14 **a** Magnitude response of the bandpass filters. **b** Message spectrum. **c** Spectrum of the final SSB signal

The message spectrum is shown in Fig. 3.14b and the spectrum of the final SSB signal must be as shown in Fig. 3.14c.

(a) Find the range of carrier frequencies that can be used by the product modulator in first stage, so that the unwanted sideband is attenuated by no less than 60 dB.

(b) Write down the expression of the SSB signal $s(t)$ at the output of the second stage, in terms of $m(t)$. Clearly specify the carrier frequency of $s(t)$ in terms of the carrier frequency of the two product modulators.

● *Solution*: Firstly, we note that if only a single stage were was employed, the required Q-factor of the BPF would be very high, since

$$Q\text{-factor} \approx \frac{\text{centre frequency}}{\text{transition bandwidth}}. \tag{3.55}$$

Secondly, we would like to align the center frequency of the bandpass filter in the center of the message band to be transmitted. The transmitted message band could be the upper or the lower sideband. Let us denote the carrier frequency of the first stage by f_1 and that of the second stage by f_2. Then, the center frequency of the BPF at the second stage is given by (see Fig. 3.15):

$$f_{c_2} = \frac{1}{2}[f_2 + f_1 - f_b + f_2 + f_1 - f_a]. \tag{3.56}$$

In the above equation:

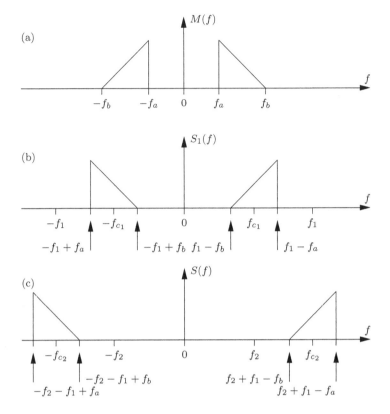

Fig. 3.15 **a** Message spectrum. **b** SSB signal at the output of the first stage. **c** SSB signal at the output of the second stage

$$f_2 = 1000\,\text{kHz}$$
$$f_a = 0.3\,\text{kHz}$$
$$f_b = 3.0\,\text{kHz}. \tag{3.57}$$

The transition bandwidth of the second BPF is $0.02 f_{c_2}$. The actual transition bandwidth at the output of the first SSB modulator is $2(f_1 - f_b)$. We require that

$$2(f_1 - f_b) \geq 0.02 f_{c_2}$$
$$\Rightarrow f_1 - f_b \geq 0.01[f_2 + f_1 - 0.5(f_a + f_b)]$$
$$f_1 \geq 13.1146\,\text{kHz}. \tag{3.58}$$

The center frequency of the first BPF is given by:

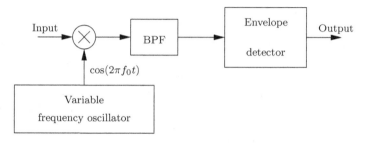

Fig. 3.16 Receiver for a frequency division multiplexing system

Fig. 3.17 Input signal to the FDM receiver in Fig. 3.16

$$f_{c_1} = \frac{1}{2}[f_1 - f_a + f_1 - f_b]$$
$$= f_1 - 1.65. \tag{3.59}$$

The transition bandwidth of the first BPF is $0.02 f_{c_1}$ and the actual transition bandwidth of the input message is $2 f_a$. We require that:

$$2 f_a \geq 0.02 f_{c_1}$$
$$\Rightarrow f_1 \leq 31.65 \,\text{kHz}. \tag{3.60}$$

Thus, the range of f_1 is given by $13.1146 \leq f_1 \leq 31.65 \,\text{kHz}$.
The final SSB signal is given by:

$$s(t) = m(t) \cos(2\pi f_c t) + \hat{m}(t) \sin(2\pi f_c t), \tag{3.61}$$

where $f_c = f_2 + f_1 - f_a + f_a = f_2 + f_1$.

11. Consider the receiver in Fig. 3.16, for a frequency division multiplexing (FDM) system. An FDM system is similar to the simultaneous radio broadcast by many stations. Assume that the BPF is ideal with a bandwidth of $2W = 10 \,\text{kHz}$.

   (a) Now consider the input signal in Fig. 3.17. Assume that the input signals are amplitude modulated, $f_0 = 1.4 \,\text{MHz}$, $f_{c_1} = 1.1 \,\text{MHz}$, and $f_{c_2} = 1.7 \,\text{MHz}$.

Assume that the BPF is centered at 300 kHz. Find out the expression for the signal at the output of the BPF. The frequency f_{c_2} is called the *image* of f_{c_1}.

(b) Now assume that $f_{c_1} = 1.1\,\text{MHz}$ is the desired carrier and $f_{c_2} = 1.7\,\text{MHz}$ is to be rejected. Assume that f_0 can only be varied between f_{c_1} and f_{c_2} and that the center frequency of the BPF cannot be less than 200 kHz (this is required for proper envelope detection) and cannot be greater than 300 kHz (so that the Q-factor of the BPF is within practical limits). Find out the permissible range of values f_0 can take. Also find out the *corresponding* permissible range of values of the center frequency of the BPF.

- *Solution*: Let the input signal be given by:

$$s(t) = A_1[1 + k_{a_1} m_1(t)]\cos(2\pi f_{c_1} t) + A_2[1 + k_{a_2} m_2(t)]\cos(2\pi f_{c_2} t).$$
$$\tag{3.62}$$

The output of the multiplier can be written as:

$$s(t)\,\frac{A_1}{2}[1 + k_{a_1} m_1(t)]\big[\cos(2\pi(f_0 - f_{c_1})t) + \cos(2\pi(f_0 + f_{c_1})t)\big]$$
$$+ \frac{A_2}{2}[1 + k_{a_2} m_2(t)]\big[\cos(2\pi(f_{c_2} - f_0)t) + \cos(2\pi(f_{c_2} + f_0)t)\big].$$
$$\tag{3.63}$$

The output of the BPF is:

$$y(t) = \frac{A_1}{2}[1 + k_{a_1} m_1(t)]\cos(2\pi(f_0 - f_{c_1})t)$$
$$+ \frac{A_2}{2}[1 + k_{a_2} m_2(t)]\cos(2\pi(f_{c_2} - f_0)t).$$
$$\tag{3.64}$$

Now consider Fig. 3.18. Denote the two difference (IF) frequencies by x and $B - x$, where $B = f_{c_2} - f_{c_1}$. To reject $B - x$, we must have:

$$B - x - x > 2W$$
$$\Rightarrow x < 295.$$
$$\tag{3.65}$$

Thus, the BPF center frequency can vary between 200 and 295 kHz. Correspondingly, f_0 can vary between 1.3 and 1.395 MHz.

12. Consider the Costas loop for AM signals as shown in Fig. 3.19. Let the received signal $r(t)$ be given by:

$$r(t) = A_c[1 + k_a m(t)]\cos(2\pi f_c t + \theta) + w(t),$$
$$\tag{3.66}$$

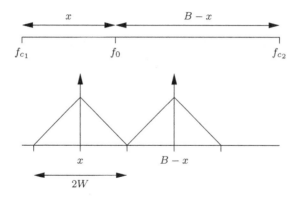

Fig. 3.18 Figure to illustrate the condition to be satisfied by the two difference frequencies, so that one of them is rejected by the BPF

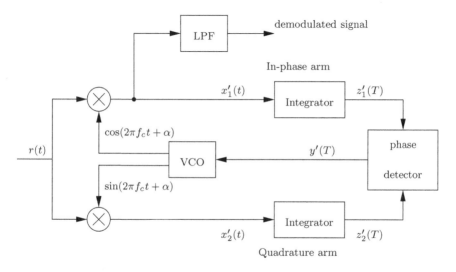

Fig. 3.19 Costas loop for AM signals

where $w(t)$ is zero-mean additive white Gaussian noise (AWGN) with psd $N_0/2$. The random variable $z_1'(T)$ is computed as:

$$z_1'(T) = \frac{1}{T} \int_0^T x_1'(t)\, dt. \tag{3.67}$$

The random variable $z_2'(T)$ is similarly computed. The random variable $y'(T)$ is computed as:

$$y'(T) = \frac{z_2'(T)}{z_1'(T)}. \tag{3.68}$$

Assume that:

(a) θ and α are uniformly distributed random variables.
(b) $\alpha - \theta$ is a constant and close to zero.
(c) $w(t)$ and α are statistically independent.
(d) $m(t)$ has zero -mean (zero dc).
(e) T is large, hence the integrator output is a dc term plus noise.
(f) The signal-to-noise ratio (SNR) at the input to the phase detector is high.

Compute the mean and the variance of the random variables $z_1'(T)$, $z_2'(T)$ and $y'(T)$.

- *Solution*: We have

$$x_1'(t) = \frac{A_c}{2}[1 + k_a m(t)][\cos(\alpha - \theta) + \cos(4\pi f_c t + \alpha + \theta)] + a_1(t)$$

$$x_2'(t) = \frac{A_c}{2}[1 + k_a m(t)][\sin(\alpha - \theta) + \sin(4\pi f_c t + \alpha + \theta)] + a_2(t)$$

$$\tag{3.69}$$

where

$$a_1(t) = w(t)\cos(2\pi f_c t + \alpha)$$
$$a_2(t) = w(t)\sin(2\pi f_c t + \alpha). \tag{3.70}$$

The autocorrelation of $a_1(t)$ and $a_2(t)$ is given by:

$$R_{a_1}(\tau) = E[a_1(t)a_1(t - \tau)]$$
$$= \frac{N_0}{4}\delta(\tau)$$
$$= R_{a_2}(\tau). \tag{3.71}$$

Since $w(t)$ and α are statistically independent

$$E[a_1(t)] = E[a_2(t)] = 0. \tag{3.72}$$

The integrator outputs are:

$$z_1'(T) = \frac{A_c}{2}\cos(\alpha - \theta) + b_1$$
$$z_2'(T) = \frac{A_c}{2}\sin(\alpha - \theta) + b_2, \tag{3.73}$$

where

$$b_1 = \frac{1}{T} \int_{t=0}^{T} a_1(t)\, dt$$

$$, b_2 = \frac{1}{T} \int_{t=0}^{T} a_2(t)\, dt. \tag{3.74}$$

Then

$$E[b_1] = \frac{1}{T} \int_{t=0}^{T} E[a_1(t)]\, dt$$
$$= 0$$
$$= E[b_2]. \tag{3.75}$$

The variance of b_1 and b_2 are given by:

$$E[b_1^2] = \frac{N_0}{4} \int_{f=-\infty}^{\infty} |H(f)|^2\, df$$
$$= \frac{N_0}{4} \int_{t=-\infty}^{\infty} h^2(t)\, dt$$
$$= \frac{N_0}{4T^2} \int_{t=0}^{T} dt$$
$$= \frac{N_0}{4T}$$
$$= E[b_2^2]. \tag{3.76}$$

In the above equation we have used the Rayleigh's energy theorem. Thus:

$$E[z_1'(T)] = \frac{A_c}{2} \cos(\alpha - \theta)$$
$$E[z_2'(T)] = \frac{A_c}{2} \sin(\alpha - \theta)$$
$$\mathrm{var}\,[z_1'(T)] = \frac{N_0}{4T}$$
$$\mathrm{var}\,[z_2'(T)] = \frac{N_0}{4T}. \tag{3.77}$$

The phase detector output is:

$$y'(T) = \frac{(A_c/2) \sin(\alpha - \theta) + b_2}{(A_c/2) \cos(\alpha - \theta) + b_1}$$
$$\approx \alpha - \theta + \frac{2b_2}{A_c}. \tag{3.78}$$

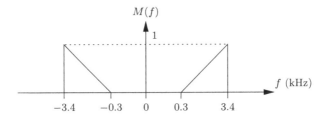

Fig. 3.20 Message spectrum

wWhere we have made the high SNR approximation. Hence, the mean and variance of $y'(T)$ is:

$$E[y'(T)]] = \alpha - \theta$$
$$\text{var}\,[y'(T)] = \frac{4}{A_c^2}\frac{N_0}{4T}. \tag{3.79}$$

13. Consider a message signal $m(t)$ whose spectrum extends over 300 Hz to 3.4 kHz, as illustrated in Fig. 3.20. This message is SSB modulated to obtain:

$$s(t) = A_c m(t) \cos(2\pi f_c t) + A_c \hat{m}(t) \sin(2\pi f_c t). \tag{3.80}$$

At the receiver, $s(t)$ is demodulated using a carrier of the form $\cos(2\pi(f_c + \Delta f)t)$. Plot the spectrum of the demodulated signal at the output of the lowpass filter when

(a) $\Delta f = 10$ Hz,
(b) $\Delta f = -10$ Hz.

Assume that the lowpass filter is ideal with unity gain and extends over $[-4, 4]$ kHz, and $f_c = 20$ kHz. Show all the steps required to arrive at the answer. Indicate all the important points on the XY-axes.

- *Solution*: The multiplier output can be written as:

$$\begin{aligned}
s_1(t) &= s(t) \cos(2\pi(f_c + \Delta f)t) \\
&= A_c \left[m(t) \cos(2\pi f_c t) + \hat{m}(t) \sin(2\pi f_c t) \right] \cos(2\pi(f_c + \Delta f)t) \\
&= \frac{A_c}{2} m(t) \left[\cos(2\pi \Delta f t) + \cos(2\pi(2f_c + \Delta f)t) \right] \\
&\quad + \frac{A_c}{2} \hat{m}(t) \left[\sin(2\pi(2f_c + \Delta f)t) - \sin(2\pi \Delta f t) \right]. \tag{3.81}
\end{aligned}$$

The output of the lowpass filter is:

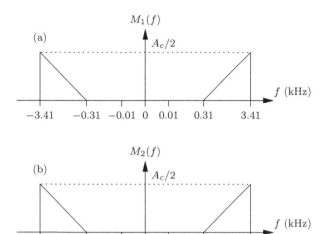

Fig. 3.21 Demodulated message spectrum in the presence of frequency offset

$$s_2(t) = \frac{A_c}{2} \left[m(t) \cos(2\pi \Delta f t) - \hat{m}(t) \sin(2\pi \Delta f t) \right]. \qquad (3.82)$$

When Δf is positive, $s_2(t)$ is an SSB signal with carrier frequency Δf and upper sideband transmitted. This is illustrated in Fig. 3.21a, for $\Delta f = 10$ Hz. When Δf is negative, $s_2(t)$ is an SSB signal with carrier frequency Δf and lower sideband transmitted. This is illustrated in Fig. 3.21b with $\Delta f = -10$.

14. Compute the envelope of the following signal:

$$s(t) = A_c[1 + 2\cos(2\pi f_m t)] \cos(2\pi f_c t + \theta). \qquad (3.83)$$

- *Solution*: The signal can be expanded as:

$$s(t) = A_c[1 + 2\cos(2\pi f_m t)] \left[\cos(2\pi f_c t) \cos(\theta) - \sin(2\pi f_c t) \sin(\theta) \right]. \qquad (3.84)$$

Therefore the envelope is:

$$\begin{aligned} a(t) &= A_c|1 + 2\cos(2\pi f_m t)|\sqrt{\cos^2(\theta) + \sin^2(\theta)} \\ &= A_c|1 + 2\cos(2\pi f_m t)|. \end{aligned} \qquad (3.85)$$

15. For the message signal shown in Fig. 3.22 compute the power efficiency (the ratio of sideband power to total power) in terms of the amplitude sensitivity k_a, A_1, A_2, and T.

The message is periodic with period T, has zero -mean, and is AM modulated

Fig. 3.22 Message signal

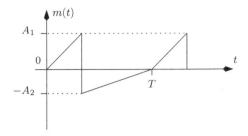

Fig. 3.23 Message signal

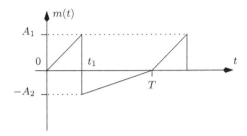

according to:

$$s(t) = A_c[1 + k_a m(t)] \cos(2\pi f_c t). \tag{3.86}$$

Assume that $T \gg 1/f_c$.

- *Solution*: Since the message is zero -mean we must have:

$$\frac{1}{2} t_1 A_1 = \frac{1}{2}(T - t_1) A_2$$
$$\Rightarrow t_1 = \frac{A_2 T}{A_1 + A_2}, \tag{3.87}$$

where t_1 is shown in Fig. 3.23. For a general AM signal given by

$$s(t) = A_c[1 + k_a m(t)] \cos(2\pi f_c t), \tag{3.88}$$

the power efficiency is:

$$\eta = \frac{A_c^2 k_a^2 P_m/2}{A_c^2/2 + A_c^2 k_a^2 P_m/2} = \frac{k_a^2 P_m}{1 + k_a^2 P_m}, \tag{3.89}$$

where P_m denotes the message power. Here we only need to compute P_m. We have

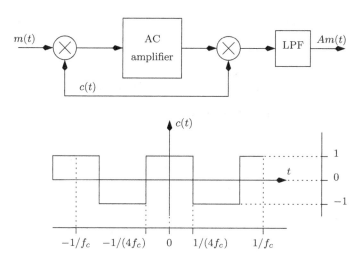

Fig. 3.24 Block diagram of a chopper-stabilized dc amplifier

$$P_m = \frac{1}{T} \int_{t=0}^{T} m^2(t)\, dt, \tag{3.90}$$

where

$$m(t) = \begin{cases} A_1 t/t_1 & \text{for } 0 < t < t_1 \\ A_2(t - T)/(T - t_1) & \text{for } t_1 < t < T. \end{cases} \tag{3.91}$$

Substituting for $m(t)$ in (3.90) we get:

$$\begin{aligned} P_m &= \frac{A_1^2}{T t_1^2} \int_{t=0}^{t_1} t^2\, dt + \frac{A_2^2}{T(T - t_1)^2} \int_{t=t_1}^{T} (t - T)^2\, dt \\ &= \frac{A_1^2 t_1}{3T} + \frac{A_2^2(T - t_1)}{3T} \\ &= \frac{A_1 A_2}{3}, \end{aligned} \tag{3.92}$$

where we have substituted for t_1 from (3.87).

16. (Haykin 1983) Figure 3.24 shows the block diagram of a chopper-stabilized dc amplifier. It uses a multiplier and an ac amplifier, which shifts the spectrum of the input signal from the vicinity of zero frequency to the vicinity of the carrier frequency (f_c). The signal at the output of the ac amplifier is then coherently demodulated. The carrier $c(t)$ is a square wave as indicated in the figure.

(a) Specify the frequency response of the ac amplifier so that there is no distortion in the LPF output, assuming that $m(t)$ is bandlimited to $-W < f < W$. Specify also the relation between f_c and W.

(b) Determine the overall gain of the system (A), assuming that the ac amplifier has a gain of K and the lowpass filter is ideal having unity gain.

• *Solution*: The square wave $c(t)$ can be written as:

$$c(t) = \frac{4}{\pi} \sum_{n=1}^{\infty} \frac{(-1)^{n-1}}{2n-1} \cos(2\pi f_c (2n-1)t). \tag{3.93}$$

The ac amplifier can have the frequency response of an ideal bandpass filter, with a gain of K and bandwidth $f_c - W < |f| < f_c + W$. Observe that for no aliasing of the spectrum centered at f_c, we require:

$$f_c + W < 3f_c - W$$
$$\Rightarrow W < f_c. \tag{3.94}$$

Thus, the output of the ac amplifier is:

$$s(t) = \frac{4K}{\pi} m(t) \cos(2\pi f_c t). \tag{3.95}$$

The output of the lowpass filter would be:

$$s(t)c(t) \xrightarrow{\text{LPF}} \frac{16K}{2\pi^2} m(t) = \frac{8K}{\pi^2} m(t). \tag{3.96}$$

Therefore, the overall gain of the system is $8K/\pi^2$.

17. (Haykin 1983) Consider the phase discrimination method of generating an SSB signal. Let the message be given by:

$$m(t) = A_m \cos(2\pi f_m t) \tag{3.97}$$

and the SSB signal be given by:

$$s(t) = m(t) \cos(2\pi f_c t) - \hat{m}(t) \sin(2\pi f_c t). \tag{3.98}$$

Determine the ratio of the amplitude of the undesired side-frequency component to that of the desired side-frequency component when the modulator deviates from the ideal condition due to the following factors considered one at a time:

(a) The Hilbert transformer introduces a phase lag of $\pi/2 - \delta$ instead of $\pi/2$.
(b) The terms $m(t)$ and $\hat{m}(t)$ are given by

$$m(t) = a A_m \cos(2\pi f_m t)$$
$$\hat{m}(t) = b A_m \sin(2\pi f_m t). \tag{3.99}$$

(c) The carrier signal applied to the product modulators are not in phase quadrature, that is:

$$c_1(t) = \cos(2\pi f_c t)$$
$$c_2(t) = \sin(2\pi f_c t + \delta). \tag{3.100}$$

• *Solution*: In the first case, we have

$$\hat{m}(t) = A_m \cos(2\pi f_m t - \pi/2 + \delta)$$
$$= A_m \sin(2\pi f_m t + \delta). \tag{3.101}$$

Thus

$$s(t) = A_m \cos(2\pi f_m t) \cos(2\pi f_c t) - A_m \sin(2\pi f_m t + \delta) \sin(2\pi f_c t)$$
$$= \frac{A_m}{2} [\cos(2\pi (f_c + f_m)t) + \cos(2\pi (f_c - f_m)t)]$$
$$- \frac{A_m}{2} [\cos(2\pi (f_c - f_m)t - \delta) - \cos(2\pi (f_c + f_m)t + \delta)] \tag{3.102}$$

The desired side frequency is $f_c + f_m$ and the undesired side frequency is $f_c - f_m$. The phasor diagram for the signal in (3.102) is shown in Fig. 3.25. Clearly, the required ratio is R_2/R_1 where

$$R_1 = \sqrt{(A_m/2)^2 (1 + \cos(\delta))^2 + (A_m \sin(\delta)/2)^2}$$
$$R_2 = \sqrt{(A_m/2)^2 (1 - \cos(\delta))^2 + (A_m \sin(\delta)/2)^2}. \tag{3.103}$$

For the second case $s(t)$ is given by:

$$s(t) = a A_m \cos(2\pi f_m t) \cos(2\pi f_c t) - b A_m \sin(2\pi f_m t) \sin(2\pi f_c t)$$
$$= \frac{a A_m}{2} [\cos(2\pi (f_c - f_m)t) + \cos(2\pi (f_c + f_m)t)]$$
$$- \frac{b A_m}{2} [\cos(2\pi (f_c - f_m)t) - \cos(2\pi (f_c + f_m)t)]. \tag{3.104}$$

Thus, the ratio of the amplitude of the undesired sideband to the desired sideband is

$$R = \frac{a - b}{a + b}. \tag{3.105}$$

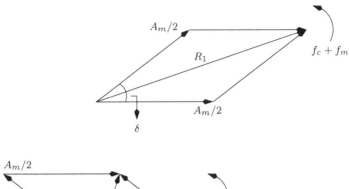

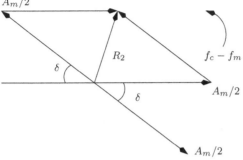

Fig. 3.25 Phasor diagram for the signal in (3.102)

For the third case $s(t)$ is given by:

$$
\begin{aligned}
s(t) &= A_m \cos(2\pi f_m t) \cos(2\pi f_c t) - A_m \sin(2\pi f_m t) \sin(2\pi f_c t + \delta) \\
&= \frac{A_m}{2} [\cos(2\pi(f_c - f_m)t) + \cos(2\pi(f_c + f_m)t)] \\
&\quad - \frac{A_m}{2} [\cos(2\pi(f_c - f_m)t + \delta) - \cos(2\pi(f_c + f_m)t + \delta)].
\end{aligned}
$$

$$(3.106)$$

Again from the phasor diagram in Fig. 3.26, we see that the ratio of the amplitude of the undesired sideband to the desired sideband is R_2/R_1 where R_1 and R_2 are given by (3.103).

18. Let the message be given by:

$$
m(t) = A_m \cos(2\pi f_m t) + A_m \sin(2\pi f_m t), \tag{3.107}
$$

and the SSB signal be given by:

$$
s(t) = m(t) \cos(2\pi f_c t) + \hat{m}(t) [\sin(2\pi f_c t) + a \sin(4\pi f_c t)]. \tag{3.108}
$$

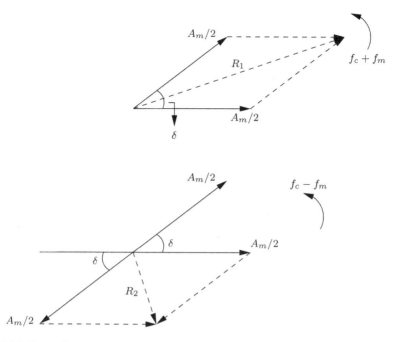

Fig. 3.26 Phasor diagram for the signal in (3.106)

Observe that the quadrature carrier contains harmonic distortion.
Determine the ratio of the power of the undesired frequency component(s) to
that of the desired frequency component(s).

• *Solution*: We have:

$$m(t) = A_m \cos(2\pi f_m t) + A_m \sin(2\pi f_m t)$$
$$= A_m \sqrt{2} \cos(2\pi f_m t - \pi/4)$$
$$\Rightarrow \hat{m}(t) = A_m \sqrt{2} \cos(2\pi f_m t - \pi/4 - \pi/2)$$
$$= A_m \sqrt{2} \sin(2\pi f_m t - \pi/4). \tag{3.109}$$

Therefore

$$s(t) = A_m \sqrt{2} \cos(2\pi f_m t - \pi/4) \cos(2\pi f_c t)$$
$$+ A_m \sqrt{2} \sin(2\pi f_m t - \pi/4) \left[\sin(2\pi f_c t) + a \sin(4\pi f_c t)\right]$$
$$= A_m \sqrt{2} \cos\left(2\pi (f_c - f_m) t - \pi/4\right)$$
$$+ \frac{a A_m}{\sqrt{2}} \cos\left(2\pi (2 f_c - f_m) t + \pi/4\right)$$
$$- \frac{a A_m}{\sqrt{2}} \cos\left(2\pi (2 f_c + f_m) t - \pi/4\right). \tag{3.110}$$

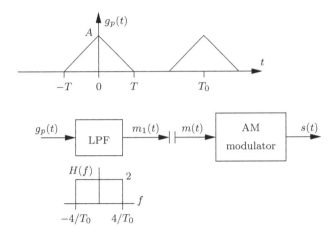

Fig. 3.27 Block diagram of an AM system

Clearly, the desired frequency is $f_c - f_m$ and the undesired frequencies are $2f_c \pm f_m$. Therefore the required power ratio is:

$$R = \frac{a^2 A_m^2 / 2}{2 A_m^2 / 2}$$

$$= \frac{a^2}{2}. \tag{3.111}$$

19. Consider the system shown in Fig. 3.27. The input signal $g_p(t)$ is periodic with a period of T_0. Assume that $T/T_0 = 0.5$. The signal $g_p(t)$ is passed through an ideal lowpass filter, that has a gain of two in the passband, as indicated in the figure. The LPF output $m_1(t)$ is further passed through a dc blocking capacitor to yield the message $m(t)$. The capacitor acts as a short for the frequencies of interest. This message is used to generate an AM signal given by

$$s(t) = A_c[1 + k_a m(t)] \cos(2\pi f_c t). \tag{3.112}$$

What is the critical value of k_a, beyond which $s(t)$ gets overmodulated?

- *Solution*: Consider the pulse:

$$g(t) = \begin{cases} g_p(t) & \text{for } -T_0/2 < t < T_0/2 \\ 0 & \text{elsewhere.} \end{cases} \tag{3.113}$$

Clearly

$$\frac{d^2 g(t)}{dt^2} = \frac{A}{T} \left(\delta(t+T) - 2\delta(t) + \delta(t-T) \right). \tag{3.114}$$

Hence

$$G(f) = -\frac{A}{4\pi^2 f^2 T} \left(\exp(\mathrm{j}\, 2\pi fT) - 2 + \exp(-\mathrm{j}\, 2\pi fT) \right)$$

$$= \frac{AT}{\pi^2 f^2 T^2} \sin^2(\pi fT)$$

$$= AT\operatorname{sinc}^2(fT). \tag{3.115}$$

We also know that

$$g_p(t) = \sum_{n=-\infty}^{\infty} c_n \exp\left(\mathrm{j}\, 2\pi nt/T_0\right), \tag{3.116}$$

where

$$c_n = \frac{1}{T_0} \int_{t=-\infty}^{\infty} g(t) \exp\left(-\mathrm{j}\, 2\pi nt/T_0\right) dt$$

$$= \frac{1}{T_0} G(n/T_0). \tag{3.117}$$

Substituting for $G(n/T_0)$ we have

$$g_p(t) = \frac{AT}{T_0} \sum_{n=-\infty}^{\infty} \operatorname{sinc}^2\left(\frac{nT}{T_0}\right) e^{\mathrm{j}\, 2\pi nt/T_0}$$

$$= \frac{A}{2} + \frac{4A}{\pi^2} \sum_{m=1}^{\infty} \frac{1}{(2m-1)^2} \cos(2\pi(2m-1)t/T_0), \tag{3.118}$$

where we have used the fact that $T/T_0 = 0.5$. The output of the dc blocking capacitor is

$$m(t) = \frac{8A}{\pi^2} \cos(2\pi t/T_0) + \frac{8A}{9\pi^2} \cos(6\pi t/T_0), . \tag{3.119}$$

where we have used the fact that the LPF has a gain of two in the frequency range $[-4/T_0, 4/T_0]$. The minimum value of $m(t)$ is

$$m_{\min}(t) = -\frac{80A}{9\pi^2} \tag{3.120}$$

and occurs at $nT_0 + T_0/2$. To prevent overmodulation we must have:

$$1 + k_a m_{\min}(t) \geq 0$$

$$\Rightarrow k_a \leq \frac{9\pi^2}{80A}. \tag{3.121}$$

Fig. 3.28 Block diagram of a scrambler

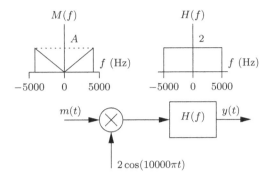

20. The system shown in Fig. 3.28 is used for scrambling audio signals. The output $y(t)$ is the scrambled version of the input $m(t)$. The spectrum (Fourier transform) of $m(t)$ and the lowpass filter $(H(f))$ are as shown in the figure.

   (a) Draw the spectrum of $y(t)$. Label all important points on the x- and y-axis.
   (b) The output $y(t)$ corresponds to a particular modulation scheme (i.e., AM, FM, SSB with lower/upper sideband transmitted, VSB, DSB-SC). Which modulation scheme is it?
   (c) Write down the precise expression for $y(t)$ in terms of $m(t)$, corresponding to the spectrum computed in part (a).
   (d) Suggest a method for recovering $m(t)$ (not $km(t)$, where k is a constant) from $y(t)$.

 • *Solution*: The multiplier output is given by:

$$y_1(t) = 2m(t)\cos(2\pi f_c t),\tag{3.122}$$

where $f_c = 5\,\text{kHz}$. The Fourier transform of $y_1(t)$ is

$$Y_1(f) = M(f - f_c) + M(f + f_c)\tag{3.123}$$

The spectrum of $y_1(t)$ and $y(t)$ is shown in Fig. 3.29.
From the spectrum of $y(t)$, we can conclude that it is an SSB signal centered at f_c, with the lower sideband transmitted. Hence we have

$$y(t) = km(t)\cos(2\pi f_c t) + k\hat{m}(t)\sin(2\pi f_c t),\tag{3.124}$$

where k is a constant to be found out. The Fourier transform of $y(t)$ is

Fig. 3.29 **a** $Y_1(f)$. **b** $Y(f)$.
c Procedure for recovering
$m(t)$

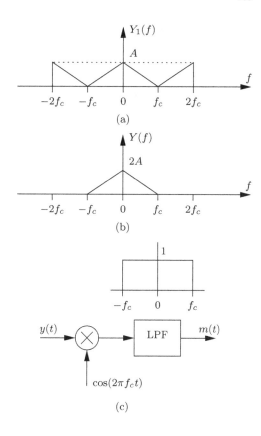

(a)

(b)

(c)

$$Y(f) = \frac{k}{2}[M(f - f_c) + M(f + f_c)]$$

$$= -\frac{k}{2}\left[\operatorname{sgn}(f - f_c)M(f - f_c) - \operatorname{sgn}(f + f_c)M(f + f_c)\right]$$

$$= \frac{k}{2}M(f - f_c)[1 - \operatorname{sgn}(f - f_c)] + \frac{k}{2}M(f + f_c)[1 + \operatorname{sgn}(f + f_c)].$$

$$(3.125)$$

Comparing the above equation with Fig. 3.29b, we conclude that $k = 2$. Therefore

$$y(t) = 2m(t)\cos(2\pi f_c t) + 2\hat{m}(t)\sin(2\pi f_c t) \qquad (3.126)$$

A method for recovering $m(t)$ is shown in Fig. 3.29c.

21. Figure 3.30 shows the block diagram of a DSB-SC modulator. Note that the oscillator generates $\cos^3(2\pi f_c t)$. The message spectrum is also shown.

    (a) Draw and label the spectrum (Fourier transform) of $y(t)$.

Fig. 3.30 Block diagram of a DSB-SC modulator

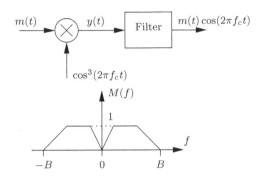

(b) Sketch the Fourier transform of a filter such that the output signal is $m(t) \cos(2\pi f_c t)$, where $m(t)$ is the message. The filter should pass only the required signal and reject all other frequencies.

(c) What is the minimum usable value of f_c?

- *Solution*: The multiplier output is given by:

$$y(t) = \frac{3}{4} m(t) \cos(2\pi f_c t) + \frac{1}{4} m(t) \cos(6\pi f_c t). \qquad (3.127)$$

Therefore

$$Y(f) = \frac{3}{8} [M(f - f_c) + M(f + f_c)] + \frac{1}{8} [M(f - 3f_c) + M(f + 3f_c)]. \qquad (3.128)$$

The spectrum of $y(t)$ and the bandpass filter is shown in Fig. 3.31. We must also have the following relationship:

$$3f_c - B \ge f_c + B$$
$$\Rightarrow f_c \ge B. \qquad (3.129)$$

Thus the minimum value of f_c is B.

22. Consider the receiver of Fig. 3.32. Both bandpass filters are assumed to be ideal with unity gain in the passband. The passband of BPF1 is in the range 500–1500 kHz. The bandwidth of BPF2 is 10 kHz and it selects only the difference frequency component. The input consists of a series of AM signals of bandwidth 10 kHz and spaced 10 kHz apart, occupying the band 500–1500 kHz as shown in the figure.
When the LO frequency is 1505 kHz, the output signal is $m_1(t)$.

(a) Compute the centere frequency of BPF2.

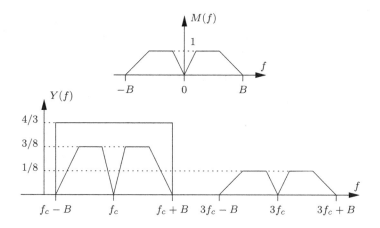

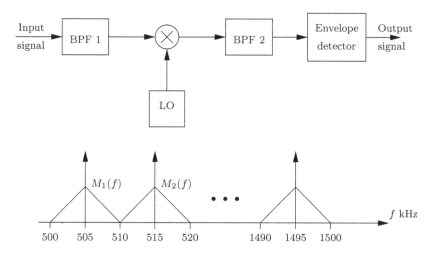

Fig. 3.31 Spectrum of $y(t)$ and the bandpass filter

Fig. 3.32 Block diagram of a radio receiver

(b) What should be the LO frequency so that the last message (occupying the band 1490–1500 kHz) is selected.

(c) Can the LO frequency be equal to 1 MHz and the center frequency of BPF2 equal to 495 kHz? Justify your answer.

• *Solution*: When the LO frequency is 1505 kHz, the difference frequency for $m_1(t)$ is $1505 - 505 = 1000$ kHz. Therefore, the center frequency of BPF2 must be 1000 kHz.

When the LO frequency is $1495 + 1000 = 2495$ kHz, the last message is selected.

When the LO frequency is 1 MHz and the center frequency of BPF2 is equal to 495 MHz, then the output signal would be the sum of the first message ($1000 - 505 = 495$ kHz) and the last message (image at $1495 - 1000 = 495$ kHz). Hence this combination should not be used.

23. (Haykin 1983) Consider a multiplex system in which four input signals $m_0(t)$, $m_1(t)$, $m_2(t)$, and $m_3(t)$ are, respectively, multiplied by the carrier waves

$$
\begin{aligned}
c_0(t) &= \cos(2\pi f_a t) + \cos(2\pi f_b t) \\
c_1(t) &= \cos(2\pi f_a t + \alpha_1) + \cos(2\pi f_b t + \beta_1) \\
c_2(t) &= \cos(2\pi f_a t + \alpha_2) + \cos(2\pi f_b t + \beta_2) \\
c_3(t) &= \cos(2\pi f_a t + \alpha_3) + \cos(2\pi f_b t + \beta_3)
\end{aligned}
\tag{3.130}
$$

to produce the multiplexed signal

$$
s(t) = \sum_{i=0}^{3} m_i(t) c_i(t).
\tag{3.131}
$$

All the messages are bandlimited to $[-W, \ W]$. At the receiver, the ith message is recovered as follows:

$$
s(t) c_i(t) \xrightarrow{LPF} m_i(t)
\tag{3.132}
$$

where the LPF is ideal with unity gain in the frequency band $[-W, \ W]$.

(a) Compute α_i and β_i, $0 \le i \le 3$ for this to be feasible.
(b) Determine the minimum separation between f_a and f_b.

- *Solution*: Note that we require:

$$
c_i(t) c_j(t) \xrightarrow{LPF} 0.5 \left[\cos(\alpha_i - \alpha_j) + \cos(\beta_i - \beta_j) \right] = \begin{cases} 1 \text{ for } i = j \\ 0 \text{ for } i \ne j \end{cases}
\tag{3.133}
$$

provided

$$
\begin{aligned}
|f_b - f_a| &\ge 2W \\
f_a &\ge W \\
f_b &\ge W.
\end{aligned}
\tag{3.134}
$$

Thus the minimum frequency separation is $2W$. Now, in order to ensure orthogonality between carriers we proceed as follows. Let us first consider $c_0(t)$ and $c_1(t)$. Orthogonality is ensured when

$$
\alpha_1 = \beta_1 = \pi/2.
\tag{3.135}
$$

Other solutions are also possible.

Let us now consider $c_2(t)$. We get the following relations:

$$c_2(t) \perp c_0(t) \Rightarrow \cos(\alpha_2) + \cos(\beta_2) = 0$$
$$c_2(t) \perp c_1(t) \Rightarrow \cos(\alpha_2 - \alpha_1) + \cos(\beta_2 - \beta_1) = 0$$
$$\Rightarrow \sin(\alpha_2) + \sin(\beta_2) = 0. \tag{3.136}$$

The two equations in (3.136) imply that if α_2 is in the 1st quadrant then β_2 must be in the 3rd quadrant. Otherwise, if α_2 is in the 2nd quadrant, then β_2 must be in the 4th quadrant. Let us take $\alpha_2 = \pi/4$. Then $\beta_2 = 5\pi/4$. Finally we consider $c_3(t)$. We get the relations:

$$c_3(t) \perp c_0(t) \Rightarrow \cos(\alpha_3) + \cos(\beta_3) = 0$$
$$c_3(t) \perp c_1(t) \Rightarrow \cos(\alpha_3 - \alpha_1) + \cos(\beta_3 - \beta_1) = 0$$
$$\Rightarrow \sin(\alpha_3) + \sin(\beta_3) = 0$$
$$c_3(t) \perp c_2(t) \Rightarrow \cos(\alpha_3 - \alpha_2) + \cos(\beta_3 - \beta_2) = 0$$
$$\Rightarrow \cos(\alpha_3) - \cos(\beta_3) + \sin(\alpha_3) - \sin(\beta_3) = 0. \tag{3.137}$$

The set of equations in (3.137) can be reduced to:

$$\cos(\alpha_3) = -\sin(\alpha_3)$$
$$\cos(\beta_3) = -\sin(\beta_3). \tag{3.138}$$

Thus we conclude that α_3 and β_3 must be an odd multiple of $\pi/4$ and must lie in the 2nd/4th quadrant and 4th/2nd quadrant, respectively. If we take $\alpha_3 = 3\pi/4$ then $\beta_3 = 7\pi/4$. Note that there are infinite solutions to α_i and β_i $(1 \leq i \leq 3)$.

24. Consider the system shown in Fig. 3.33a. The signal $s(t)$ is given by:

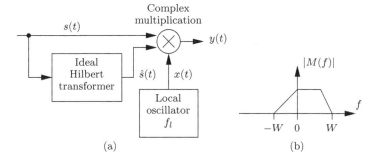

Fig. 3.33 Demodulation using a Hilbert transformer

$$s(t) = A_1 m_1(t) \cos(2\pi f_c t) - A_2 m_2(t) \sin(2\pi f_c t). \qquad (3.139)$$

Both $m_1(t)$ and $m_2(t)$ are real-valued and bandlimited to $[-W,\ W]$. The carrier frequency $f_c \gg W$. The inputs to the complex multiplier are $s(t) + j\hat{s}(t)$ and $x(t)$ (which could be real or complex-valued and depends on the local oscillator frequency f_l). The multiplier output is the complex-valued signal $y(t)$. Let

$$m(t) = A_1 m_1(t) + j A_2 m_2(t) \rightleftharpoons M(f). \qquad (3.140)$$

(a) Find f_l and $x(t)$ such that $y(t) = m(t)$.
(b) The local oscillator that generates $x(t)$ has a frequency $f_l = f_c + \Delta f$. Find $y(t)$. Sketch $|Y(f)|$ when $|M(f)|$ is depicted in Fig. 3.33b.

• *Solution*: We know that

$$\hat{s}(t) = A_1 m_1(t) \cos(2\pi f_c t - \pi/2) - A_2 m_2(t) \sin(2\pi f_c t - \pi/2)$$
$$= A_1 m_1(t) \sin(2\pi f_c t) + A_2 m_2(t) \cos(2\pi f_c t). \qquad (3.141)$$

Therefore

$$s(t) + j\hat{s}(t) = m(t) e^{j 2\pi f_c t}. \qquad (3.142)$$

If $y(t) = m(t)$, then we must have $f_l = f_c$ and

$$x(t) = e^{-j 2\pi f_c t}. \qquad (3.143)$$

If

$$x(t) = e^{-j 2\pi (f_c + \Delta f) t}. \qquad (3.144)$$

then

$$y(t) = m(t) e^{-j 2\pi \Delta f t}$$
$$\Rightarrow Y(f) = M(f + \Delta f)$$
$$\Rightarrow |Y(f)| = |M(f + \Delta f)|. \qquad (3.145)$$

The spectrum of $|Y(f)|$ is shown in Fig. 3.34.

25. An AM signal is given by:

$$s(t) = A_c [1 + k_a m(t)] \sin(2\pi f_c t + \theta). \tag{3.146}$$

The message $m(t)$ is real-valued, does not have a dc component and its one-sided bandwidth is $W \ll f_c$. Explain how $km(t)$, where k is a constant, can be recovered from $s(t)$ using a Hilbert transformer and other components.

• *Solution*: When $s(t)$ is passed through a Hilbert transformer, its output is:

$$\begin{aligned}\hat{s}(t) &= A_c [1 + k_a m(t)] \sin(2\pi f_c t + \theta - \pi/2) \\ &= -A_c [1 + k_a m(t)] \cos(2\pi f_c t + \theta).\end{aligned} \tag{3.147}$$

Now

$$s(t) + j\hat{s}(t) = A_c [1 + k_a m(t)] e^{j(2\pi f_c t + \theta - \pi/2)}. \tag{3.148}$$

The message $m(t)$ can be recovered using the block diagram in Fig. 3.35. Note that

$$y(t) = A_c [1 + k_a m(t)]. \tag{3.149}$$

26. Consider the Costas loop shown in Fig. 3.36. The input signal $s(t)$ is

$$s(t) = m(t) \cos(2\pi f_c t), \tag{3.150}$$

Fig. 3.34 Spectrum of $|Y(f)|$

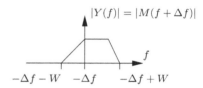

$$|Y(f)| = |M(f + \Delta f)|$$

$$-\Delta f - W \quad -\Delta f \qquad -\Delta f + W$$

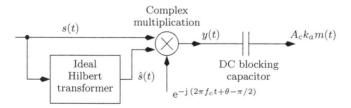

Complex multiplication

$s(t)$

$y(t)$

$A_c k_a m(t)$

Ideal Hilbert transformer

$\hat{s}(t)$

DC blocking capacitor

$e^{-j(2\pi f_c t + \theta - \pi/2)}$

Fig. 3.35 Demodulating an AM signal using a Hilbert transformer

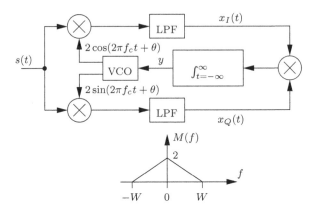

Fig. 3.36 Costas loop

where $m(t)$ is a real-valued message signal. The spectrum of $m(t)$ is shown in Fig. 3.36. The LPFs are ideal with a gain of two in the passband $[-W, W]$. It is given that $f_c \gg W$.

(a) Determine the signals $x_I(t)$, $x_Q(t)$, and y.
(b) For what values of θ will y be equal to zero?

- *Solution*: Clearly

$$2m(t)\cos(2\pi f_c t)\cos(2\pi f_c t + \theta) = m(t)\cos(\theta) + 2f_c \text{ term}$$
$$2m(t)\cos(2\pi f_c t)\sin(2\pi f_c t + \theta) = m(t)\sin(\theta) + 2f_c \text{ term}. \quad (3.151)$$

Since the gain of the LPF is two in the passband

$$x_I(t) = 2m(t)\cos(\theta)$$
$$x_Q(t) = 2m(t)\sin(\theta). \quad (3.152)$$

Hence

$$y = 2\sin(2\theta)\int_{t=-\infty}^{\infty} m^2(t)\,dt$$
$$= 2\sin(2\theta)\int_{f=-\infty}^{\infty} |M(f)|^2\,df$$
$$= 4\sin(2\theta)\int_{f=0}^{W}\left(\frac{-2f}{W}+2\right)^2\,df$$
$$= \frac{16}{3}W\sin(2\theta) \quad (3.153)$$

Fig. 3.37 Spectrum of each
voice signal

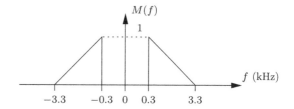

where we have used the Rayleigh's energy theorem. It is clear that $y = 0$ when
$\theta = k\pi/2$, where k is an integer.

27. It is desired to transmit 400 voice signals using frequency division multiplexing
   (FDM). The voice signals are SSB modulated with lower sideband transmitted.
   The spectrum of each of the voice signals is illustrated in Fig. 3.37.

   (a) What is the minimum carrier spacing required to multiplex all the signals,
       such that there is no overlap of spectra.
   (b) Using a single- stage approach and assuming a carrier spacing of 7 kHz and
       a minimum carrier frequency of 200 kHz, determine the lower and upper
       frequency limits occupied by the FDM signal containing 400 voice signals.
   (c) Determine the spectrum (lower and upper frequency limits) occupied by the
       100th voice signal.
   (d) In the two-stage approach, the first stage uses SSB with lower sideband
       transmitted. In the first stage, the voice signals are grouped into L blocks,
       with each block containing K voice signals. The carrier frequencies in each
       block in the first stage is given by $7n$ kHz, for $1 \leq n \leq K$. Note that the
       FDM signal using the two-stage approach must be identical to the single-
       stage approach in (b).
       i. Sketch the spectrum of each block at the output of the first stage.
       ii. Specify the modulation required for each of the blocks in the second
           stage.
       iii. How many carriers are required to generate the composite FDM signal?
            Express your answer in terms of L and K.
       iv. Determine L and K such that the number of carriers is minimized.
       v. Give the expression for the carrier frequencies required to modulate
          each block in the second stage.

   • *Solution*: The minimum carrier spacing required is 3 kHz.
     If the minimum carrier frequency is 200 kHz, then the spectrum of the first
     voice signal starts at $200 - 3.3 = 196.7$ kHz. The carrier for the 400th voice
     signal is at

$$200 + 399 \times 7 = 2993 \text{ kHz.} \tag{3.154}$$

The spectrum of the 400th voice signal ends at $2993 - 0.3 = 2992.7\,\text{kHz}$. Therefore, the spectrum of the FDM signal extends from 196.7 to 2992.7 kHz, that is, $196.7 \leq |f| \leq 2992.7\,\text{kHz}$.
The carrier for the 100th voice signal is at

$$200 + 99 \times 7 = 893\,\text{kHz}. \tag{3.155}$$

Hence, the spectrum of the 100th voice signal extends from $893 - 3.3 = 889.7\,\text{kHz}$ to $893 - 0.3 = 892.7\,\text{kHz}$, that is, $889.7 \leq |f| \leq 892.7\,\text{kHz}$.
The spectrum of each block at the output of the first stage is shown in Fig. 3.38b. The modulation required for each of the blocks in the second stage is SSB with upper sideband transmitted.
For the second approach, let us first evaluate the number of carriers required to obtain each block in the first stage. Clearly, the number of carriers required is K. In the next stage, L carriers are required to translate each block to the appropriate frequency band. Thus, the total number of carriers required is $L + K$.
Now, we need to minimize $L + K$ subject to the constraint $LK = 400$. The problem can be restated as

$$\min_{K} \frac{400}{K} + K. \tag{3.156}$$

Differentiating with respect to K and setting the result to zero, we get the solution as $K = 20$, $L = 20$.
The carrier frequencies required in the second stage is given by $200 - 7 + 7K(l - 1)$ for $1 \leq l \leq L$.
The block diagram of the two-stage approach is given in Fig. 3.38a.

28. It is desired to transmit 500 voice signals using frequency division multiplexing (FDM). The voice signals are DSB-SC modulated. The spectrum of each of the voice signals is illustrated in Fig. 3.39.

(a) How many carrier signals are required to multiplex all the 500 voice signals?
(b) What is the minimum carrier spacing required to multiplex all the signals, such that there is no overlap of spectra.
(c) Using a single- stage approach and assuming a carrier spacing of 10 kHz and a minimum carrier frequency of 300 kHz, determine the lower and upper frequency limits occupied by the FDM signal containing 500 voice signals.
(d) Determine the spectrum (lower and upper frequency limits) occupied by the 150th voice signal.
(e) It is desired to reduce the number of carriers using a two-stage approach. The first stage uses DSB-SC modulation. In the first stage, the voice signals are grouped into L blocks, with each block containing K voice signals. The carrier frequencies in each block in the first stage is given by $10n$ kHz, for

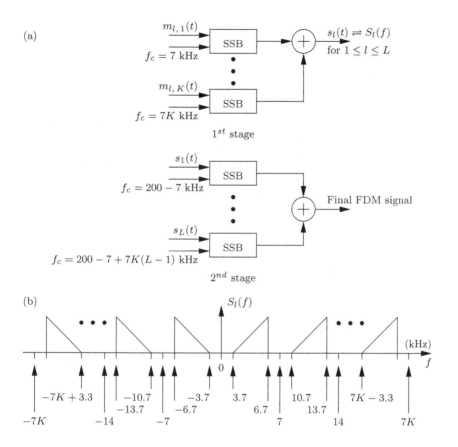

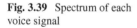

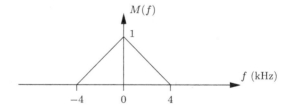

Fig. 3.38   a Two-stage approach for obtaining the final FDM signal. **b** Spectrum of each block at the output of the first stage

Fig. 3.39 Spectrum of each voice signal

$1 \leq n \leq K$. Note that the FDM signal using the two-stage approach must be identical to the single-stage approach in (c).

    i. Sketch the spectrum of each block at the output of the first stage.

    ii. Specify the modulation required for each of the blocks in the second stage, such that, the minimum carrier frequency used in the second stage, is closest to 300 kHz.

    iii. How many carriers are required to generate the composite FDM signal? Express your answer in terms of L and K.

    iv. Determine L and K such that the number of carriers is minimized.

    v. Give the expression for the carrier frequencies required to modulate each block in the second stage.

- *Solution*: The number of carriers required to multiplex all the voice signals is 500.

The minimum carrier spacing required is 8 kHz.

If the minimum carrier frequency is 300 kHz, then the spectrum of the first voice signal starts at $300 - 4 = 296$ kHz. The carrier for the 500th voice signal is at

$$300 + 499 \times 10 = 5290 \, \text{kHz}. \tag{3.157}$$

The spectrum of the 500th voice signal ends at $5290 + 4 = 5294$ kHz. Therefore, the spectrum of the FDM signal extends from 296 to 5294 kHz, that is $296 \leq |f| \leq 5294$ kHz.

The carrier for the 150th voice signal is at

$$300 + 149 \times 10 = 1790 \, \text{kHz}. \tag{3.158}$$

Hence the spectrum of the 150th voice signal extends from $1790 - 4 = 1786$ kHz to $1790 + 4 = 1794$ kHz, that is $1786 \leq |f| \leq 1794$ kHz.

The spectrum of each block at the output of the first stage is shown in Fig. 3.40b. The modulation required for each of the blocks in the second stage is SSB. In order to ensure that the minimum carrier frequency in the second stage is closest to 300 kHz, the upper sideband must be transmitted. Thus, the minimum carrier frequency in the second stage is $300 - 10 = 290$ kHz.

For the second approach, let us first evaluate the number of carriers required to obtain each block in the first stage. Clearly, the number of carriers required is K. In the next stage, L carriers are required to translate each block to the appropriate frequency band. Thus, the total number of carriers required is $L + K$.

Now, we need to minimize $L + K$ subject to the constraint $LK = 500$. The problem can be restated as

$$\min_{K} \frac{500}{K} + K. \tag{3.159}$$

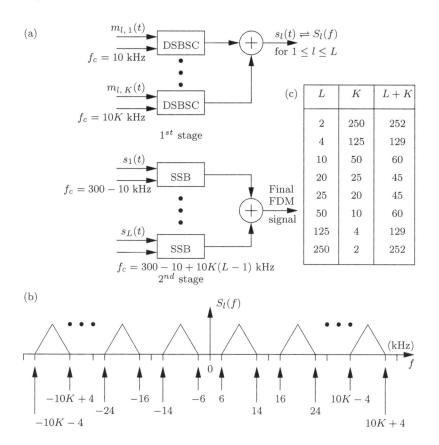

Fig. 3.40 **a** Two-stage approach for obtaining the final FDM signal. **b** Spectrum of each block at the output of the first stage. **c** Various possible values of L and K

Differentiating with respect to K and setting the result to zero, we get the solution as $K = 22.36$, which is not an integer. Hence, we need to obtain the solution manually, as given in Fig. 3.40c. We find that, there are two sets of solutions, $L = 20$, $K = 25$, and $L = 25$, $K = 20$.

The carrier frequencies required in the second stage is given by $300 - 10 + 10K(l - 1)$ for $1 \leq l \leq L$.

The block diagram of the two-stage approach is given in Fig. 3.40a.

29. Let $m(t)$ be a signal having Fourier transform $M(f)$, with $M(0) = 0$. When $m(t)$ is passed through a filter with impulse response $h(t)$, the Fourier transform of the output can be written as

$$Y(f) = \begin{cases} M(f)\,e^{-j\theta} & \text{for } f > 0 \\ 0 & \text{for } f = 0 \\ M(f)\,e^{j\theta} & \text{for } f < 0 \end{cases} \qquad (3.160)$$

Determine $h(t)$.

• *Solution*: We know that if (this was done in class)

$$y(t) = m(t)\cos(\theta) + \hat{m}(t)\sin(\theta) \qquad (3.161)$$

the Fourier transform of $y(t)$ is

$$Y(f) = M(f)\cos(\theta) - j\,\mathrm{sgn}(f)M(f)\sin(\theta)$$
$$= \begin{cases} M(f)\,e^{-j\theta} & \text{for } f > 0 \\ 0 & \text{for } f = 0 \\ M(f)\,e^{j\theta} & \text{for } f < 0. \end{cases} \qquad (3.162)$$

From (3.162) it is clear that

$$h(t) = \cos(\theta)\delta(t) + \frac{\sin(\theta)}{\pi t}. \qquad (3.163)$$

30. Consider the DSB-SC signal

$$s(t) = A_c \cos(2\pi f_c t)m(t). \qquad (3.164)$$

Assume that the energy signal $m(t)$ occupies the frequency band $[-W,\ W]$. Now, $s(t)$ is applied to a square law device given by

$$y(t) = s^2(t). \qquad (3.165)$$

The output $y(t)$ is applied to an ideal bandpass filter with a passband transfer function equal to $1/(\Delta f)$, midband frequency of $\pm 2 f_c$ and bandwidth Δf. Assume that $\Delta f \to 0$. All signals are real-valued.

(a) Determine the spectrum of $y(t)$.
(b) Find the relation between f_c and W for no aliasing in the spectrum of $y(t)$.
(c) Find the expression for the signal $v(t)$ at the BPF output.

• *Solution*: We have

$$y(t) = \frac{A_c^2}{2}[1 + \cos(4\pi f_c t)]\,m^2(t)$$
$$\Rightarrow Y(f) = \frac{A_c^2}{2}G(f) + \frac{A_c^2}{4}[G(f - 2f_c) + G(f + 2f_c)], \quad (3.166)$$

where

$$G(f) = M(f) \star M(f)$$
$$M(f) \rightleftharpoons m(t). \tag{3.167}$$

In the above equation "$\star$" denotes convolution and $G(f)$ extends over $[-2W,\, 2W]$. For no aliasing in $Y(f)$, we require

$$2f_c - 2W > 2W$$
$$\Rightarrow f_c > 2W. \tag{3.168}$$

Observe that, as $\Delta f \to 0$, the transfer function of the BPF can be expressed as two delta functions at $\pm 2f_c$. Therefore, the spectrum at the BPF output is given by:

$$V(f) = \frac{A_c^2}{4} G(0) \left[\delta(f - 2f_c) + \delta(f + 2f_c) \right]. \tag{3.169}$$

Therefore

$$v(t) = \frac{A_c^2}{2} G(0) \cos(4\pi f_c t). \tag{3.170}$$

Now

$$G(f) = \int_{x=-\infty}^{\infty} M(x)M(f - x)\, dx$$
$$\Rightarrow G(0) = \int_{x=-\infty}^{\infty} M(x)M(-x)\, dx. \tag{3.171}$$

Since $m(t)$ is real-valued

$$M(-x) = M^*(x)$$
$$\Rightarrow G(0) = \int_{x=-\infty}^{\infty} |M(x)|^2\, dx$$
$$= E, \tag{3.172}$$

where we have used the Rayleigh's energy theorem. Thus (3.170) reduces to:

$$v(t) = \frac{A_c^2}{2} E \cos(4\pi f_c t). \tag{3.173}$$

31. (Haykin 1983) Consider the quadrature -carrier multiplex system shown in Fig. 3.41. The multiplexed signal $s(t)$ is applied to a communication channel of frequency response $H(f)$. The channel output is then applied to the receiver input. Here f_c denotes the carrier frequency and the message spectra extends over $[-W,\, W]$. Find

Fig. 3.41 Quadrature carrier multiplexing system

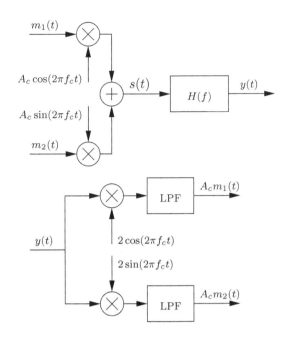

(a) the relation between f_c and W,

(b) the condition on $H(f)$, and

(c) the frequency response of the LPF

that is necessary for recovery of the message signals $m_1(t)$ and $m_2(t)$ at the receiver output. Assume a real-valued channel impulse response and $A_c = 1$.

- *Solution*: Assuming $A_c = 1$ we have

$$s(t) = m_1(t) \cos(2\pi f_c t) + m_2(t) \sin(2\pi f_c t)$$
$$\Rightarrow S(f) = \frac{M_1(f - f_c) + M_1(f + f_c)}{2} + \frac{M_2(f - f_c) - M_2(f + f_c)}{2j}$$

$$(3.174)$$

Let

$$Y(f) = S(f)H(f) \rightleftharpoons y(t).$$ (3.175)

Now at the receiver:

$$2y(t) \cos(2\pi f_c t) \rightleftharpoons Y(f - f_c) + Y(f + f_c).$$ (3.176)

Now

$$Y(f - f_c) = H(f - f_c)$$
$$\left[\frac{M_1(f - 2f_c) + M_1(f)}{2} + \frac{M_2(f - 2f_c) - M_2(f)}{2j}\right]$$
$$Y(f + f_c) = H(f + f_c)$$
$$\left[\frac{M_1(f) + M_1(f + 2f_c)}{2} + \frac{M_2(f) - M_2(f + 2f_c)}{2j}\right]$$

$$\text{(3.177)}$$

Since the LPF eliminates frequency components beyond $\pm W$, the output of the upper LPF is given by:

$$G_1(f) = \left[\frac{M_1(f)}{2} - \frac{M_2(f)}{2j}\right] H(f - f_c)$$
$$+ \left[\frac{M_1(f)}{2} + \frac{M_2(f)}{2j}\right] H(f + f_c). \qquad \text{(3.178)}$$

From the above equation, it is clear that to recover $M_1(f)$ from the upper LPF $(G_1(f) = M_1(f))$ we require

$$H(f - f_c) = H(f + f_c) \qquad \text{for } -W \le f \le W$$
$$\Rightarrow H^*(f_c - f) = H(f + f_c) \qquad \text{for } -W \le f \le W \qquad \text{(3.179)}$$

where it is assumed that the channel impulse response is real-valued. It can be shown that the above condition is also required for the recovery of $M_2(f)$ from the lower LPF.

Note that for distortionless recovery of the message signals, the frequency response of both the LPFs must be:

$$H_{\mathrm{LPF}}(f) = 2/(H(f - f_c) + H(f + f_c)) \qquad \text{for } -W \le f \le W \text{(3.180)}$$

Finally, for no aliasing we require $f_c > W$.

32. (Haykin 2001) A particular version of AM stereo uses quadrature multiplexing. Specifically, the carrier $A_c \cos(2\pi f_c t)$ is used to modulate the sum signal

$$m_1(t) = V_0 + m_l(t) + m_r(t), \qquad \text{(3.181)}$$

where V_0 is a dc offset included for the purpose of transmitting the carrier component, $m_l(t)$ is the left hand audio signal, and $m_r(t)$ is the right hand audio signal. The quadrature carrier $A_c \sin(2\pi f_c t)$ is used to modulate the difference signal

$$m_2(t) = m_l(t) - m_r(t). \qquad \text{(3.182)}$$

(a) Show that an envelope detector may be used to recover the sum signal $m_r(t) + m_l(t)$ from the quadrature multiplexed signal. How would you minimize the signal distortion produced by the envelope detector.

(b) Show that a coherent detector can recover the difference $m_l(t) - m_r(t)$.

- *Solution*: The quadrature -multiplexed signal can be written as:

$$s(t) = A_c m_1(t) \cos(2\pi f_c t) + A_c m_2(t) \sin(2\pi f_c t). \qquad (3.183)$$

Ideally, the output of the envelope detector is given by:

$$y(t) = A_c \sqrt{m_1^2(t) + m_2^2(t)}$$

$$= A_c m_1(t) \sqrt{1 + \frac{m_2^2(t)}{m_1^2(t)}}. \qquad (3.184)$$

The desired signal is $A_c m_1(t)$ and the distortion term is

$$D(t) = \sqrt{1 + \frac{m_2^2(t)}{m_1^2(t)}}. \qquad (3.185)$$

The distortion can be minimized by increasing the dc offset V_0. Note, however, that increasing V_0 results in increasing the carrier power, which makes the system more power inefficient.

Multiplying $s(t)$ by $2\cos(2\pi f_c t)$ and passing the output through a lowpass filter yields:

$$z_1(t) = A_c m_1(t)$$

$$= A_c(V_0 + m_l(t) + m_r(t)). \qquad (3.186)$$

Similarly, multiplying $s(t)$ by $2\sin(2\pi f_c t)$ and passing the output through a lowpass filter yields:

$$z_2(t) = A_c m_2(t)$$

$$= A_c(m_l(t) - m_r(t)). \qquad (3.187)$$

Adding $z_1(t)$ and $z_2(t)$ we get

$$z_1(t) + z_2(t) = A_c V_0 + 2 A_c m_l(t). \qquad (3.188)$$

Subtracting $z_2(t)$ from $z_1(t)$ yields:

$$z_1(t) - z_2(t) = A_c V_0 + 2 A_c m_r(t). \qquad (3.189)$$

The message signals $m_l(t)$ and $m_r(t)$ typically have zero dc, hence $A_c V_0$ can be removed by a dc blocking capacitor.

33. (Haykin 1983) Using the message signal

$$m(t) = \frac{1}{1 + t^2}, \tag{3.190}$$

determine the modulated signal for the following methods of modulation:

(a) AM with 50% modulation. Assume a cosine carrier.
(b) DSB-SC. Assume a cosine carrier.
(c) SSB with upper sideband transmitted.

In all cases, the area under the Fourier transform of the modulated signal must be unity.

• *Solution*: Note that the maximum value of $m(t)$ occurs at $t = 0$. Moreover

$$m(0) = 1$$

$$= \int_{f=-\infty}^{\infty}, M(f) \, df \tag{3.191}$$

where $M(f)$ is the Fourier transform of $m(t)$. Hence the AM signal with 50% modulation is given by:

$$s(t) = A_{c1} \left(1 + \frac{0.5}{1 + t^2} \right) \cos(2\pi f_c t). \tag{3.192}$$

The Fourier transform of $s(t)$ is:

$$S(f) = \frac{A_{c1}}{2} [\delta(f - f_c) + \delta(f + f_c)] + \frac{A_{c1}}{4} [M(f - f_c) + M(f + f_c)]. \tag{3.193}$$

Since

$$\int_{f=-\infty}^{\infty} S(f) \, df = 1, \tag{3.194}$$

we require

$$\frac{A_{c1}}{2} [1 + 1] + \frac{A_{c1}}{4} [1 + 1] = 1$$

$$\Rightarrow A_{c1} = 2/3, \tag{3.195}$$

where we have used (3.191). Note that shifting the spectrum of $M(f)$ does not change the area. The DSB-SC modulated signal is given by:

$$s(t) = \frac{A_{c2}}{1+t^2} \cos(2\pi f_c t). \tag{3.196}$$

The Fourier transform of $s(t)$ is

$$S(f) = \frac{A_{c2}}{2} \left[M(f - f_c) + M(f + f_c) \right]. \tag{3.197}$$

Again, due to (3.191) and (3.194) we have

$$\frac{A_{c2}}{2} [1 + 1] = 1$$
$$\Rightarrow A_{c2} = 1. \tag{3.198}$$

The Hilbert transform of $m(t)$ is given by:

$$\frac{1}{1+t^2} \overset{HT}{\rightleftharpoons} \frac{t}{1+t^2}. \tag{3.199}$$

Therefore, the SSB modulated signal with upper sideband transmitted is given by:

$$s(t) = A_{c3} \left[\frac{1}{1+t^2} \cos(2\pi f_c t) - \frac{t}{1+t^2} \sin(2\pi f_c t) \right]. \tag{3.200}$$

The spectrum of $s(t)$ is

$$\begin{aligned} S(f) &= \frac{A_{c3}}{2} M(f - f_c) \left[1 + \mathrm{sgn}(f - f_c) \right] \\ &+ \frac{A_{c3}}{2} M(f + f_c) \left[1 - \mathrm{sgn}(f + f_c) \right]. \end{aligned} \tag{3.201}$$

Note that $m(t)$ is real-valued and an even function of time. Hence, $M(f)$ is also real-valued and an even function of frequency. Therefore from (3.191)

$$\int_{f=0}^{\infty} M(f) \, df = 1/2. \tag{3.202}$$

Hence, from (3.194) and (3.202) we again have

$$\frac{2 \times 0.5 A_{c3}}{2} + \frac{2 \times 0.5 A_{c3}}{2} = 1$$
$$\Rightarrow A_{c3} = 1. \tag{3.203}$$

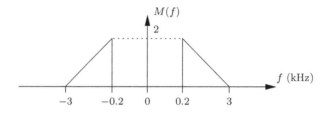

Fig. 3.42 Message spectrum

Note that we can also use the relation

$$\int_{f=-\infty}^{\infty} S(f)\,df = s(0) = 1 \qquad (3.204)$$

to obtain A_{c1}, A_{c2}, and A_{c3}.

34. Consider a message signal $m(t)$ whose spectrum extends over 200 Hz to 3 kHz, as illustrated in Fig. 3.42. This message is SSB modulated to obtain:

$$s(t) = A_c m(t) \cos(2\pi f_c t) + A_c \hat{m}(t) \sin(2\pi f_c t). \qquad (3.205)$$

At the receiver, $s(t)$ is demodulated using a carrier of the form $\cos(2\pi(f_c + \Delta f)t)$. Plot the spectrum of the demodulated signal at the output of the lowpass filter when

(a) $\Delta f = 20$ Hz,
(b) $\Delta f = -10$ Hz.

Assume that the lowpass filter is ideal with unity gain and extends over $[-4, 4]$ kHz, and $f_c = 20$ kHz.
Show all the steps required to arrive at the answer. In the sketch, indicate all the important points on the XY-axes.

- *Solution*: The multiplier output can be written as:

$$\begin{aligned}
s_1(t) &= s(t)\cos(2\pi(f_c + \Delta f)t) \\
&= A_c\big[m(t)\cos(2\pi f_c t) + \hat{m}(t)\sin(2\pi f_c t)\big]\cos(2\pi(f_c + \Delta f)t) \\
&= \frac{A_c}{2}m(t)\big[\cos(2\pi\Delta f t) + \cos(2\pi(2f_c + \Delta f)t)\big] \\
&\quad + \frac{A_c}{2}\hat{m}(t)\big[\sin(2\pi(2f_c + \Delta f)t) - \sin(2\pi\Delta f t)\big]. \qquad (3.206)
\end{aligned}$$

The output of the lowpass filter is:

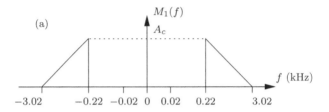

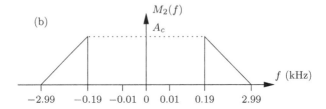

Fig. 3.43 Demodulated message spectrum in the presence of frequency offset

Fig. 3.44 An AM signal
transmitted through a series
RLC circuit

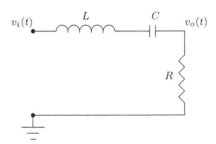

$$s_2(t) = \frac{A_c}{2}\left[m(t)\cos(2\pi\Delta ft) - \hat{m}(t)\sin(2\pi\Delta ft)\right]. \qquad (3.207)$$

When Δf is positive, $s_2(t)$ is an SSB signal with carrier frequency Δf and upper sideband transmitted. This is illustrated in Fig. 3.43a, for $\Delta f = 20$ Hz. When Δf is negative, $s_2(t)$ is an SSB signal with carrier frequency Δf and lower sideband transmitted. This is illustrated in Fig. 3.43b with $\Delta f = -10$ Hz.

35. Consider the series RLC circuit in Fig. 3.44. The resonant frequency of the circuit is 1 MHz and the Q-factor is 100. The input signal $v_i(t)$ is given by:

$$v_i(t) = A_c\left[1 + \mu\cos(2\pi f_m t)\right]\cos(2\pi f_c t), \qquad (3.208)$$

where $\mu < 1$, $f_c = 1$ MHz, $f_m = 5$ kHz.
Determine $v_o(t)$. Note that the Q-factor of the circuit is defined as

$$Q = 2\pi f_c L / R$$

$$= \frac{f_c}{-3 \text{ dB bandwidth}} \qquad (3.209)$$

where f_c denotes the resonant frequency of the RLC circuit.

- *Solution*: The input signal $v_i(t)$ can be written as:

$$v_i(t) = A_c \cos(2\pi f_c t) + \frac{\mu A_c}{2} [\cos(2\pi(f_c - f_m)t) + \cos(2\pi(f_c + f_m)t)] \qquad (3.210)$$

The transfer function of the circuit is given by:

$$\frac{V_o(\omega)}{V_i(\omega)} = H(\omega) = \frac{R}{R + j\omega L - j/(\omega C)}, \qquad (3.211)$$

where $\omega = 2\pi f$. Let us now consider a frequency

$$\omega = \omega_c + \Delta\omega \qquad (3.212)$$

where $\Delta\omega \ll \omega_c = 2\pi f_c$. Therefore $H(\omega)$ becomes:

$$
\begin{aligned}
H(\omega) &= \frac{R}{R + j(\omega_c + \Delta\omega)L - j/((\omega_c + \Delta\omega)C)} \\
&\approx \frac{R}{R + j\Delta\omega L + j\Delta\omega/(\omega_c^2 C)} \\
&= \frac{1}{1 + j\Delta\omega L/R + j\Delta\omega/(\omega_c^2 R C)} \\
&= \frac{1}{1 + j(2\Delta\omega)L/R},
\end{aligned}
\qquad (3.213)
$$

where we have used the following relationships:

$$\omega_c L = 1/(\omega_c C)$$

$$\frac{1}{1 + \Delta\omega/\omega_c} \approx 1 - \Delta\omega/\omega_c \qquad \text{when } \Delta\omega \ll \omega_c. \qquad (3.214)$$

Since the Q-factor of the filter is high, we can assume that $2\Delta\omega$ is the 3-dB bandwidth of the filter. Hence

$$\Delta\omega = \frac{\omega_c}{2Q} = 2\pi \times 5000 \, \text{rad/s}, \qquad (3.215)$$

which is also equal to the message frequency. Thus

$$H(\omega_c + \Delta\omega) = \frac{1}{\sqrt{2}}e^{-j\pi/4}$$

$$H(\omega_c - \Delta\omega) = \frac{1}{\sqrt{2}}e^{j\pi/4}$$

$$H(\omega_c) = 1. \tag{3.216}$$

The output signal can be written as:

$$v_o(t) = A_c \cos(2\pi f_c t) + \frac{\mu A_c}{2\sqrt{2}} \{\cos[2\pi(f_c - f_m)t + \pi/4]$$
$$+ \cos[2\pi(f_c + f_m)t - \pi/4]\}$$
$$= A_c \left[1 + \frac{\mu}{\sqrt{2}} \cos(2\pi f_m t - \pi/4)\right] \cos(2\pi f_c t). \tag{3.217}$$

36. Let $s_u(t)$ denote the SSB wave obtained by transmitting only the upper sideband, that is

$$s_u(t) = A_c \left[m(t) \cos(2\pi f_c t) - \hat{m}(t) \sin(2\pi f_c t)\right], \tag{3.218}$$

where $m(t)$ is bandlimited to $|f| < W \ll f_c$.
Explain how $m(t)$ can be recovered from $s_u(t)$ using only multipliers, adders/subtracters and Hilbert transformers. Lowpass filters should not be used.

- *Solution*: We have

$$s_u(t) = A_c \left[m(t) \cos(2\pi f_c t) - \hat{m}(t) \sin(2\pi f_c t)\right]$$
$$\hat{s}_u(t) = A_c \left[m(t) \sin(2\pi f_c t) + \hat{m}(t) \cos(2\pi f_c t)\right]. \tag{3.219}$$

From (3.219), we get

$$m(t) = \frac{1}{A_c} \left[s_u(t) \cos(2\pi f_c t) + \hat{s}_u(t) \sin(2\pi f_c t)\right]. \tag{3.220}$$

37. (Haykin 1983) A method that is used for carrier recovery in SSB modulation systems involves transmitting two pilot frequencies that are appropriately positioned with respect to the transmitted sideband. This is shown in Fig. 3.45a for the case where the lower sideband is transmitted. Here, the two pilot frequencies are defined by:

$$f_1 = f_c - W - \Delta f$$
$$f_2 = f_c + \Delta f, \tag{3.221}$$

where f_c is the carrier frequency, W is the message bandwidth, and Δf is chosen such that

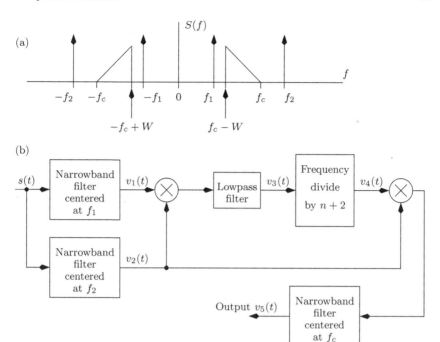

Fig. 3.45 Carrier recovery for SSB signals with lower sideband transmitted

$$n = \frac{W}{\Delta f} \tag{3.222}$$

where n is an integer. Carrier recovery is accomplished using the scheme shown in Fig. 3.45b. The outputs of the two narrowband filters centered at f_1 and f_2 are given by:

$$v_1(t) = A_1 \cos(2\pi f_1 t + \phi_1)$$
$$v_2(t) = A_2 \cos(2\pi f_2 t + \phi_2). \tag{3.223}$$

The lowpass filter is designed to select the difference frequency component at the first multiplier output, due to $v_1(t)$ and $v_2(t)$.

(a) Show that the output signal in Fig. 3.45b is proportional to the carrier wave $A_c \cos(2\pi f_c t)$ if the phase angles satisfy:

$$\phi_2 = \frac{-\phi_1}{1+n}. \tag{3.224}$$

(b) For the case when only the upper sideband is transmitted, the two pilot frequencies are

$$f_1 = f_c - \Delta f$$
$$f_2 = f_c + W + \Delta f. \tag{3.225}$$

How would you modify the carrier recovery scheme in order to deal with this case. What is the corresponding relation between ϕ_1 and ϕ_2 for the output to be proportional to the carrier signal?

- *Solution*: The output of the lowpass filter can be written as:

$$v_3(t) = \frac{A_1 A_2}{2} \cos(2\pi(f_2 - f_1)t + \phi_2 - \phi_1). \tag{3.226}$$

Substituting for f_1 and f_2 we get:

$$\begin{aligned} v_3(t) &= \frac{A_1 A_2}{2} \cos(2\pi(n+2)Wt/n + \phi_2 - \phi_1) \\ &= \frac{A_1 A_2}{2} \cos(2\pi(n+2)Wt/n + \phi_2 - \phi_1 + 2\pi k) \end{aligned} \tag{3.227}$$

for $0 \le k < n+2$, where k is an integer. However, $v_3(t)$ in (3.227) can be written as:

$$v_3(t) = \frac{A_1 A_2}{2} \cos(2\pi(n+2)W(t - t_0)/n) \tag{3.228}$$

where

$$\frac{-2\pi(n+2)Wt_0}{n} = \phi_2 - \phi_1 + 2\pi k \tag{3.229}$$

for some value of t_0. The output of the frequency divider is:

$$\begin{aligned} v_4(t) &= \frac{A_1 A_2}{2} \cos(2\pi W(t - t_0)/n) \\ &= \frac{A_1 A_2}{2} \cos(2\pi Wt/n + (\phi_2 - \phi_1)/(n+2) + 2\pi k/(n+2)) \\ &= \frac{A_1 A_2}{2} \cos(2\pi \Delta f t + (\phi_2 - \phi_1)/(n+2) + 2\pi k/(n+2)). \end{aligned} \tag{3.230}$$

Note that frequency division by $n+2$ results in a phase ambiguity of $2\pi k/(n+2)$. We need to choose the phase corresponding to $k = 0$. How this can be done will be explained later. The output of the second multiplier is given by:

$$v_4(t)v_2(t) = \frac{A_1 A_2^2}{2} \cos(2\pi \Delta f t + (\phi_2 - \phi_1)/(n+2))$$
$$\times \cos(2\pi(f_c + \Delta f)t + \phi_2). \tag{3.231}$$

The output of the narrowband filter centered at f_c is

$$v_5(t) = \frac{A_1 A_2^2}{4} \cos(2\pi f_c t + \phi_2 - (\phi_2 - \phi_1)/(n+2)), \tag{3.232}$$

which is proportional to the carrier frequency when

$$\phi_2 - (\phi_2 - \phi_1)/(n+2) = 0$$
$$\Rightarrow \phi_2 = \frac{-\phi_1}{n+1}. \tag{3.233}$$

Note that $\phi_1 = \phi_2 = 0$ is a trivial solution.

One possible method to choose the correct value of $k (= 0)$ could be as follows. Obtain $v_5(t)$ using each value of k, for $0 \leq k < n+2$, and use it to demodulate the SSB signal. The value of k that maximizes the message energy/power, is selected.

For the case where the upper sideband is transmitted, the carrier recovery scheme is shown in Fig. 3.46. Here we again have

$$v_3(t) = \frac{A_1 A_2}{2} \cos(2\pi(f_2 - f_1)t + \phi_2 - \phi_1)$$
$$= \frac{A_1 A_2}{2} \cos(2\pi(n+2)Wt/n + \phi_2 - \phi_1 + 2\pi k), \tag{3.234}$$

for $0 \leq k < n+2$. Using similar arguments, the output of the frequency divider is (for $k = 0$):

$$v_4(t) = \frac{A_1 A_2}{2} \cos(2\pi Wt/n + (\phi_2 - \phi_1)/(n+2))$$
$$= \frac{A_1 A_2}{2} \cos(2\pi \Delta f t + (\phi_2 - \phi_1)/(n+2)). \tag{3.235}$$

The output of the second multiplier is given by:

$$v_4(t)v_1(t) = \frac{A_1^2 A_2}{2} \cos(2\pi \Delta f t + (\phi_2 - \phi_1)/(n+2))$$
$$\times \cos(2\pi(f_c - \Delta f)t + \phi_1). \tag{3.236}$$

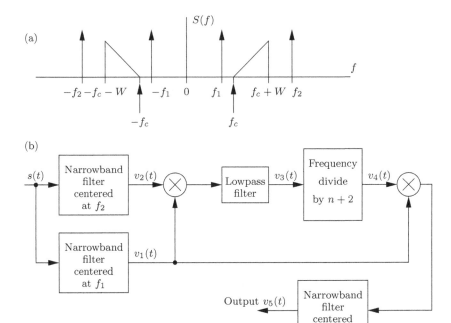

Fig. 3.46 Carrier recovery for SSB signals with upper sideband transmitted

The output of the narrowband filter centered at f_c is

$$v_5(t) = \frac{A_1^2 A_2}{4} \cos(2\pi f_c t), \tag{3.237}$$

provided

$$\frac{\phi_2 - \phi_1}{n + 2} + \phi_1 = 0$$
$$\Rightarrow \phi_2 = -(n + 1)\phi_1. \tag{3.238}$$

38. Consider a signal given by:

$$s(t) = \hat{m}(t) \cos(2\pi f_c t) - m(t) \sin(2\pi f_c t), \tag{3.239}$$

where $\hat{m}(t)$ is the Hilbert transform of $m(t)$.

(a) Sketch the spectrum of $s(t)$ when the spectrum of $m(t)$ is shown in Fig. 3.47a. Label all the important points along the axes.
(b) It is desired to recover $m(t)$ from $s(t)$ using the scheme shown in Fig. 3.47b. Give the specifications of filter1, filter2, and the value of x.

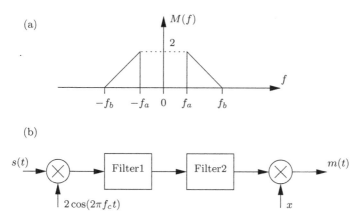

Fig. 3.47 a Spectrum of the message signal. b Scheme to recover the message

- *Solution*: We know that the Fourier transform of $\hat{m}(t)$ is given by

$$\hat{m}(t) \rightleftharpoons -j \operatorname{sgn}(f)M(f). \tag{3.240}$$

Then

$$
\begin{aligned}
\hat{m}(t)\cos(2\pi f_c t) &\rightleftharpoons \frac{-j}{2}\Big[\operatorname{sgn}(f - f_c)M(f - f_c) \\
&\quad + \operatorname{sgn}(f + f_c)M(f + f_c)\Big] \\
&\stackrel{\Delta}{=} S_1(f),
\end{aligned} \tag{3.241}
$$

which is plotted in Fig. 3.48b. Similarly

$$
\begin{aligned}
m(t)\sin(2\pi f_c t) &\rightleftharpoons \frac{-j}{2}[M(f - f_c) - M(f + f_c)] \\
&\stackrel{\Delta}{=} S_2(f)
\end{aligned} \tag{3.242}
$$

which is plotted in Fig. 3.48c.
Finally, we have

$$S(f) = S_1(f) - S_2(f), \tag{3.243}$$

which is plotted in Fig. 3.48d.
One possible implementation of the receiver is shown in Fig. 3.48e.

39. It is desired to transmit 100 voice signals using frequency division multiplexing (FDM). The voice signals are SSB modulated, with the upper sideband transmitted. The spectrum of each of the voice signals is illustrated in Fig. 3.49.

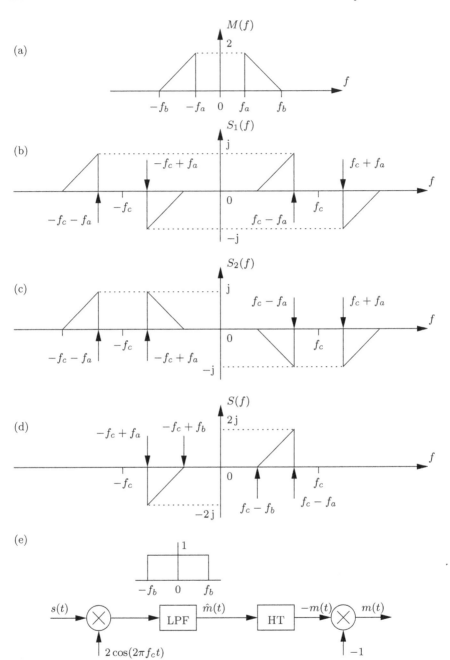

Fig. 3.48 **a** $M(f)$. **b** $S_1(f)$. **c** $S_2(f)$. **d** $S(f)$

Fig. 3.49 Spectrum of the message signal

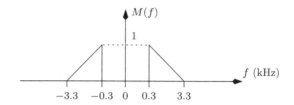

(a) What is the minimum carrier spacing required to multiplex all the signals, such that there is no overlap of spectra.

(b) Using a single- stage approach and assuming a carrier spacing of 4 kHz and a minimum carrier frequency of 100 kHz, determine the lower and upper frequency limits occupied by the FDM signal containing 100 voice signals.

(c) In the two-stage approach, both stages use SSB with upper sideband transmitted. In the first stage, the voice signals are grouped into L blocks, each containing K voice signals. The carrier frequencies in each block in the first stage is given by $4n$ kHz, for $0 \leq n \leq K - 1$. Note that the FDM signal using the two-stage approach must be identical to the single-stage approach in (b).

　i. Sketch the spectrum of each block in the first stage.

　ii. How many carriers are required to generate the FDM signal? Express your answer in terms of L and K.

　iii. Determine L and K such that the number of carriers is minimized.

　iv. Give the expression for the carrier frequencies required to modulate each block in the 2nd stage.

- *Solution*: The minimum carrier spacing required is 3 kHz. If the minimum carrier frequency is 100 kHz, then the spectrum of the first voice signal starts at 100.3 kHz. The carrier for the 100th voice signal is at

$$100 + 99 \times 4 = 496 \text{ kHz}. \tag{3.244}$$

The spectrum of the 100th voice signal starts at 496.3 kHz and ends at 499.3 kHz. Therefore, the spectrum of the composite FDM signal extends from 100.3 to 499.3 kHz, that is $100.3 \leq |f| \leq 499.3$ kHz.

The spectrum of each block in the first stage is shown in Fig. 3.50b.

For the second approach, let us first evaluate the number of carriers required to obtain each block in the first stage. Clearly, the number of carriers required is $K - 1$, since the first message in each block is not modulated at all. In the next stage, L carriers are required to translate each block to the appropriate frequency band. Thus, the total number of carriers required is $L + K - 1$.

Now, we need to minimize $L + K - 1$ subject to the constraint $LK = 100$. The problem can be restated as

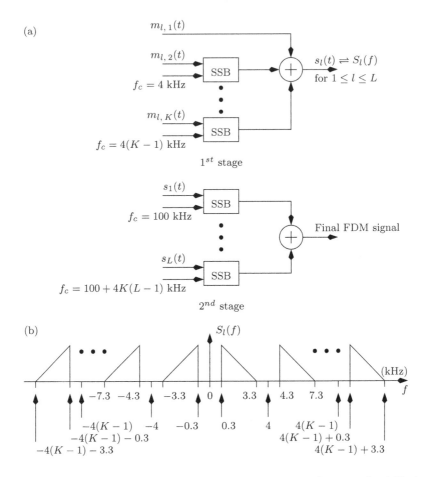

Fig. 3.50 a Two-stage approach for obtaining the final FDM signal. **b** Spectrum of each block at the output of the first stage

$$\min_{K} \frac{100}{K} + K - 1. \tag{3.245}$$

Differentiating with respect to K and setting the result to zero, we get the solution as $K = 10, L = 10$.

The carrier frequencies required to modulate each block in the second stage is given by $100 + 4Kl$ for $0 \leq l \leq L - 1$.

The block diagram of the two-stage approach is given in Fig. 3.50a.

40. Consider the modified switching modulator shown in Fig. 3.51. Note that

$$v_1(t) = A_c \cos(2\pi f_c t) + m(t)$$
$$v_2(t) = v_1(t) g_p(t), \tag{3.246}$$

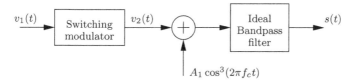

Fig. 3.51 Block diagram of a modified switching modulator

where

$$g_p(t) = \frac{1}{2} + \frac{2}{\pi} \sum_{n=1}^{\infty} \frac{(-1)^{n-1}}{2n-1} \cos(2\pi f_c (2n-1)t). \qquad (3.247)$$

It is desired to obtain an AM signal $s(t)$ centered at $3 f_c$.

(a) Draw the spectrum of the ideal BPF required to obtain $s(t)$. Assume a BPF gain of k. The bandwidth of $m(t)$ extends over $[-W, W]$.
(b) Write down the expression for $s(t)$, assuming a BPF gain of k.
(c) It is given that the carrier power is $10\,W$ with 100% modulation. The absolute maximum value of $m(t)$ is $2\,V$. Compute A_1 and k.

- *Solution*: We have

$$v_2(t) = \frac{A_c}{2} \cos(2\pi f_c t)$$
$$+ \frac{A_c}{\pi} \sum_{n=1}^{\infty} \frac{(-1)^{n-1}}{2n-1} (\cos(4\pi(n-1)f_c t) + \cos(4\pi n f_c t))$$
$$+ \frac{m(t)}{2} + \frac{2}{\pi} \sum_{n=1}^{\infty} \frac{(-1)^{n-1}}{2n-1} \cos(2\pi(2n-1) f_c t) m(t). \quad (3.248)$$

In (3.248), the only term centered at $3 f_c$ is

$$- \frac{2}{3\pi} \cos(6\pi f_c t) m(t), \qquad (3.249)$$

which occurs at $n = 2$. Similarly

$$A_1 \cos^3(2\pi f_c t) = \frac{A_1}{4} [3 \cos(2\pi f_c t) + \cos(6\pi f_c t)], \qquad (3.250)$$

which has a component at f_c and $3 f_c$. Thus it is clear that in order to extract the components at $3 f_c$, the spectrum of the BPF must be as indicated in Fig. 3.52. The BPF output is given by:

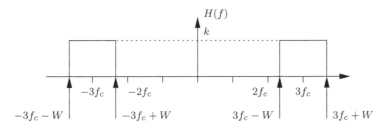

Fig. 3.52 Spectrum of the BPF

$$s(t) = \frac{A_1 k}{4} \cos(6\pi f_c t) - \frac{2k}{3\pi} m(t) \cos(6\pi f_c t)$$

$$= \frac{kA_1}{4} \left[1 - \frac{8}{3\pi A_1} m(t) \right] \cos(6\pi f_c t). \qquad (3.251)$$

For 100% modulation we require

$$\max \frac{8}{3\pi A_1} |m(t)| = 1$$

$$\Rightarrow \frac{16}{3\pi A_1} = 1$$

$$\Rightarrow A_1 = 1.7. \qquad (3.252)$$

The carrier power is given by

$$\frac{k^2 A_1^2}{32} = 10$$

$$\Rightarrow k = 10.5. \qquad (3.253)$$

41. Consider an AM signal given by:

$$s(t) = A_c \left[1 + \mu \cos(2\pi f_m t) \right] \cos(2\pi f_c t) \quad |\mu| < 1. \qquad (3.254)$$

Assume that the envelope detector is implemented using a diode and an RC filter, as shown in Fig. 3.53. Determine the upper limit on RC such that the capacitor voltage follows the envelope. Assume that $RC \gg 1/f_c$, $e^{-x} \approx 1 - x$ for small values of x and f_c to be very large.

- *Solution*: The envelope of $s(t)$ is (see Fig. 3.54)

$$a(t) = A_c \left| 1 + \mu \cos(2\pi f_m t) \right|$$

$$= A_c \left[1 + \mu \cos(2\pi f_m t) \right], \qquad (3.255)$$

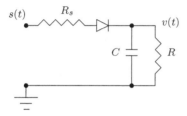

Fig. 3.53 Envelope detector using an RC filter

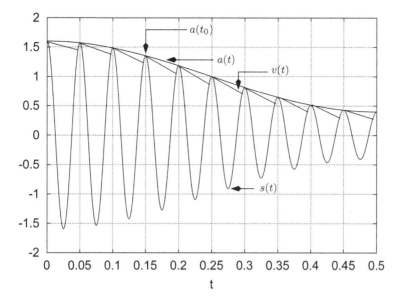

Fig. 3.54 Plots of $s(t)$, the envelope $a(t)$ and the capacitor voltage $v(t)$

since $|\mu| < 1$. Consider any point $a(t_0)$ on the envelope at time t_0, which coincides with the carrier peak. The capacitor starts discharging from $a(t_0)$ upto approximately the next carrier peak. The capacitor voltage $v(\tau)$ can be written as:

$$v(\tau) = a(t_0)e^{-\tau/RC}$$
$$\approx a(t_0)\left[1 - \frac{\tau}{RC}\right] \quad \text{for } 0 < \tau < 1/f_c, \quad (3.256)$$

where it is assumed that the new time variable $\tau = 0$ at t_0. The absolute value of the slope of the capacitor voltage at t_0 is given by

$$\left|\frac{dv(\tau)}{d\tau}\right|_{\tau=0} = \frac{a(t_0)}{RC}. \quad (3.257)$$

We require that the absolute value of the slope of the capacitor voltage must exceed that of the envelope at time t_0. The absolute value of the slope of the envelope at t_0 is

$$\left. \frac{da(t)}{dt} \right|_{t=t_0} = \mu A_c (2\pi f_m) \left| \sin(2\pi f_m t_0) \right| . \tag{3.258}$$

Note that since t_0 must coincide with the carrier peak, we must have

$$t_0 = \frac{k}{f_c} \tag{3.259}$$

for integer values of k. From (3.257) and (3.258) we get the required condition as

$$\frac{a(t_0)}{RC} \geq \mu A_c (2\pi f_m) \left| \sin(2\pi f_m t_0) \right|$$

$$\Rightarrow RC \leq \frac{1 + \mu \cos(2\pi f_m t_0)}{2\pi \mu f_m \left| \sin(2\pi f_m t_0) \right|} \tag{3.260}$$

for every t_0 given by (3.259). However, as f_c becomes very large $t_0 \rightarrow t$. Thus the upper limit of RC can be found by computing the minimum value of the RHS of (3.260). Assuming that $\sin(2\pi f_m t)$ is positive, we have

$$\frac{d}{dt} \left[\frac{1 + \mu \cos(2\pi f_m t)}{2\pi \mu f_m \sin(2\pi f_m t)} \right] = 0$$

$$\Rightarrow \cos(2\pi f_m t) = -\mu. \tag{3.261}$$

We get the same result when $\sin(2\pi f_m t)$ is negative. Substituting (3.261) in (3.260) we get

$$RC \leq \frac{\sqrt{1 - \mu^2}}{2\pi \mu f_m}. \tag{3.262}$$

Recall that for proper functioning of the envelope detector, we require:

$$RC \ll \frac{1}{W}, \tag{3.263}$$

where W is the one-sided bandwidth of the message. We find that (3.263) is valid when μ is close to unity and f_m is replaced by W in (3.262).

42. Consider a VSB signal $s(t)$ obtained by passing

$$s_1(t) = A_c m(t) \cos(2\pi f_c t) \tag{3.264}$$

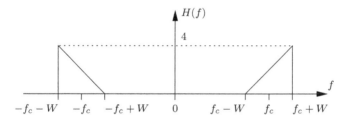

Fig. 3.55 Spectrum of the VSB filter

through a VSB filter $H(f)$ as illustrated in Fig. 3.55. Assume $m(t)$ to be ban-dlimited to $[-W, W]$ and $f_c \gg W$. Let

$$s(t) = s_c(t) \cos(2\pi f_c t) - s_s(t) \sin(2\pi f_c t). \tag{3.265}$$

(a) Determine the bandwidth of $s(t)$.
(b) Determine $s_c(t)$ and $s_s(t)$.
(c) Draw the block diagram of the phase discrimination method of generating the VSB signal $s(t)$ for this particular problem. Sketch the frequency response of $H_s(f)/j$. Label all the important points.
(d) In this problem, most of the energy in $s(t)$ is concentrated on the upper side-band. How should the block diagram in (c) be modified (without changing $H_s(f)$) such that most of the energy is concentrated in the lower sideband.
(e) How should $H_s(f)$ in (c) be modified so that $s(t)$ becomes an SSB signal with upper sideband transmitted? Sketch the new response of $H_s(f)/j$. Label all the important points. Note that in this part only $H_s(f)$ is to be modified.

There is no need to derive any formula. All symbols have their usual meaning.

- *Solution*: The bandwidth of $s(t)$ is $2W$.
  We know that

$$S_c(f) = \begin{cases} A_c M(f) H(f_c) & \text{for } -W \le f \le W \\ 0 & \text{otherwise.} \end{cases} \tag{3.266}$$

In the given problem $H(f_c) = 2$. Therefore

$$s_c(t) = 2A_c m(t). \tag{3.267}$$

Similarly

$$S_s(f) = A_c M(f) H_s(f)/2, \tag{3.268}$$

where

$$H_s(f) = \begin{cases} j[H(f - f_c) - H(f + f_c)] & \text{for } -W \le f \le W \\ 0 & \text{otherwise.} \end{cases} \tag{3.269}$$

From Fig. 3.56a–c and (3.269), it is clear that

$$H_s(f) = \begin{cases} -j(4f/W) & \text{for } -W \le f \le W \\ 0 & \text{otherwise.} \end{cases} \tag{3.270}$$

$H_s(f)/j$ is depicted in Fig. 3.56d. Hence

$$S_s(f) = (-A_c/(\pi W))j\, 2\pi f M(f). \tag{3.271}$$

Taking the inverse Fourier transform of (3.271) we get

$$s_s(t) = -\frac{A_c}{\pi W}\frac{dm(t)}{dt}. \tag{3.272}$$

The block diagram of the phase discrimination method of generating $s(t)$ is given in Fig. 3.57.
If most of the energy is to be concentrated in the lower sideband, $z(t)$ in Fig. 3.57 must be subtracted instead of being added.
If $s(t)$ is to be an SSB signal with upper sideband transmitted, then $H_s(f)/j$ must be as shown in Fig. 3.56e.

43. The signal $m_1(t) = B\cos(2\pi f_m t)$, $B > 0$, is passed through a first-order RC-filter as shown in Fig. 3.58. It is given that $1/(2\pi RC) = f_m/2$. The output $m(t)$ is amplitude modulated to obtain:

$$s(t) = A_c[1 + k_a m(t)]\cos(2\pi f_c t). \tag{3.273}$$

(a) Find $m(t)$.
(b) Find the limits of k_a for no envelope distortion.
(c) Compute the power of $s(t)$.

• *Solution*: The transfer function of the first-order lowpass filter is

$$H(f) = \frac{1}{1 + j\, f/f_0}, \tag{3.274}$$

where f_0 is the –3 dB frequency given by:

$$f_0 = \frac{1}{2\pi RC} = \frac{f_m}{2}. \tag{3.275}$$

Clearly

$$m(t) = B|H(f_m)|\cos(2\pi f_m t + \theta), \tag{3.276}$$

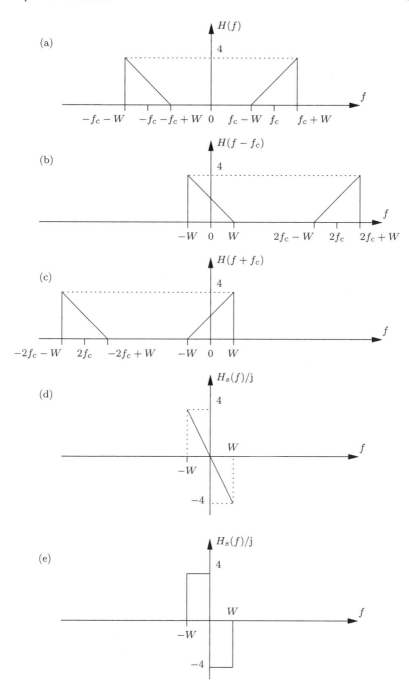

Fig. 3.56 Spectrum of the VSB filter and $H_s(f)/\mathrm{j}$

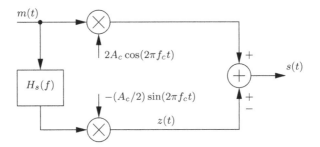

Fig. 3.57 Block diagram for generating $s(t)$

Fig. 3.58 First-order RC
filter

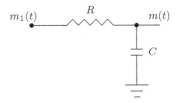

where

$$\theta = -\tan^{-1}(2)$$

$$|H(f_m)| = \frac{1}{\sqrt{5}}. \qquad (3.277)$$

For no envelope distortion we require

$$|k_a|B|H(f_m)| \leq 1$$

$$\Rightarrow |k_a| \leq \frac{\sqrt{5}}{B} \qquad (3.278)$$

with the constraint that $k_a \neq 0$. In order to compute the power of $s(t)$ we note that

$$s(t) = A_c \cos(2\pi f_c t)$$
$$+ A_c k_a \frac{B}{2\sqrt{5}} \cos(2\pi (f_c - f_m)t - \theta)$$
$$+ A_c k_a \frac{B}{2\sqrt{5}} \cos(2\pi (f_c + f_m)t + \theta). \qquad (3.279)$$

Thus the power of $s(t)$ is just the sum of the power of the individual sinusoids and is equal to:

Fig. 3.59 a Message
spectrum. **b** Recovery of the
message

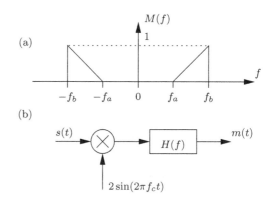

$$P = \frac{A_c^2}{2} + \frac{A_c^2 k_a^2 B^2}{20}. \tag{3.280}$$

44. An AM signal

$$s_1(t) = A_c[1 + k_a m(t)]\cos(2\pi f_c t) \tag{3.281}$$

is passed through an ideal bandpass filter (BPF) having a frequency response

$$H(f) = \begin{cases} 3 \text{ for } f_c - f_b \le |f| \le f_c - f_a \\ 0 \text{ otherwise.} \end{cases} \tag{3.282}$$

The message spectrum is shown in Fig. 3.59a.

(a) Write down the expression of the signal $s(t)$, at the BPF output, in terms of $m(t)$.

(b) It is desired to recover $m(t)$ (not $cm(t)$ where c is a constant) from $s(t)$ using the block diagram in Fig. 3.59b.
Give the specifications of $H(f)$. Assume ideal filter characteristics.

- *Solution*: Clearly, the signal at the BPF output is SSB modulated with lower sideband transmitted. Therefore, the expression for $s(t)$ is

$$s(t) = km(t)\cos(2\pi f_c t) + k\hat{m}(t)\sin(2\pi f_c t), \tag{3.283}$$

where k is a constant that needs to be found out.
Now, the Fourier transform of $s(t)$ is

$$S(f) = \frac{k}{2}M(f - f_c)\left[1 - \mathrm{sgn}\,(f - f_c)\right]$$
$$+ \frac{k}{2}M(f + f_c)\left[1 + \mathrm{sgn}\,(f + f_c)\right]. \tag{3.284}$$

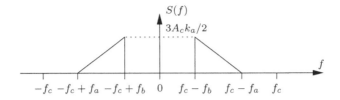

Fig. 3.60 Spectrum at BPF output

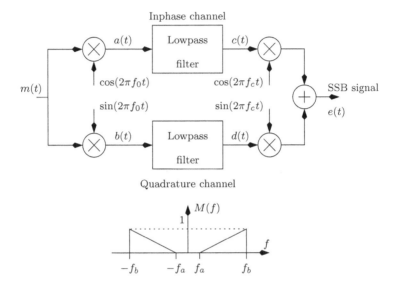

Fig. 3.61 Weaver's method of generating SSB signals transmitting the upper sideband

Comparing $S(f)$ in (3.284) with the spectrum in Fig. 3.60, we get

$$k = \frac{3A_c k_a}{2}. \tag{3.285}$$

In order to recover the message, we need a lowpass filter with unity gain and passband $[-f_b, \ f_b]$ in cascade with an ideal Hilbert transformer followed by a gain of $-1/k$. These three elements can be combined into a single filter whose frequency response is given by

$$H(f) = \begin{cases} \mathrm{j}\,(1/k)\mathrm{sgn}(f) & \text{for } |f| < f_b \\ 0 & \text{otherwise.} \end{cases} \tag{3.286}$$

45. (Haykin 1983) Consider the block diagram in Fig. 3.61. The message $m(t)$ is bandlimited to $f_a \leq |f| \leq f_b$. The auxiliary carrier applied to the first pair of

product modulators is given by:

$$f_0 = \frac{f_a + f_b}{2}.$$

(3.287)

The lowpass filters are identical and can be assumed to be ideal with unity gain and cutoff frequency equal to $(f_b - f_a)/2$. The carrier frequency $f_c > (f_b - f_a)/2$.

(a) Plot the spectra of the complex signals $a(t) + j b(t)$, $c(t) + j d(t)$ and the real-valued signal $e(t)$.
(b) Write down the expression for $e(t)$ in terms of $m(t)$.
(c) How would you modify Fig. 3.61 so that only the lower sideband is transmitted.

• *Solution*: We have

$$a(t) + j b(t) = m(t) \exp (j 2\pi f_0 t) = m_1(t) \quad \text{(say)}.$$

(3.288)

Thus

$$m_1(t) \rightleftharpoons M_1(f) = M(f - f_0).$$

(3.289)

The plot of $M_1(f)$ is given in Fig. 3.62b. Note that $m_1(t)$ cannot be regarded as a pre-envelope, since it has non-zero negative frequencies. The various edge frequencies in $M_1(f)$ are given by:

$$f_1 = (f_b - f_a)/2$$
$$f_2 = (f_b + 3 f_a)/2$$
$$f_3 = (f_a + 3 f_b)/2.$$

(3.290)

Now the complex-valued signal $m_1(t)$ gets convolved with a *real-valued* low-pass filter (the lowpass filters have been stated to be identical, hence they can be considered to be a single real-valued lowpass filter), resulting in a complex-valued output $m_2(t)$. The plot of $M_2(f)$ is shown in Fig. 3.62c. Let

$$m_2(t) = m_{2,c}(t) + j m_{2,s}(t) = c(t) + j d(t).$$

(3.291)

Now $e(t)$ is given by

$$
\begin{aligned}
e(t) &= \Re \{m_2(t) \exp (-j 2\pi f_c t)\} \\
&= \frac{1}{2} \left[m_2(t) \exp (-j 2\pi f_c t) + m_2^*(t) \exp (j 2\pi f_c t) \right] \\
&\rightleftharpoons \frac{1}{2} \left[M_2(f + f_c) + M_2^*(-f + f_c) \right].
\end{aligned}
$$

(3.292)

Fig. 3.62 Plot of the signal
spectra at various points

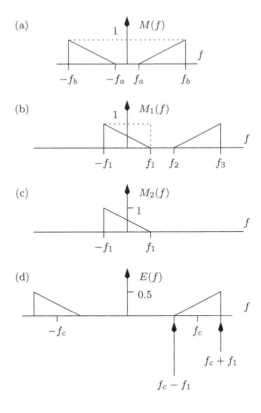

Since $M_2(f)$ is real-valued, we have

$$e(t) = \frac{1}{2}[M_2(f + f_c) + M_2(-f + f_c)] = E(f) \quad \text{(say)}. \quad (3.293)$$

$E(f)$ is plotted in Fig. 3.62d.
Let

$$f_{c1} = f_c - f_1 - f_a. \quad (3.294)$$

Then

$$e(t) = m(t)\cos(2\pi f_{c1}t) - \hat{m}(t)\sin(2\pi f_{c1}t). \quad (3.295)$$

In order to transmit the lower sideband, the first product modulator in the quadrature arm must be fed with $-\sin(2\pi f_0 t)$.

46. (Haykin 2001) The spectrum of a voice signal $m(t)$ is zero outside the interval $f_a < |f| < f_b$. To ensure communication privacy, this signal is applied to a

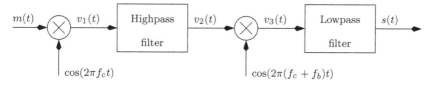

Fig. 3.63 Block diagram of a scrambler

scrambler that consists of the following components in cascade: product modulator, highpass filter, second product modulator, and lowpass filter as illustrated
in Fig. 3.63. The carrier wave applied to the first product modulator has frequency equal to f_c, whereas that applied to the second product modulator has
frequency equal to $f_b + f_c$. Both the carriers have unit amplitude. The highpass
and lowpass filters are ideal with unity gain and have the same cutoff frequency
at f_c. Assume that $f_c > f_b$.

(a) Derive an expression for the scrambler output $s(t)$.
(b) Show that the original voice signal $m(t)$ may be recovered from $s(t)$ by
using an unscrambler that is identical to the scrambler.

- *Solution*: The block diagram of the system is shown in Fig. 3.63. The output
of the first product modulator is given by:

$$V_1(f) = \frac{1}{2}\left[M(f - f_c) + M(f + f_c)\right]. \tag{3.296}$$

The output of the highpass filter is an SSB signal with the upper sideband
transmitted. This is illustrated in Fig. 3.64. Hence:

$$v_2(t) = \frac{1}{2}\left[m(t)\cos(2\pi f_c t) - \hat{m}(t)\sin(2\pi f_c t)\right]. \tag{3.297}$$

The output of the second product modulator is given by:

$$\begin{aligned}
v_3(t) &= v_2(t)\cos(2\pi(f_c + f_b)t)\\
&= \frac{1}{4}\left[m(t)\left(\cos(2\pi(2f_c + f_b)t) + \cos(2\pi f_b t)\right)\right.\\
&\quad \left. - \hat{m}(t)\left(\sin(2\pi(2f_c + f_b)t) - \sin(2\pi f_b t)\right)\right], \tag{3.298}
\end{aligned}$$

which after lowpass filtering becomes:

$$s(t) = \frac{1}{4}\left[m(t)\cos(2\pi f_b t) + \hat{m}(t)\sin(2\pi f_b t)\right]. \tag{3.299}$$

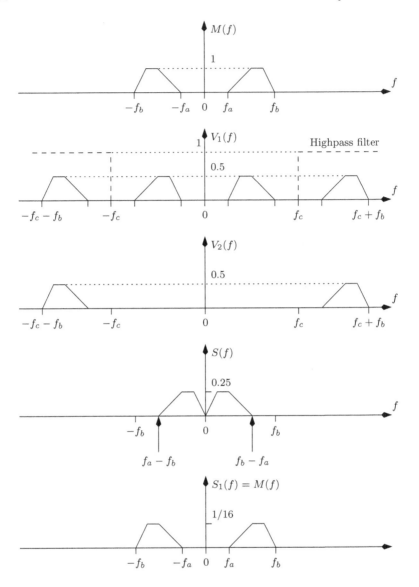

Fig. 3.64 Illustrating the spectra of various signals in the scrambler

It is clear that $s(t)$ is an SSB signal with lower sideband transmitted and carrier frequency f_b. This is illustrated in Fig. 3.64.

If $s(t)$ is fed to the scrambler, the output would be similar to (3.299), that is:

$$s_1(t) = \frac{1}{4}\left[s(t)\cos(2\pi f_b t) + \hat{s}(t)\sin(2\pi f_b t)\right]. \tag{3.300}$$

In other words, we transmit the lower sideband of $s(t)$ with carrier frequency f_b. Thus

$$s_1(t) = \frac{1}{16}m(t) \tag{3.301}$$

which is again shown in Fig. 3.64.

47. (Haykin 1983) The single-tone modulating signal

$$m(t) = A_m \cos(2\pi f_m t) \tag{3.302}$$

is used to generate the VSB signal

$$s(t) = \frac{A_m A_c}{2}\left[a\cos(2\pi(f_c + f_m)t) + (1 - a)\cos(2\pi(f_c - f_m)t)\right](3.303)$$

where $0 \le a \le 1$ is a constant, representing the attenuation of the upper side frequency.

(a) From the canonical representation of a bandpass signal, find the quadrature component of $s(t)$.
(b) The VSB signal plus a carrier $A_c \cos(2\pi f_c t)$, is passed through an envelope detector. Determine the distortion produced by the quadrature component.
(c) What is the value of a for which the distortion is maximum.
(d) What is the value of a for which the distortion is minimum.

• *Solution*: The VSB signal $s(t)$ can be simplified to:

$$s(t) = \frac{A_m A_c}{2}\left[\cos(2\pi f_c t)\cos(2\pi f_m t) + (1 - 2a)\sin(2\pi f_c t)\sin(2\pi f_m t)\right]. \tag{3.304}$$

By comparing with the canonical representation of a bandpass signal, we see that the quadrature component is:

$$-\frac{A_m A_c}{2}(1 - 2a)\sin(2\pi f_m t). \tag{3.305}$$

After the addition of the carrier, the modified signal is:

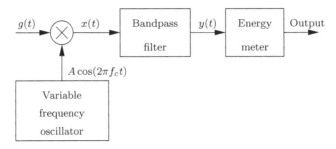

Fig. 3.65 Block diagram of a heterodyne spectrum analyzer

$$s(t) = A_c \cos(2\pi f_c t) \left[1 + \frac{A_m}{2} \cos(2\pi f_m t)\right]$$

$$+ \frac{A_m A_c}{2} (1 - 2a) \sin(2\pi f_c t) \sin(2\pi f_m t). \qquad (3.306)$$

The envelope is given by:

$$E(t) = A_c \left[1 + \frac{A_m}{2} \cos(2\pi f_m t)\right] \sqrt{1 + D(t)}, \qquad (3.307)$$

where $D(t)$ is the distortion defined by:

$$D(t) = \left[\frac{(A_m/2)(1 - 2a) \sin(2\pi f_m t)}{1 + (A_m/2) \cos(2\pi f_m t)}\right]^2. \qquad (3.308)$$

Clearly, $D(t)$ is minimum when $a = 1/2$ and maximum when $a = 0, 1$.

48. (Haykin 1983) Figure 3.65 shows the block diagram of a heterodyne spectrum analyzer. The oscillator has an amplitude A and operates over the range f_0 to $f_0 + W$ where $\pm f_0$ is the midband frequency of the BPF and $g(t)$ extends over the frequency band $[-W, W]$. Assume that the BPF bandwidth $\Delta f \ll W$ and $f_0 \gg W$ and that the passband response of the BPF is unity. Determine the value of the energy meter output for an input signal $g(t)$, for a particular value of the oscillator frequency, say f_c. Assume that $g(t)$ is a real-valued energy signal.

- *Solution*: Let us denote the output of the product modulator by $x(t)$. Hence

$$x(t) = A \cos(2\pi f_c t) g(t)$$

$$\rightleftharpoons \frac{A}{2} [G(f - f_c) + G(f + f_c)] = X(f), \qquad (3.309)$$

where f_c denotes the oscillator frequency. The BPF output can be written as:

$$Y(f) = X(f)\left[\text{rect}\left(\frac{f - f_0}{\Delta f}\right) + \text{rect}\left(\frac{f + f_0}{\Delta f}\right)\right]$$

$$= \frac{A}{2}\left[X(f_0)\text{rect}\left(\frac{f - f_0}{\Delta f}\right) + X(-f_0)\text{rect}\left(\frac{f + f_0}{\Delta f}\right)\right]$$

$$= \frac{A}{2}\left[G(f_0 - f_c)\text{rect}\left(\frac{f - f_0}{\Delta f}\right) + G(-f_0 + f_c)\text{rect}\left(\frac{f + f_0}{\Delta f}\right)\right]$$

, (3.310)

where we have assumed that Δf is small enough so that $X(f)$ can be assumed to be constant in the bandwidth of the filter. The energy of $y(t)$ is given by:

$$E = \int_{t=-\infty}^{\infty} y^2(t)\, dt$$

$$= \int_{f=-\infty}^{\infty} |Y(f)|^2\, df \tag{3.311}$$

where we have used the Rayleigh's energy theorem. Hence

$$E = \frac{A^2}{2}|G(f_0 - f_c)|^2 \Delta f \tag{3.312}$$

where we have used the fact that $g(t)$ is real-valued, hence

$$|G(f_0 - f_c)| = |G(-f_0 + f_c)|. \tag{3.313}$$

Note that as f_c varies from f_0 to $f_0 + W$, we obtain the magnitude spectrum of $g(t)$.

49. An AM signal is generated using a switching modulator as shown in Fig. 3.66. The signal $v_1(t) = A_c \cos(2\pi f_c t) + m(t)$, where $A_c \gg |m(t)|$. The diode D has

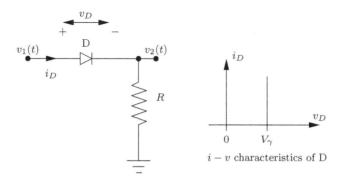

Fig. 3.66 Generation of AM signal using a switching modulator

a cut-in voltage equal to V_γ ($< A_c$). The $i - v$ characteristics of the diode is also shown.

(a) Assuming that $V_\gamma/A_c = 1/\sqrt{2}$, derive the expression for the AM signal centered at f_c and explain how it can be obtained. Assume also that $f_c \gg W$, where W is the two-sided bandwidth of $m(t)$.
(b) Find the minimum value of A_c for no envelope distortion, given that $|m(t)| \le 1$ for all t.

- *Solution*: We have

$$v_2(t) = \begin{cases} v_1(t) - V_\gamma & \text{when D is ON} \\ 0 & \text{when D is OFF} \end{cases}$$

$$\Rightarrow v_2(t) = \begin{cases} v_1(t) - V_\gamma & \text{when } v_1(t) \ge V_\gamma \\ 0 & \text{when } v_1(t) < V_\gamma \end{cases}$$

$$\Rightarrow v_2(t) = \begin{cases} v_1(t) - V_\gamma & \text{when } A_c \cos(2\pi f_c t) \ge V_\gamma \\ 0 & \text{when } A_c \cos(2\pi f_c t) < V_\gamma \end{cases} \qquad (3.314)$$

where we have assumed that $v_1(t) \approx A_c \cos(2\pi f_c t)$. The output voltage $v_2(t)$ can also be written as:

$$v_2(t) = (v_1(t) - V_\gamma)g_p(t), \qquad (3.315)$$

where

$$g_p(t) = \begin{cases} 1 \text{ when } A_c \cos(2\pi f_c t) \ge V_\gamma \\ 0 \text{ when } A_c \cos(2\pi f_c t) < V_\gamma \end{cases} \qquad (3.316)$$

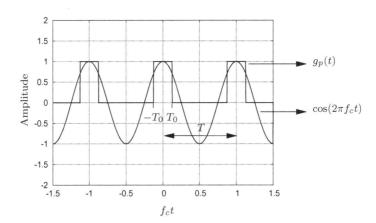

Fig. 3.67 Plot of $g_p(t)$ and $\cos(2\pi f_c t)$ for $V_\gamma/A_c = 1/\sqrt{2}$

as illustrated in Fig. 3.67. Note that:

$$A_c \cos(2\pi f_c T_0) = V_\gamma$$
$$\Rightarrow \cos(2\pi f_c T_0) = 1/\sqrt{2}$$
$$\Rightarrow 2\pi f_c T_0 = \pi/4$$
$$\Rightarrow T_0/T = 1/8, \tag{3.317}$$

where $f_c = 1/T$. The Fourier series expansion for $g_p(t)$ is:

$$g_p(t) = a_0 + 2\sum_{n=1}^{\infty} \left[a_n \cos\left(\frac{2\pi nt}{T}\right) + b_n \sin\left(\frac{2\pi nt}{T}\right) \right], \tag{3.318}$$

where

$$a_0 = \frac{1}{T} \int_{-T/2}^{T/2} g_p(t)\, dt$$
$$= \frac{1}{T} \int_{-T_0}^{T_0} dt$$
$$= 1/4. \tag{3.319}$$

Since $g_p(t)$ is an even function, $b_n = 0$. Now

$$a_n = \frac{1}{T} \int_{-T/2}^{T/2} \cos(2\pi nt/T)\, dt$$
$$= \frac{1}{T} \int_{-T_0}^{T_0} \cos(2\pi nt/T)\, dt$$
$$= \frac{1}{n\pi} \sin(n\pi/4). \tag{3.320}$$

Therefore

$$g_p(t) = \frac{1}{4} + 2\sum_{n=1}^{\infty} \frac{1}{n\pi} \sin(n\pi/4) \cos(2\pi nt/T)$$
$$= \frac{1}{4} + \frac{\sqrt{2}}{\pi} \cos(2\pi f_c t) + \frac{1}{\pi} \cos(4\pi f_c t) + \cdots \tag{3.321}$$

Substituting (3.321) in (3.315) and collecting terms corresponding to the AM signal centered at f_c, we obtain:

$$v_2(t) = \frac{A_c}{4} \cos(2\pi f_c t) + \frac{\sqrt{2}}{\pi} m(t) \cos(2\pi f_c t)$$

$$- \frac{\sqrt{2}}{\pi} V_\gamma \cos(2\pi f_c t) + \frac{A_c}{2\pi} \cos(2\pi f_c t) + \cdots \quad (3.322)$$

Using a bandpass signal centered at f_c and two-sided bandwidth equal to W we obtain the desired AM signal as:

$$s(t) = \left[\frac{A_c}{4} - \frac{V_\gamma \sqrt{2}}{\pi} + \frac{A_c}{2\pi} + \frac{\sqrt{2}}{\pi} m(t) \right] \cos(2\pi f_c t)$$

$$= A_c \left[\frac{1}{4} - \frac{1}{\pi} + \frac{1}{2\pi} + \frac{\sqrt{2}}{\pi A_c} m(t) \right] \cos(2\pi f_c t)$$

$$= A_c \left[\frac{1}{4} - \frac{1}{2\pi} + \frac{\sqrt{2}}{\pi A_c} m(t) \right] \cos(2\pi f_c t)$$

$$= A_c \left[0.0908451 + \frac{0.4501582}{A_c} m(t) \right] \cos(2\pi f_c t)$$

$$= 0.0908451 A_c \left[1 + \frac{4.9552301}{A_c} m(t) \right] \cos(2\pi f_c t). \quad (3.323)$$

For no envelope distortion, the minimum value of A_c is 4.9552301.

50. A signal $x(t) = A \sin(2000\pi t)$ is a full wave rectified and passed through an ideal lowpass filter (LPF) having a bandwidth $[-4.5,\ 4.5]$ kHz and a gain of 2 in the passband. Let the LPF output be denoted by $m(t)$. Find $y(t) = m(t) + j\,\hat{m}(t)$ and sketch its Fourier transform. Compute the power of $y(t)$.

- *Solution*: Note that the period of $x(t)$ is 1 ms. The full wave rectified sine wave can be represented by the complex Fourier series coefficients as follows:

$$|x(t)| = \sum_{n=-\infty}^{\infty} c_n e^{j 2\pi n t / T_1}, \quad (3.324)$$

where $T_1 = 0.5$ ms and

$$c_n = \frac{A}{T_1} \int_{t=0}^{T_1} \sin(\omega_0 t) e^{-j 2\pi n t / T_1} \, dt \quad (3.325)$$

where $\omega_0 = 2\pi / T_0$, with $T_0 = 2T_1 = 1$ ms. The above equation can be rewritten as:

$$c_n = \frac{A}{2j T_1} \int_{t=0}^{T_1} \left(e^{j \omega_0 t} - e^{-j \omega_0 t} \right) e^{-j 2\pi n t / T_1} \, dt$$

Fig. 3.68 Fourier transform of $y(t)$

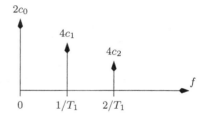

$$= \frac{A}{2jT_1} \left. \frac{e^{j(\omega_0 - 2\pi n/T_1)t}}{j(\omega_0 - 2\pi n/T_1)} \right|_{t=0}^{T_1}$$

$$- \frac{A}{2jT_1} \left. \frac{e^{-j(\omega_0 + 2\pi n/T_1)t}}{-j(\omega_0 + 2\pi n/T_1)} \right|_{t=0}^{T_1}$$

$$= \frac{2A}{\pi(1 - 4n^2)}. \tag{3.326}$$

The output of the LPF is (gain of the LPF is 2):

$$m(t) = 2 \sum_{n=-2}^{2} c_n e^{j2\pi nt/T_1}$$

$$= 2c_0 + 4c_1 \cos(\omega_1 t) + 4c_2 \cos(2\omega_1 t), \tag{3.327}$$

where $\omega_1 = 2\pi/T_1 = 2\omega_0$. The output of the Hilbert transformer is:

$$\hat{m}(t) = 4c_1 \cos(\omega_1 t - \pi/2) + 4c_2 \cos(2\omega_1 t - \pi/2)$$

$$= 4c_1 \sin(\omega_1 t) + 4c_2 \sin(2\omega_1 t). \tag{3.328}$$

Note the absence of c_0 in (3.328), since the Hilbert transformer blocks the dc component. Therefore

$$y(t) = m(t) + j\hat{m}(t)$$

$$= 2c_0 + 4c_1 e^{j\omega_1 t} + 4c_2 e^{2\omega_1 t}. \tag{3.329}$$

The Fourier transform of $y(t)$ is given by:

$$Y(f) = 2c_0 \delta(f) + 4c_1 \delta(f - 1/T_1) + 4c_2 \delta(f - 2/T_1), \tag{3.330}$$

which is shown in Fig. 3.68. The power of $y(t)$ is

$$P = 4c_0^2 + 16c_1^2 + 16c_2^2. \tag{3.331}$$

51. Explain the principle of operation of the Costas loop for DSB-SC signals given by $s(t) = A_c m(t) \cos(2\pi f_c t)$. Draw the block diagram. Clearly identify the phase discriminator. Explain phase ambiguity.

- *Solution:* See Fig. 3.69. We assume that $m(t)$ is real-valued with energy E and Fourier transform $M(f)$. Clearly

$$x_1(t) = A_c m(t) \cos(\phi)$$
$$x_2(t) = A_c m(t) \sin(\phi). \tag{3.332}$$

and

$$x_3(t) = \frac{1}{2} A_c^2 m^2(t) \sin(2\phi). \tag{3.333}$$

Note that

$$m^2(t) \rightleftharpoons \int_{\alpha=-\infty}^{\infty} M(\alpha) M(f - \alpha)\, d\alpha$$
$$\stackrel{\Delta}{=} G(f). \tag{3.334}$$

Therefore

$$G(0) = \int_{\alpha=-\infty}^{\infty} |M(\alpha)|^2\, d\alpha$$
$$= E \tag{3.335}$$

since $M(-\alpha) = M^*(\alpha)$ ($m(t)$ is real-valued), and we have used Rayleigh's energy theorem. We also have

$$\lim_{\Delta f \to 0} H_1(f) = \delta(f). \tag{3.336}$$

Hence

$$X_4(f) = X_3(f)\delta(f)$$
$$= \frac{1}{2} A_c^2 \sin(2\phi) E \delta(f)$$
$$\rightleftharpoons \frac{1}{2} A_c^2 \sin(2\phi) E$$
$$= x_4(t). \tag{3.337}$$

Thus, the (dc) control signal $x_4(t)$ determines the phase of the voltage controlled oscillator (VCO). Clearly, $x_4(t) = 0$ when

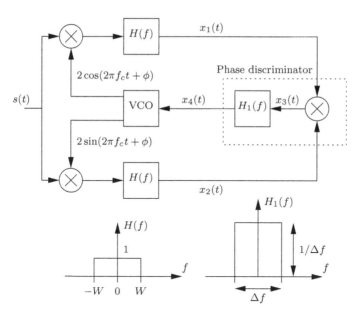

Fig. 3.69 Block diagram of the Costas loop

Fig. 3.70 Message signal

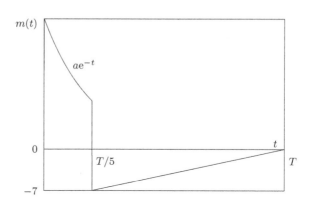

$$2\phi = n\pi$$
$$\Rightarrow \phi = n\pi/2. \tag{3.338}$$

Therefore, the Costas loop exhibits 90° phase ambiguity.

52. For the message signal shown in Fig. 3.70 compute the power efficiency (the ratio of sideband power to total power) in terms of the amplitude sensitivity k_a, and T. Find a in terms of T.

The message is periodic with period T, has zero -mean and is AM modulated according to:

$$s(t) = A_c[1 + k_a m(t)] \cos(2\pi f_c t). \tag{3.339}$$

Assume that $T \gg 1/f_c$.

- *Solution*: Since the message is zero- mean we must have:

$$\int_{t=0}^{T} m(t)\, dt = 0$$

$$\Rightarrow \int_{t=0}^{T/5} a e^{-t}\, dt = \frac{1}{2} \times 7 \times \left(T - \frac{T}{5}\right)$$

$$\Rightarrow a = \frac{14T}{5(1 - e^{-T/5})}. \tag{3.340}$$

For a general AM signal given by

$$s(t) = A_c[1 + k_a m(t)] \cos(2\pi f_c t), \tag{3.341}$$

the power efficiency is:

$$\eta = \frac{A_c^2 k_a^2 P_m / 2}{A_c^2/2 + A_c^2 k_a^2 P_m / 2} = \frac{k_a^2 P_m}{1 + k_a^2 P_m} \tag{3.342}$$

where P_m denotes the message power. Here we only need to compute P_m. We have

$$P_m = \frac{1}{T} \int_{t=0}^{T} m^2(t)\, dt. \tag{3.343}$$

Let

$$P_1 = \frac{1}{T} \int_{t=0}^{T/5} a^2 e^{-2t}\, dt$$

$$= \frac{a^2}{2T}\left[1 - e^{-2T/5}\right]. \tag{3.344}$$

Let

$$P_2 = \frac{1}{T} \int_{t=T/5}^{T} (mt + c)^2\, dt$$

$$= \frac{1}{T} \int_{t=T/5}^{T} \left(m^2 t^2 + 2mct + c^2\right) dt$$

$$= \frac{1}{T}\left[\frac{m^2 t^3}{3} + mct^2 + c^2 t\right]_{t=T/5}^{T}$$

$$= \left[\frac{m^2 T^2}{3} \left(1 - \frac{1}{125} \right) + mTc \left(1 - \frac{1}{25} \right) + c^2 \left(1 - \frac{1}{5} \right) \right]$$

$$= \left(\frac{35}{4} \right)^2 \left[\frac{124}{3 \times 125} - \frac{24}{25} + \frac{4}{5} \right]$$

$$= 13.0667, \tag{3.345}$$

where

$$m = \frac{35}{4T}$$

$$c = -mT. \tag{3.346}$$

Now

$$P_m = P_1 + P_2. \tag{3.347}$$

The power efficiency can be obtained by substituting (3.340), (3.344), (3.345), and (3.347) in (3.342).

References

Simon Haykin. *Communication Systems.* Wiley Eastern, second edition, 1983.
Simon Haykin. *Communication Systems.* Wiley Eastern, fourth edition, 2001.
Rodger E. Ziemer and William H. Tranter. *Principles of Communications.* John Wiley, fifth edition, 2002.

Chapter 4
Frequency Modulation

1. (Haykin 1983) In a frequency-modulated radar, the instantaneous frequency of the transmitted carrier $f_t(t)$ is varied as given in Fig. 4.1. The instantaneous frequency of the received echo $f_r(t)$ is also shown, where τ is the round-trip delay time. Assuming that $f_0\tau \ll 1$ determine the number of beat (difference frequency) cycles in one second, in terms of the frequency deviation (Δf) of the carrier frequency, the delay τ, and the repetition frequency f_0. Assume that f_0 is an integer.

   • *Solution*: Let the transmitted signal be given by

$$s(t) = A_1 \cos\left(2\pi \int_{\tau=0}^{t} f_t(\tau)\, d\tau\right). \tag{4.1}$$

   Let the received signal be given by

$$r(t) = A_2 \cos\left(2\pi \int_{\tau=0}^{t} f_r(\tau)\, d\tau\right). \tag{4.2}$$

   The variation of the beat (difference) frequency $(f_t(t) - f_r(t))$ with time is plotted in Fig. 4.2b. Note that the number of beat cycles over the time duration $1/f_0$ is given by

$$N = \int_{t=t_1}^{t_1+1/f_0} |f_t(t) - f_r(t)|\, dt$$
$$= |\text{area of ABCD}| + |\text{area of DEFG}|. \tag{4.3}$$

   Note that

$$f_1 = f_c + \Delta f - f_2. \tag{4.4}$$

© The Editor(s) (if applicable) and The Author(s), under exclusive license
to Springer Nature Switzerland AG 2021
K. Vasudevan, *Analog Communications*,
https://doi.org/10.1007/978-3-030-50337-6_4

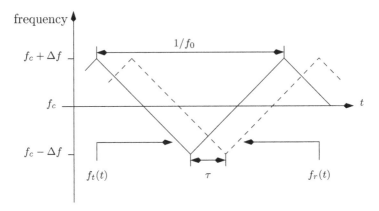

Fig. 4.1 Variation of the instantaneous frequency with time in an FM radar

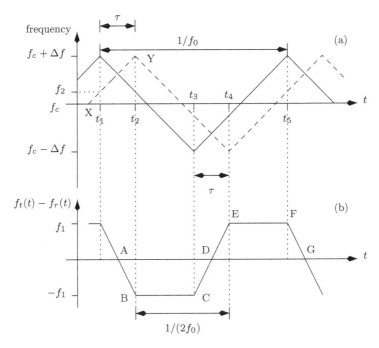

Fig. 4.2 Variation of the instantaneous difference frequency with time in an FM radar

The equation of the line XY in Fig. 4.2a is

$$y = mt + c, \qquad (4.5)$$

where

$$m = 4 f_0 \Delta f. \tag{4.6}$$

At $t = t_1 + \tau$ we have

$$f_c + \Delta f = 4 f_0 \Delta f (t_1 + \tau) + c. \tag{4.7}$$

At $t = t_1$ we have

$$\begin{aligned} f_2 &= 4 f_0 \Delta f t_1 + c \\ &= f_c + \Delta f - 4 f_0 \Delta f \tau. \end{aligned} \tag{4.8}$$

Therefore

$$\begin{aligned} f_1 &= f_c + \Delta f - f_2 \\ &= 4 f_0 \Delta f \tau. \end{aligned} \tag{4.9}$$

Now

$$\begin{aligned} N &= 4 \frac{1}{2} \frac{\tau}{2} f_1 + 2 f_1 \left(\frac{1}{2 f_0} - \tau \right) \\ &= \tau f_1 + f_1 \frac{1 - 2\tau f_0}{f_0} \\ &\approx \tau f_1 + f_1 / f_0. \end{aligned} \tag{4.10}$$

Since f_0 is an integer, the number of cycles in one second is

$$\begin{aligned} N f_0 &= f_1 (1 + \tau f_0) \\ &\approx f_1 = 4 f_0 \Delta f \tau, \end{aligned} \tag{4.11}$$

which is proportional to τ and hence twice the distance between the target and the radar. In other words,

$$\tau = 2x/c, \tag{4.12}$$

where x is the distance between the target and the radar and c is the velocity of light. Thus the FM radar can be used for *ranging*.

2. (Haykin 1983) The instantaneous frequency of a cosine wave is equal to $f_c - \Delta f$ for $|t| < T/2$ and f_c for $|t| > T/2$. Determine the spectrum of this signal.

- *Solution*: The Fourier transform of this signal is given by

$$S(f) = \int_{t=-\infty}^{-T/2} \cos(2\pi f_c t) e^{-j2\pi ft}\, dt$$

$$+ \int_{t=-T/2}^{T/2} \cos(2\pi (f_c + \Delta f)t) e^{-j2\pi ft}\, dt$$

$$+ \int_{t=T/2}^{\infty} \cos(2\pi f_c t) e^{-j2\pi ft}\, dt$$

$$= \int_{t=-\infty}^{\infty} \cos(2\pi f_c t) e^{-j2\pi ft}\, dt$$

$$+ \int_{t=-T/2}^{T/2} (\cos(2\pi (f_c + \Delta f)t) - \cos(2\pi f_c t)) e^{-j2\pi ft}\, dt$$

$$= \frac{1}{2} [\delta(f - f_c) + \delta(f + f_c)]$$

$$+ \frac{T}{2} [\text{sinc}\,((f - f_c - \Delta f)T) + \text{sinc}\,((f + f_c + \Delta f)T)]$$

$$- \frac{T}{2} [\text{sinc}\,((f - f_c)T) + \text{sinc}\,((f + f_c)T)]. \tag{4.13}$$

3. (Haykin 1983) Single sideband modulation may be viewed as a hybrid form of amplitude modulation and frequency modulation. Evaluate the envelope and the instantaneous frequency of an SSB wave, in terms of the message signal and its Hilbert transform, for the two cases:

(a) When only the upper sideband is transmitted.
(b) When only the lower sideband is transmitted.

Assume that the message signal is $m(t)$, the carrier amplitude is $A_c/2$ and carrier frequency is f_c.

- *Solution*: The SSB signal can be written as

$$s(t) = \frac{A_c}{2} \left[m(t) \cos(2\pi f_c t) \pm \hat{m}(t) \sin(2\pi f_c t) \right]. \tag{4.14}$$

When the upper sideband is to be transmitted, the minus sign is used and when the lower sideband is to be transmitted, the plus sign is used. The envelope of $s(t)$ is given by

$$a(t) = \frac{A_c}{2} \sqrt{m^2(t) + \hat{m}^2(t)} \tag{4.15}$$

is independent of whether the upper or lower sideband is transmitted. Note that $s(t)$ can be written as

$$s(t) = a(t) \cos(2\pi f_c t + \theta(t)), \tag{4.16}$$

where $\theta(t)$ denotes the instantaneous phase which is given by

$$\theta(t) = \pm \tan^{-1} \left(\frac{\hat{m}(t)}{m(t)} \right). \tag{4.17}$$

The plus sign in the above equation is used when the upper sideband is transmitted and the minus sign is used when the lower sideband is transmitted. The total instantaneous phase is given by

$$\theta_{\text{tot}}(t) = 2\pi f_c t + \theta(t). \tag{4.18}$$

The total instantaneous frequency is given by

$$f_{\text{tot}}(t) = \frac{1}{2\pi} \frac{d\theta_{\text{tot}}(t)}{dt}. \tag{4.19}$$

When the upper sideband is transmitted, the total instantaneous frequency is given by

$$f_{\text{tot}}(t) = f_c + \frac{1}{2\pi} \frac{\hat{m}'(t)m(t) - \hat{m}(t)m'(t)}{(m^2(t) + \hat{m}^2(t))}, \tag{4.20}$$

where $m'(t)$ denotes the derivative of $m(t)$ and $\hat{m}'(t)$ denotes the derivative of $\hat{m}(t)$. When the lower sideband is transmitted, the total instantaneous frequency is given by

$$f_{\text{tot}}(t) = f_c - \frac{1}{2\pi} \frac{\hat{m}'(t)m(t) - \hat{m}(t)m'(t)}{(m^2(t) + \hat{m}^2(t))}. \tag{4.21}$$

4. (Haykin 1983) Consider a narrowband FM signal approximately defined by

$$s(t) \approx A_c \cos(2\pi f_c t) - \beta A_c \sin(2\pi f_m t) \sin(2\pi f_c t). \tag{4.22}$$

(a) Determine the envelope of $s(t)$. What is the ratio of the maximum to the minimum value of this envelope.

(b) Determine the total average power of the narrowband FM signal. Determine the total average power in the sidebands.

(c) Assuming that $s(t)$ in (4.22) can be written as

$$s(t) = a(t) \cos(2\pi f_c t + \theta(t)) \tag{4.23}$$

expand $\theta(t)$ in the form of a Maclaurin series. Assume that $\beta < 0.3$. What is the power ratio of the third harmonic to the fundamental component.

• *Solution*: The envelope is given by

$$a(t) = A_c\sqrt{1 + \beta^2 \sin^2(2\pi f_m t)}. \tag{4.24}$$

Therefore, the maximum and the minimum values of the envelope are given by

$$A_{max} = A_c\sqrt{1 + \beta^2}$$
$$A_{min} = A_c$$
$$\Rightarrow \frac{A_{max}}{A_{min}} = \sqrt{1 + \beta^2}. \tag{4.25}$$

The total average power of narrowband FM is equal to

$$P_{tot} = \frac{A_c^2}{2} + \frac{2\beta^2 A_c^2}{8}. \tag{4.26}$$

The total average power in the sidebands is equal to

$$P_{mes} = \frac{2\beta^2 A_c^2}{8}. \tag{4.27}$$

Assuming that $\beta \ll 1$, the narrowband FM signal can be written as

$$s(t) = a(t)\cos(2\pi f_c t + \theta(t)), \tag{4.28}$$

where $\theta(t)$ denotes the instantaneous phase of the message component and is given by

$$\theta(t) \approx \tan^{-1}(\beta \sin(2\pi f_m t)). \tag{4.29}$$

Now, the Maclaurin series expansion of $\tan^{-1}(x)$ is (ignoring higher terms)

$$\tan^{-1}(x) \approx x - \frac{x^3}{3}. \tag{4.30}$$

Thus

$$\theta(t) = \beta \sin(2\pi f_m t) - \frac{1}{3}\beta^3 \sin^3(2\pi f_m t). \tag{4.31}$$

Using the fact that

$$\sin^3(\theta) = \frac{3\sin(\theta) - \sin(3\theta)}{4} \tag{4.32}$$

(4.31) becomes

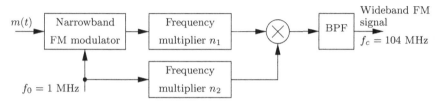

Fig. 4.3 Armstrong-type FM modulator

$$\theta(t) \approx \beta \sin(2\pi f_m t)$$
$$-\frac{1}{12}\beta^3 (3 \sin(2\pi f_m t) - \sin(2\pi (3 f_m)t))$$
$$= \left(\beta - \frac{\beta^3}{4}\right) \sin(2\pi f_m t) + \frac{\beta^3}{12} \sin(2\pi (3 f_m)t). \qquad (4.33)$$

Therefore, the power ratio of the third harmonic to the fundamental is

$$R = \left[\frac{\beta^3}{12} \times \frac{4}{(4\beta - \beta^3)}\right]^2. \qquad (4.34)$$

5. (Proakis and Salehi 2005) To generate wideband FM, we can first generate a narrowband FM signal, and then use frequency multiplication to spread the signal bandwidth. This is illustrated in Fig. 4.3, which is called the Armstrong-type FM modulator. The narrowband FM has a frequency deviation of 1 kHz.

   (a) If the frequency of the first oscillator is 1 MHz, determine n_1 and n_2 that is necessary to generate an FM signal at a carrier frequency of 104 MHz and a maximum frequency deviation of 75 kHz. The BPF allows only the difference frequency component.
   (b) If the error in the carrier frequency f_c for the wideband FM signal is to be within ± 200 Hz, determine the maximum allowable error in the 1 MHz oscillator.

   • *Solution*: The maximum frequency deviation of the output wideband FM signal is 75 kHz. The maximum frequency deviation of the narrowband FM (NBFM) signal is 1 kHz. Therefore

$$n_1 = \frac{75}{1} = 75. \qquad (4.35)$$

   Consequently, the carrier frequency at the output of the first frequency multiplier is 75 MHz. However, the required carrier frequency is 104 MHz. Hence, what we now require is a frequency translation. Therefore, we must have

Fig. 4.4 Fourier transform
of the frequency
discriminator for negative
frequencies

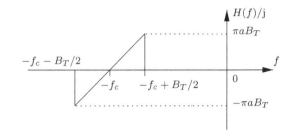

$$1 \times n_2 - 75 = 104\,\text{MHz}$$
$$\Rightarrow n_2 = 179. \tag{4.36}$$

Let us assume that the error in f_0 is x Hz. Hence, the error in the final output carrier frequency is

$$n_2 x - n_1 x = \pm 200\,\text{Hz}$$
$$\Rightarrow x = \pm 1.92\,\text{Hz}. \tag{4.37}$$

6. Figure 4.4 shows the Fourier transform of a frequency discriminator for negative frequencies. Here B_T denotes the transmission bandwidth of the FM signal and f_c denotes the carrier frequency.
   The block diagram of the system proposed for frequency demodulation is also shown, where

$$s(t) = A_c \cos\left(2\pi f_c t + 2\pi k_f \int_{\tau=0}^{t} m(\tau)\,d\tau\right), \tag{4.38}$$

where $m(t)$ denotes the message signal.

   (a) Sketch the frequency response of $H(f)$ for positive frequencies given that $h(t)$ is real-valued.
   (b) Find the output $z(t)$.
   (c) Can the proposed system be used for frequency demodulation? Justify your answer.

• *Solution*: Since $h(t)$ is real-valued we must have

$$H(-f) = H^*(f), \tag{4.39}$$

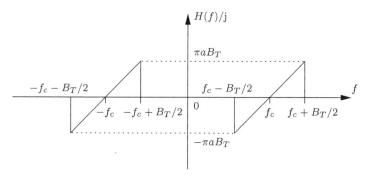

Fig. 4.5 Frequency response of the proposed discriminator

which is illustrated in Fig. 4.5. Therefore

$$H(f) = \begin{cases} j\,2\pi a(f - f_c) \text{ for } f_c - B_T/2 < f < f_c + B_T/2 \\ j\,2\pi a(f + f_c) \text{ for } -f_c - B_T/2 < f < -f_c + B_T/2. \end{cases} \quad (4.40)$$

In order to compute $y(t)$, we use the complex lowpass equivalent model. We assume that

$$\begin{aligned} h(t) &= 2h_c(t)\cos(2\pi f_c t) - 2h_s(t)\sin(2\pi f_c t) \\ &= \Re\left\{2\tilde{h}(t)\mathrm{e}^{j\,2\pi f_c t}\right\}, \end{aligned} \quad (4.41)$$

where

$$\tilde{h}(t) = h_c(t) + j\,h_s(t). \quad (4.42)$$

We know that the Fourier transform of the complex envelope of $h(t)$ is given by

$$\tilde{H}(f) = \begin{cases} H(f + f_c) \text{ for } -B_T/2 < f < B_T/2 \\ 0 \qquad\qquad \text{elsewhere} \end{cases}, \quad (4.43)$$

which reduces to

$$\tilde{H}(f) = \begin{cases} j\,2\pi a f \text{ for } -B_T/2 < f < B_T/2 \\ 0 \qquad \text{elsewhere} \end{cases}. \quad (4.44)$$

Let us denote the Fourier transform of the complex envelope of the output by $\tilde{Y}(f)$. Then

$$\tilde{Y}(f) = \tilde{H}(f)\tilde{S}(f)$$
$$= j2\pi f a \tilde{S}(f), \tag{4.45}$$

where $\tilde{S}(f)$ is the Fourier transform of the complex envelope of $s(t)$. Note that

$$\tilde{s}(t) = A_c \exp\left(j2\pi k_f \int_{\tau=0}^{t} m(\tau)\,d\tau\right). \tag{4.46}$$

From (4.45) it is clear that

$$\tilde{y}(t) = a\frac{d\tilde{s}(t)}{dt}$$

$$= jaA_c 2\pi k_f m(t) \exp\left(j2\pi k_f \int_{\tau=0}^{t} m(\tau)\,d\tau\right)$$

$$= 2aA_c \pi k_f m(t) \exp\left(j2\pi k_f \int_{\tau=0}^{t} m(\tau)\,d\tau + j\pi/2\right). \tag{4.47}$$

Therefore

$$y(t) = 2\pi a A_c k_f m(t) \cos\left(2\pi f_c t + 2\pi k_f \int_{\tau=0}^{t} m(\tau)\,d\tau + \pi/2\right). \tag{4.48}$$

The output is

$$z(t) = |\tilde{y}(t)| = 2\pi a A_c k_f |m(t)|. \tag{4.49}$$

Thus, the proposed system cannot be used for frequency demodulation.

7. Figure 4.6 shows the Fourier transform of a frequency discriminator for negative frequencies. Here B_T denotes the transmission bandwidth of the FM signal and f_c denotes the carrier frequency.
The block diagram of the system proposed for frequency demodulation is also shown, where

$$s(t) = A_c \cos\left(2\pi f_c t + 2\pi k_f \int_{\tau=0}^{t} m(\tau)\,d\tau\right), \tag{4.50}$$

where $m(t)$ denotes the message signal.

(a) Sketch the frequency response of $H(f)$ for positive frequencies given that $h(t)$ is real-valued.
(b) Find the output $z(t)$.
(c) Can the proposed system be used for frequency demodulation? Justify your answer.

Fig. 4.6 Fourier transform of the frequency discriminator for negative frequencies

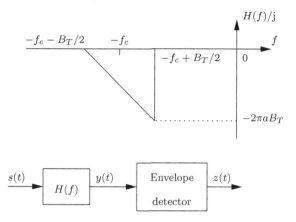

- *Solution*: Since $h(t)$ is real-valued we must have

$$H(-f) = H^*(f),\qquad(4.51)$$

which is illustrated in Fig. 4.7. Therefore

$$H(f) = \begin{cases} -\mathrm{j}\,2\pi a(f - f_c - B_T/2) \text{ for } f_c - B_T/2 < f < f_c + B_T/2 \\ -\mathrm{j}\,2\pi a(f + f_c + B_T/2) \text{ for } - f_c - B_T/2 < f < -f_c + B_T/2. \end{cases}\qquad(4.52)$$

In order to compute $y(t)$, we use the complex lowpass equivalent model. We assume that

$$\begin{aligned} h(t) &= 2h_c(t)\cos(2\pi f_c t) - 2h_s(t)\sin(2\pi f_c t) \\ &= \Re\left\{ 2\tilde{h}(t)\mathrm{e}^{\mathrm{j}2\pi f_c t} \right\}, \end{aligned}\qquad(4.53)$$

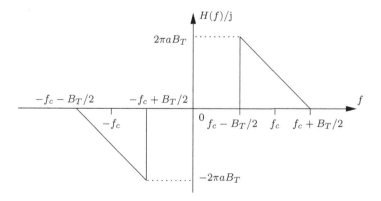

Fig. 4.7 Frequency response of the proposed discriminator

where

$$\tilde{h}(t) = h_c(t) + j h_s(t).\tag{4.54}$$

We know that the Fourier transform of the complex envelope of $h(t)$ is given by

$$\tilde{H}(f) = \begin{cases} H(f + f_c) & \text{for } - B_T/2 < f < B_T/2 \\ 0 & \text{elsewhere} \end{cases},\tag{4.55}$$

which reduces to

$$\tilde{H}(f) = \begin{cases} -j\,2\pi a(f - B_T/2) & \text{for } - B_T/2 < f < B_T/2 \\ 0 & \text{elsewhere} \end{cases}.\tag{4.56}$$

Let us denote the Fourier transform of the complex envelope of the output by $\tilde{Y}(f)$. Then

$$\begin{aligned} \tilde{Y}(f) &= \tilde{H}(f)\tilde{S}(f) \\ &= -j\,2\pi a(f - B_T/2)\tilde{S}(f),\end{aligned}\tag{4.57}$$

where $\tilde{S}(f)$ is the Fourier transform of the complex envelope of $s(t)$. Note that

$$\tilde{s}(t) = A_c \exp\left(j\,2\pi k_f \int_{\tau=0}^{t} m(\tau)\,d\tau \right).\tag{4.58}$$

From (4.45) it is clear that

$$\begin{aligned} \tilde{y}(t) &= -a\frac{d\tilde{s}(t)}{dt} + j\pi a B_T \tilde{s}(t) \\ &= -j\,a A_c 2\pi k_f m(t) \exp\left(j\,2\pi k_f \int_{\tau=0}^{t} m(\tau)\,d\tau \right) \\ &\quad + j\,a A_c \pi B_T \exp\left(j\,2\pi k_f \int_{\tau=0}^{t} m(\tau)\,d\tau \right) \\ &= j\,\pi a A_c \left[B_T - 2k_f m(t) \right] \exp\left(j\,2\pi k_f \int_{\tau=0}^{t} m(\tau)\,d\tau \right).\end{aligned}\tag{4.59}$$

Therefore

$$y(t) = \pi a A_c [B_T - 2k_f m(t)] \cos\left(2\pi f_c t + 2\pi k_f \int_{\tau=0}^{t} m(\tau)\,d\tau + \pi/2 \right).\tag{4.60}$$

The output is

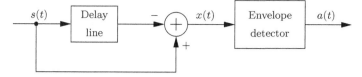

Fig. 4.8 Delay-line method of demodulating FM signals

$$z(t) = |\tilde{y}(t)| = \pi a A_c B_T [1 - 2k_f m(t)/B_T]. \tag{4.61}$$

Therefore, the proposed system can be used for frequency demodulation provided

$$|2k_f m(t)/B_T| < 1. \tag{4.62}$$

8. (Haykin 1983) Consider the frequency demodulation scheme shown in Fig. 4.8 in which the incoming FM signal is passed through a delay line that produces a delay of T such that $2\pi f_c T = \pi/2$. The delay-line output is subtracted from the incoming FM signal and the resulting output is envelope detected. This demodulator finds wide application in demodulating microwave FM waves. Assuming that

$$s(t) = A_c \cos(2\pi f_c t + \beta \sin(2\pi f_m t)) \tag{4.63}$$

compute $a(t)$ when $\beta < 1$ and the delay T is such that

$$2\pi f_m T \ll 1. \tag{4.64}$$

- *Solution*: The signal $x(t)$ can be written as

$$
\begin{aligned}
x(t) &= s(t) - s(t - T) \\
&= A_c \cos(2\pi f_c t + \beta \sin(2\pi f_m t)) \\
&\quad - A_c \cos(2\pi f_c (t - T) + \beta \sin(2\pi f_m (t - T))) \\
&\approx A_c \cos(2\pi f_c t + \beta \sin(2\pi f_m t)) \\
&\quad - A_c \cos(2\pi f_c t + \beta \sin(2\pi f_m t) - 2\pi f_m T \beta \cos(2\pi f_m t) - \pi/2) \\
&= A_c \cos(2\pi f_c t + \beta \sin(2\pi f_m t)) \\
&\quad - A_c \sin(2\pi f_c t + \beta \sin(2\pi f_m t) - 2\pi f_m T \beta \cos(2\pi f_m t)). \tag{4.65}
\end{aligned}
$$

Let

$$
\begin{aligned}
\theta(t) &= \beta \sin(2\pi f_m t) \\
\alpha(t) &= \theta(t) - 2\pi f_m T \beta \cos(2\pi f_m t). \tag{4.66}
\end{aligned}
$$

Then $x(t)$ can be written as

$$x(t) = A_c \cos(2\pi f_c t + \theta(t)) - A_c \sin(2\pi f_c t + \alpha(t))$$
$$= A_c \cos(2\pi f_c t) \left[\cos(\theta(t)) - \sin(\alpha(t))\right]$$
$$\quad - A_c \sin(2\pi f_c t) \left[\sin(\theta(t)) + \cos(\alpha(t))\right]$$
$$= a(t) \cos(2\pi f_c t + \phi(t)), \tag{4.67}$$

where

$$\tan(\phi(t)) = \frac{\sin(\theta(t)) + \cos(\alpha(t))}{\cos(\theta(t)) - \sin(\alpha(t))} \tag{4.68}$$

and the envelope of $x(t)$ is

$$a(t) = A_c \sqrt{2 + 2\sin(\theta(t) - \alpha(t))}$$
$$\approx A_c \sqrt{2} \left[1 + (\theta(t) - \alpha(t))/2\right]$$
$$= A_c \sqrt{2} \left[1 + \pi f_m T \beta \cos(2\pi f_m t)\right], \tag{4.69}$$

where we have made use of the fact that

$$2\pi f_m T \ll 1$$
$$\beta < 1$$
$$\Rightarrow \theta(t) - \alpha(t) \ll 1$$
$$\Rightarrow \sin(\theta(t) - \alpha(t)) \approx \theta(t) - \alpha(t). \tag{4.70}$$

Note that $a(t) > 0$ for all t. The message signal $\cos(2\pi f_m t)$ can be recovered by passing $a(t)$ through a capacitor.

9. Consider the frequency demodulation scheme shown in Fig. 4.9 in which the incoming FM signal is passed through a delay line that produces a delay of T such that $2\pi f_c T = \pi/2$. The delay-line output is subtracted from the incoming FM signal and the resulting output is envelope detected. This demodulator finds wide application in demodulating microwave FM waves. Assuming that

$$s(t) = A_c \cos\left(2\pi f_c t + 2\pi k_f \int_{\tau=0}^{t} m(\tau)\, d\tau\right) \tag{4.71}$$

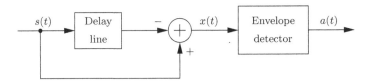

Fig. 4.9 Delay-line method of demodulating FM signals

compute $a(t)$ when the delay T is such that

$$2\pi \Delta f T \ll 1 \tag{4.72}$$

where Δf denotes the frequency deviation. Assume also that $m(t)$ is constant over any interval T.

- *Solution*: The signal $x(t)$ can be written as

$$x(t) = s(t) - s(t - T), \tag{4.73}$$

where

$$
\begin{aligned}
s(t - T) &= A_c \cos \left(2\pi f_c(t - T) + 2\pi k_f \int_{\tau=0}^{t-T} m(\tau) \, d\tau \right) \\
&= A_c \cos \left(2\pi f_c t - \pi/2 + 2\pi k_f \int_{\tau=0}^{t-T} m(\tau) \, d\tau \right) \\
&= A_c \sin \left(2\pi f_c t + 2\pi k_f \int_{\tau=0}^{t-T} m(\tau) \, d\tau \right).
\end{aligned}
\tag{4.74}
$$

Now

$$
\begin{aligned}
2\pi k_f \int_{\tau=0}^{t-T} m(\tau) \, d\tau &= 2\pi k_f \int_{\tau=0}^{t} m(\tau) \, d\tau - 2\pi k_f \int_{\tau=t-T}^{t} m(\tau) \, d\tau \\
&= 2\pi k_f \int_{\tau=0}^{t} m(\tau) \, d\tau - 2\pi k_f T m(t) \\
&= \theta(t) - 2\pi k_f T m(t) \\
&= \alpha(t) \quad \text{(say)},
\end{aligned}
\tag{4.75}
$$

where

$$2\pi k_f \int_{\tau=0}^{t} m(\tau) \, d\tau = \theta(t). \tag{4.76}$$

Therefore $x(t)$ in (4.73) becomes

$$
\begin{aligned}
x(t) &= A_c \cos(2\pi f_c t + \theta(t)) - A_c \sin(2\pi f_c t + \alpha(t)) \\
&= A_c \cos(2\pi f_c t) \left[\cos(\theta(t)) - \sin(\alpha(t)) \right] \\
&\quad - A_c \sin(2\pi f_c t) \left[\sin(\theta(t)) + \cos(\alpha(t)) \right] \\
&= a(t) \cos(2\pi f_c t + \phi(t)),
\end{aligned}
\tag{4.77}
$$

where

$$\tan(\phi(t)) = \frac{\sin(\theta(t)) + \cos(\alpha(t))}{\cos(\theta(t)) - \sin(\alpha(t))} \tag{4.78}$$

and the envelope of $x(t)$ is

$$
\begin{aligned}
a(t) &= A_c\sqrt{2 + 2\sin(\theta(t) - \alpha(t))} \\
&\approx A_c\sqrt{2}\left[1 + (\theta(t) - \alpha(t))/2\right] \\
&= A_c\sqrt{2}\left[1 + \pi k_f T m(t)\right], \tag{4.79}
\end{aligned}
$$

where we have made use of the fact that

$$
\begin{aligned}
\Delta f &= \max|k_f m(t)| \\
2\pi \Delta f T &\ll 1 \quad \text{(given)} \\
\Rightarrow |\theta(t) - \alpha(t)| &\ll 1 \\
\Rightarrow \sin(\theta(t) - \alpha(t)) &\approx \theta(t) - \alpha(t). \tag{4.80}
\end{aligned}
$$

Note that $a(t) > 0$ for all t. The message signal $m(t)$ can be recovered by passing $a(t)$ through a capacitor.

10. A tone-modulated FM signal of the form:

$$s(t) = A_c \sin(2\pi f_c t + \beta \cos(2\pi f_m t)) \tag{4.81}$$

is passed through an ideal unity gain BPF with center frequency equal to the carrier frequency and bandwidth equal to $3 f_m$ ($\pm 1.5 f_m$ on either side of the carrier), yielding the signal $z(t)$.

(a) Derive the expression for $z(t)$.
(b) Assuming that $z(t)$ is of the form

$$z(t) = a(t)\cos(2\pi f_c t + \theta(t)) \tag{4.82}$$

compute $a(t)$ and $\theta(t)$.

Use the relations

$$
\begin{aligned}
J_n(\beta) &= \frac{1}{2\pi}\int_{x=-\pi}^{x=\pi} e^{j(\beta\sin(x) - nx)}\, dx \\
J_n(\beta) &= (-1)^n J_{-n}(\beta), \tag{4.83}
\end{aligned}
$$

where $J_n(\beta)$ denotes the nth order Bessel function of the first kind and argument β.

• *Solution*: The input FM signal can be written as

$$s(t) = A_c \cos(2\pi f_c t + \beta \cos(2\pi f_m t) - \pi/2)$$
$$= \Re\left\{\tilde{s}(t)e^{j 2\pi f_c t}\right\}, \tag{4.84}$$

where

$$\tilde{s}(t) = A_c e^{j\beta \cos(2\pi f_m t) - j\pi/2}$$
$$= -j A_c e^{j\beta \cos(2\pi f_m t)}, \tag{4.85}$$

which is periodic with a period $1/f_m$. Hence, $\tilde{s}(t)$ can be expanded in the form of a complex Fourier series given by

$$\tilde{s}(t) = \sum_{n=-\infty}^{\infty} c_n e^{j 2\pi n f_m t}, \tag{4.86}$$

where

$$c_n = f_m \int_{t=-1/(2 f_m)}^{1/(2 f_m)} \tilde{s}(t) e^{-j 2\pi n f_m t}\, dt$$
$$= -j A_c f_m \int_{t=-1/(2 f_m)}^{1/(2 f_m)} e^{j\beta \cos(2\pi f_m t)} e^{-j 2\pi n f_m t}\, dt. \tag{4.87}$$

Let

$$2\pi f_m t = \pi/2 - x$$
$$\Rightarrow 2\pi f_m dt = -dx. \tag{4.88}$$

Thus

$$c_n = \frac{j A_c}{2\pi} \int_{x=3\pi/2}^{-\pi/2} e^{j(\beta \sin(x) - n(\pi/2 - x))}\, dx. \tag{4.89}$$

Noting that the integrand is periodic with respect to x with a period 2π we can interchange the limits and integrate from $-\pi$ to π to get

$$c_n = \frac{-j A_c}{2\pi} \int_{x=-\pi}^{\pi} e^{j(\beta \sin(x) - n(\pi/2 - x))}\, dx$$
$$= \frac{A_c}{2\pi} e^{-j(n+1)\pi/2} \int_{x=-\pi}^{\pi} e^{j(\beta \sin(x) + nx)}\, dx$$
$$= A_c J_{-n}(\beta) e^{-j(n+1)\pi/2}. \tag{4.90}$$

Therefore

$$\tilde{s}(t) = A_c \sum_{n=-\infty}^{\infty} J_{-n}(\beta) e^{j(2\pi n f_m t - (n+1)\pi/2)}. \tag{4.91}$$

Thus

$$s(t) = A_c \sum_{n=-\infty}^{\infty} J_{-n}(\beta) \cos(2\pi f_c t + 2\pi n f_m t - (n+1)\pi/2). \tag{4.92}$$

The output of the BPF is

$$\begin{aligned}
z(t) &= A_c J_0(\beta) \cos(2\pi f_c t - \pi/2) + A_c J_{-1}(\beta) \cos(2\pi f_c t + 2\pi f_m t - \pi) \\
&\quad + A_c J_1(\beta) \cos(2\pi f_c t - 2\pi f_m t) \\
&= A_c J_0(\beta) \sin(2\pi f_c t) + A_c J_1(\beta) \cos(2\pi f_c t + 2\pi f_m t) \\
&\quad + A_c J_1(\beta) \cos(2\pi f_c t - 2\pi f_m t) \\
&= A_c J_0(\beta) \sin(2\pi f_c t) + 2 A_c J_1(\beta) \cos(2\pi f_c t) \cos(2\pi f_m t). \tag{4.93}
\end{aligned}$$

Assuming that $z(t)$ is of the form

$$\begin{aligned}
z(t) &= a(t) \cos(2\pi f_c t + \theta(t)) \\
&= a(t) \cos(2\pi f_c t) \cos(\theta(t)) - a(t) \sin(2\pi f_c t) \sin(\theta(t)) \tag{4.94}
\end{aligned}$$

we have

$$\begin{aligned}
a(t) \cos(\theta(t)) &= 2 A_c J_1(\beta) \cos(2\pi f_m t) \\
a(t) \sin(\theta(t)) &= -A_c J_0(\beta). \tag{4.95}
\end{aligned}$$

Thus, the envelope of the output is

$$a(t) = A_c \sqrt{(2 J_1(\beta) \cos(2\pi f_m t))^2 + J_0^2(\beta)}. \tag{4.96}$$

The phase is

$$\theta(t) = -\tan^{-1}\left(\frac{J_0(\beta)}{2 J_1(\beta) \cos(2\pi f_m t)}\right). \tag{4.97}$$

11. (Haykin 1983) The bandwidth of an FM signal extends over both sides of the carrier frequency. However, in the single sideband version of FM, it is possible to transmit either the upper or the lower sideband.

   (a) Assuming that the FM signal is given by

$$s(t) = A_c \cos(2\pi f_c t + \phi(t)) \tag{4.98}$$

explain how we can transmit only the upper sideband. Express your result in terms of complex envelope of $s(t)$ and Hilbert transforms. Assume that for all practical purposes, $s(t)$ is bandlimited to $f_c - B_T/2 < |f| < f_c + B_T/2$ and $f_c \gg B_T$.

(b) Verify your answer for single-tone FM modulation when

$$\phi(t) = \beta \sin(2\pi f_m t). \tag{4.99}$$

Use the Fourier series representation for this FM signal given by

$$s(t) = A_c \sum_{n=-\infty}^{\infty} J_n(\beta) \cos(2\pi(f_c + nf_m)t). \tag{4.100}$$

- *Solution*: Recall that in SSB modulation with upper sideband transmitted, the signal is given by

$$
\begin{aligned}
s(t) &= m(t) \cos(2\pi f_c t) - \hat{m}(t) \sin(2\pi f_c t) \\
&= \Re\left\{ \left(m(t) + j\hat{m}(t)\right) e^{j2\pi f_c t} \right\},
\end{aligned} \tag{4.101}
$$

where $\hat{m}(t)$ is the Hilbert transform of the message $m(t)$, which is typically bandlimited between $[-W, W]$ and $f_c \gg W$. Observe that $m(t)$ in (4.101) can be complex. For the case of the FM signal given by

$$
\begin{aligned}
s(t) &= A_c \cos(2\pi f_c t + \phi(t)) \\
&= A_c \Re\left\{ e^{j(2\pi f_c t + \phi(t))} \right\}
\end{aligned} \tag{4.102}
$$

the complex envelope is given by

$$\tilde{s}(t) = A_c e^{j\phi(t)}. \tag{4.103}$$

Clearly, $\tilde{s}(t)$ is bandlimited to $-B_T/2 < |f| < B_T/2$.

Using the concept given in (4.101) the required equation for the FM signal with only the upper sideband transmitted is given by

$$s_1(t) = \Re\left\{ \left(\tilde{s}(t) + j\widehat{\tilde{s}}(t)\right) e^{j2\pi f_c t} \right\}, \tag{4.104}$$

where $\widehat{\tilde{s}}(t)$ is the Hilbert transform of $\tilde{s}(t)$.

Now in the given example

$$
\begin{aligned}
s(t) &= A_c \cos(2\pi f_c t + \beta \sin(2\pi f_m t)) \\
&= A_c \sum_{n=-\infty}^{\infty} J_n(\beta) \cos(2\pi(f_c + nf_m)t).
\end{aligned} \tag{4.105}
$$

The complex envelope can be written as

$$\tilde{s}(t) = A_c \sum_{n=-\infty}^{\infty} J_n(\beta) e^{j 2\pi n f_m t}$$

$$= A_c \sum_{n=-\infty}^{\infty} J_n(\beta) \left(\cos(2\pi n f_m t) + j \, \sin(2\pi n f_m t)\right). \quad (4.106)$$

The Hilbert transform of $\tilde{s}(t)$ in (4.106) is

$$\widehat{\tilde{s}}(t) = A_c \sum_{n=-\infty}^{-1} J_n(\beta) \left(\cos(2\pi n f_m t + \pi/2) + j \, \sin(2\pi n f_m t + \pi/2)\right)$$

$$+ A_c \sum_{n=1}^{\infty} J_n(\beta) \left(\cos(2\pi n f_m t - \pi/2) + j \, \sin(2\pi n f_m t - \pi/2)\right)$$

$$= A_c \sum_{n=-\infty}^{-1} J_n(\beta) \left(-\sin(2\pi n f_m t) + j \, \cos(2\pi n f_m t)\right)$$

$$+ A_c \sum_{n=1}^{\infty} J_n(\beta) \left(\sin(2\pi n f_m t) - j \, \cos(2\pi n f_m t)\right) \quad (4.107)$$

since the Hilbert transformer removes the dc component ($n = 0$), introduces a phase shift of $\pi/2$ for negative frequencies ($n < 0$) and a phase shift of $-\pi/2$ for positive frequencies ($n > 0$). Hence

$$\tilde{s}(t) + j \widehat{\tilde{s}}(t) = A_c J_0(\beta) + 2 A_c \sum_{n=1}^{\infty} J_n(\beta) e^{j 2\pi n f_m t} \quad (4.108)$$

and $s_1(t)$ in (4.104) becomes

$$s_1(t) = A_c J_0(\beta) \cos(2\pi f_c t) + 2 A_c \sum_{n=1}^{\infty} J_n(\beta) \cos(2\pi (f_c + n f_m) t) \quad (4.109)$$

Thus, we find that only the upper sideband is transmitted, which verifies our result.

12. (Haykin 1983) Consider the message signal as shown in Fig. 4.10, which is used to frequency modulate a carrier. Assume a frequency sensitivity of k_f Hz/V and that the FM signal is given by

$$s(t) = A_c \cos(2\pi f_c t + \phi(t)). \quad (4.110)$$

Fig. 4.10 A periodic square
wave corresponding to the
message signal

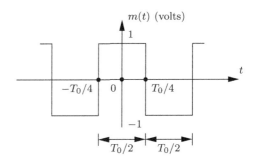

(a) Sketch the waveform corresponding to the total instantaneous frequency
$f(t)$ of the FM signal for $-T_0/4 \leq t \leq 3T_0/4$. Label all the important points
on the axes.
(b) Sketch $\phi(t)$ for $-T_0/4 \leq t \leq 3T_0/4$. Label all the important points on the
axes. Assume that $\phi(t)$ has zero mean.
(c) Write down the expression for the complex envelope of $s(t)$ in terms of $\phi(t)$.
(d) The FM signal $s(t)$ can be written as

$$s(t) = \sum_{n=-\infty}^{\infty} c_n \cos(2\pi f_c t + 2n\pi t/T_0), \qquad (4.111)$$

where

$$c_n = \left[a_n \operatorname{sinc}\left(\frac{\beta - n}{2}\right) + b_n \operatorname{sinc}\left(\frac{\beta + n}{2}\right) \right], \qquad (4.112)$$

where $\beta = k_f T_0$. Compute a_n and b_n. Assume the limits of integration (for
computing c_n) to be from $-T_0/4$ to $3T_0/4$.

• *Solution*: The total instantaneous frequency is depicted in Fig. 4.11b. Note
that

$$\phi(t) = 2\pi k_f \int m(\tau) \, d\tau. \qquad (4.113)$$

The variation of $\phi(t)$ is shown in Fig. 4.11c where

$$- A + 2\pi k_f \frac{T_0}{2} = A$$

$$\Rightarrow A = \pi k_f \frac{T_0}{2}. \qquad (4.114)$$

The complex envelope of $s(t)$ is

Fig. 4.11 Plot of the total instantaneous frequency and $\phi(t)$

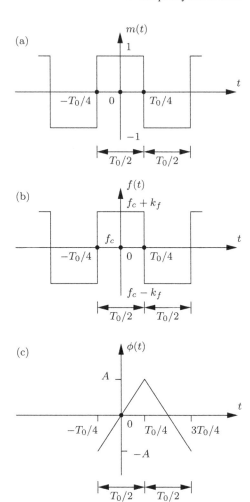

$$\tilde{s}(t) = A_c e^{j\phi(t)}. \tag{4.115}$$

In order to compute the Fourier coefficients c_n we note that

$$\phi(t) = \begin{cases} 2\pi k_f t & \text{for } -T_0/4 \le t \le T_0/4 \\ 2\pi k_f(-t + T_0/2) & \text{for } T_0/4 \le t \le 3T_0/4 \end{cases}. \tag{4.116}$$

Therefore

$$c_n = \frac{1}{T_0} \int_{t=-T_0/4}^{3T_0/4} \tilde{s}(t)e^{-j\,2\pi n t/T_0} \, dt$$

$$= \frac{A_c}{T_0} \int_{t=-T_0/4}^{T_0/4} e^{j\,2\pi t (k_f - n/T_0)}$$

$$+ \frac{A_c}{T_0} e^{j\pi\beta} \int_{t=T_0/4}^{3T_0/4} e^{-j\,2\pi t (k_f + n/T_0)}. \tag{4.117}$$

The first integral in (4.117) evaluates to

$$I = \frac{A_c}{2} \, \mathrm{sinc} \left(\frac{\beta - n}{2} \right). \tag{4.118}$$

The second integral in (4.117) evaluates to

$$II = A_c \left[\frac{e^{-j\beta\pi/2} e^{-j\,3n\pi/2} - e^{j\beta\pi/2} e^{-jn\pi/2}}{-j\,2\pi(\beta + n)} \right]. \tag{4.119}$$

Now

$$-\frac{3n\pi}{2} = \left(-\frac{3\pi}{2} + \pi \right) n - n\pi = -\frac{n\pi}{2} - n\pi. \tag{4.120}$$

Similarly

$$-\frac{n\pi}{2} = \left(-\frac{\pi}{2} + \pi \right) n - n\pi = \frac{n\pi}{2} - n\pi. \tag{4.121}$$

Substituting (4.120) and (4.121) in (4.119), we get

$$II = \frac{A_c}{2} e^{-jn\pi} \mathrm{sinc} \left(\frac{\beta + n}{2} \right)$$

$$= \frac{A_c}{2} (-1)^n \, \mathrm{sinc} \left(\frac{\beta + n}{2} \right). \tag{4.122}$$

Therefore

$$c_n = \left[\frac{A_c}{2} \, \mathrm{sinc} \left(\frac{\beta - n}{2} \right) + \frac{A_c}{2} (-1)^n \, \mathrm{sinc} \left(\frac{\beta + n}{2} \right) \right] \tag{4.123}$$

, which implies that

$$a_n = \frac{A_c}{2}$$

$$b_n = \frac{A_c}{2} (-1)^n. \tag{4.124}$$

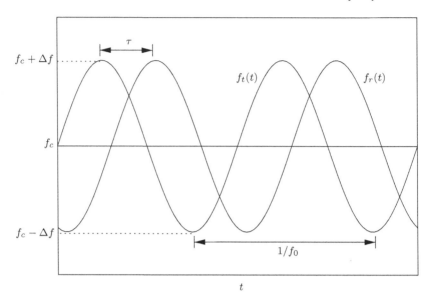

Fig. 4.12 Variation of the instantaneous frequency with time in an FM radar

13. In a frequency-modulated radar, the instantaneous frequency of the transmitted carrier $f_t(t)$ is varied as given in Fig. 4.12. The instantaneous frequency of the received echo $f_r(t)$ is also shown, where τ is the round-trip delay time. Assuming that $f_0\tau \ll 1$ so that $\cos(2\pi f_0\tau) \approx 1$ and $\sin(2\pi f_0\tau) \approx 2\pi f_0\tau$, determine the number of beat (difference frequency) cycles in one second, in terms of the frequency deviation (Δf) of the carrier frequency, the delay τ, and the repetition frequency f_0. Assume that f_0 is an integer. Note that in Fig. 4.12

$$f_t(t) = f_c + \Delta f \sin(2\pi f_0 t). \qquad (4.125)$$

- *Solution*: Let the transmitted signal be given by

$$s(t) = A_1 \cos\left(2\pi \int_{\tau=0}^{t} f_t(\tau)\, d\tau\right). \qquad (4.126)$$

Let the received signal be given by

$$r(t) = A_2 \cos\left(2\pi \int_{\tau=0}^{t} f_r(\tau)\, d\tau\right). \qquad (4.127)$$

The beat (difference) frequency is $f_t(t) - f_r(t)$. The number of beat cycles over the time duration $1/f_0$ is given by

$$N = \int_{t=0}^{1/f_0} |f_t(t) - f_r(t)| \, dt. \tag{4.128}$$

Note that

$$\begin{aligned}
f_r(t) &= f_c + \Delta f \sin(2\pi f_0(t - \tau)) \\
&= f_c + \Delta f \left[\sin(2\pi f_0 t) \cos(2\pi f_0 \tau) - \cos(2\pi f_0 t) \sin(2\pi f_0 \tau) \right] \\
&\approx f_c + \Delta f \left[\sin(2\pi f_0 t) - 2\pi f_0 \tau \cos(2\pi f_0 t) \right]. \tag{4.129}
\end{aligned}$$

Therefore

$$f_t(t) - f_r(t) = 2\pi f_0 \tau \Delta f \cos(2\pi f_0 t). \tag{4.130}$$

Hence (4.128) becomes

$$\begin{aligned}
N &= 2\pi f_0 \tau \Delta f \times 4 \int_{t=0}^{1/(4f_0)} \cos(2\pi f_0 t) \, dt \\
&= 8\pi f_0 \tau \Delta f \left| \frac{\sin(2\pi f_0 t)}{2\pi f_0} \right|_{t=0}^{1/(4f_0)} \\
&= 4\tau \Delta f. \tag{4.131}
\end{aligned}$$

Since f_0 is an integer, the number of beat cycles per second is

$$N f_0 = 4\tau \Delta f f_0. \tag{4.132}$$

14. (Haykin 1983) The sinusoidal modulating wave

$$m(t) = A_m \cos(2\pi f_m t) \tag{4.133}$$

is applied to a phase modulator with phase sensitivity k_p. The modulated signal $s(t)$ is of the form $A_c \cos(2\pi f_c t + \phi(t))$.

(a) Determine the spectrum of $s(t)$, assuming that the maximum phase deviation $\beta_p = k_p A_m$ does not exceed 0.3 rad.
(b) Construct a phasor diagram for $s(t)$.

• *Solution*: The PM signal is given by

Fig. 4.13 Phasor diagram
for narrowband phase
modulation

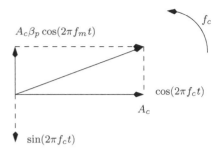

$$s(t) = A_c \cos(2\pi f_c t + k_p A_m \cos(2\pi f_m t))$$
$$= A_c \cos(2\pi f_c t) \cos(k_p A_m \cos(2\pi f_m t))$$
$$- A_c \sin(2\pi f_c t) \sin(k_p A_m \cos(2\pi f_m t))$$
$$\approx A_c \cos(2\pi f_c t) - A_c \beta_p \sin(2\pi f_c t) \cos(2\pi f_m t)$$
$$= A_c \cos(2\pi f_c t) - \frac{A_c \beta_p}{2} [\sin(2\pi (f_c - f_m)t) + \sin(2\pi (f_c + f_m)t)].$$

$$(4.134)$$

Hence, the spectrum is given by

$$S(f) = \frac{A_c}{2} [\delta(f - f_c) + \delta(f + f_c)]$$
$$- \frac{A_c \beta_p}{4j} [\delta(f - f_c + f_m) - \delta(f + f_c - f_m)]$$
$$- \frac{A_c \beta_p}{4j} [\delta(f - f_c - f_m) - \delta(f + f_c + f_m)]. \quad (4.135)$$

The phasor diagram is shown in Fig. 4.13.

15. Consider a tone-modulated PM signal of the form

$$s(t) = A_c \cos(2\pi f_c t + \beta_p \cos(2\pi f_m t)), \quad (4.136)$$

where $\beta_p = k_p A_m$. This modulated signal is applied to an ideal BPF with
unity gain, midband frequency f_c, and passband extending from $f_c - 1.5 f_m$
to $f_c + 1.5 f_m$. Determine the envelope, phase, and instantaneous frequency of
the modulated signal at the filter output as functions of time.

• *Solution*: The PM signal is given by

$$s(t) = A_c \cos(2\pi f_c t + k_p A_m \cos(2\pi f_m t)). \quad (4.137)$$

The complex envelope is

$$\tilde{s}(t) = A_c \exp(j k_p A_m \cos(2\pi f_m t)), \tag{4.138}$$

which is periodic with period $1/f_m$ and hence can be represented in the form of a Fourier series as follows:

$$\tilde{s}(t) = \sum_{n=-\infty}^{\infty} c_n \exp(j 2\pi n f_m t). \tag{4.139}$$

The coefficients c_n are given by

$$c_n = A_c f_m \int_{t=-1/(2f_m)}^{1/(2f_m)} \exp(j k_p A_m \cos(2\pi f_m t) - j 2\pi n f_m t) \, dt. \tag{4.140}$$

Let

$$2\pi f_m t = \pi/2 - x. \tag{4.141}$$

Then

$$c_n = \frac{-A_c}{2\pi} \int_{x=3\pi/2}^{-\pi/2} \exp(j k_p A_m \sin(x) - j(n\pi/2 - nx)) \, dx. \tag{4.142}$$

Since the above integrand is periodic with a period of 2π, we can write

$$c_n = \frac{A_c}{2\pi} \exp(-j n\pi/2) \int_{x=-\pi}^{\pi} \exp(j k_p A_m \sin(x) + j nx) \, dx. \tag{4.143}$$

We know that

$$J_n(\beta) = \frac{1}{2\pi} \int_{x=-\pi}^{\pi} \exp(j \beta \sin(x) - j nx) \, dx, \tag{4.144}$$

where $J_n(\beta)$ is the nth-order Bessel function of the first kind and argument β. Thus

$$c_n = A_c \exp(-j n\pi/2) J_{-n}(\beta_p). \tag{4.145}$$

The transmitted PM signal can be written as

$$s(t) = \Re\left\{\tilde{s}(t)\exp(j\,2\pi f_c t)\right\}$$

$$= A_c\Re\left\{\sum_{n=-\infty}^{\infty}\exp(-j\,n\pi/2)J_{-n}(\beta_p)\exp(j\,2\pi n f_m t)\exp(j\,2\pi f_c t)\right\}$$

$$= A_c\sum_{n=-\infty}^{\infty}J_{-n}(\beta_p)\cos(2\pi(f_c + n f_m)t - n\pi/2). \tag{4.146}$$

The BPF allows only the components at f_c, $f_c - f_m$ and $f_c + f_m$. Thus, the BPF output is

$$\begin{aligned}x(t) &= A_c J_0(\beta_p)\cos(2\pi f_c t)\\ &\quad + A_c J_{-1}(\beta_p)\cos(2\pi(f_c + f_m)t - \pi/2)\\ &\quad + A_c J_1(\beta_p)\cos(2\pi(f_c - f_m)t + \pi/2)\\ &= A_c J_0(\beta_p)\cos(2\pi f_c t)\\ &\quad + A_c J_{-1}(\beta_p)\sin(2\pi(f_c + f_m)t)\\ &\quad - A_c J_1(\beta_p)\sin(2\pi(f_c - f_m)t). \end{aligned}\tag{4.147}$$

However

$$J_{-1}(\beta_p) = -J_1(\beta_p). \tag{4.148}$$

Hence

$$\begin{aligned}x(t) &= A_c J_0(\beta_p)\cos(2\pi f_c t)\\ &\quad - A_c J_1(\beta_p)\sin(2\pi(f_c + f_m)t)\\ &\quad - A_c J_1(\beta_p)\sin(2\pi(f_c - f_m)t)\\ &= A_c J_0(\beta_p)\cos(2\pi f_c t)\\ &\quad - 2A_c J_1(\beta_p)\sin(2\pi f_c t)\cos(2\pi f_m t)\\ &= a(t)\cos(2\pi f_c t + \theta_i(t)). \end{aligned}\tag{4.149}$$

The envelope of $x(t)$ is

$$a(t) = A_c\sqrt{J_0^2(\beta_p) + 4J_1^2(\beta_p)\cos^2(2\pi f_m t)}. \tag{4.150}$$

The total instantaneous phase of $x(t)$ is

$$\phi_i(t) = 2\pi f_c t + \tan^{-1}\left[\frac{2J_1(\beta_p)}{J_0(\beta_p)}\cos(2\pi f_m t)\right]. \tag{4.151}$$

The total instantaneous frequency of $x(t)$ is

$$f_i(t) = f_c + \frac{1}{2\pi}\frac{d}{dt}\left[\tan^{-1}\left(\frac{2J_1(\beta_p)}{J_0(\beta_p)}\cos(2\pi f_m t)\right)\right]$$

$$= f_c - \frac{2J_1(\beta_p)/J_0(\beta_p)}{1 + (2J_1(\beta_p)/J_0(\beta_p))^2\cos^2(2\pi f_m t)}f_m\sin(2\pi f_m t)$$

$$= f_c - \frac{2J_1(\beta_p)J_0(\beta_p)}{J_0^2(\beta_p) + (2J_1(\beta_p))^2\cos^2(2\pi f_m t)}f_m\sin(2\pi f_m t).$$

$$(4.152)$$

16. (Haykin 1983) A carrier wave is frequency modulated using a sinusoidal signal of frequency f_m and amplitude A_m.

   (a) Determine the values of the modulation index β for which the carrier component of the FM signal is reduced to zero.
   (b) In a certain experiment conducted with $f_m = 1\,\text{kHz}$ and increasing A_m starting from zero volts, it is found that the carrier component of the FM signal is reduced to zero for the first time when $A_m = 2\,\text{V}$. What is the frequency sensitivity of the modulator? What is the value of A_m for which the carrier component is reduced to zero for the second time?

   • *Solution*: From the table of Bessel functions, we see that $J_0(\beta)$ is equal to zero for

$$\beta = 2.44$$
$$\beta = 5.52$$
$$\beta = 8.65$$
$$\beta = 11.8.$$

$$(4.153)$$

For tone modulation

$$\beta = \frac{k_f A_m}{f_m}$$
$$\Rightarrow k_f = \frac{\beta f_m}{A_m}$$
$$= 1.22\,\text{kHz/V}.$$

$$(4.154)$$

The value of A_m for which the carrier component goes to zero for the second time is equal to

$$A_m = \frac{\beta f_m}{k_f}$$
$$\Rightarrow A_m = \frac{5.52}{1.22}$$
$$= 4.52\,\text{V}.$$

$$(4.155)$$

17. An FM signal with modulation index $\beta = 2$ is transmitted through an ideal bandpass filter with midband frequency f_c and bandwidth $7 f_m$, where f_c is the carrier frequency and f_m is the frequency of the sinusoidal modulating wave. Determine the spectrum of the filter output.

- *Solution*: We know that the spectrum of the tone-modulated FM signal is given by

$$S(f) = \frac{A_c}{2} \sum_{n=-\infty}^{\infty} J_n(\beta) \left[\delta(f - f_c - nf_m) + \delta(f + f_c + nf_m) \right] \quad (4.156)$$

The spectrum at the output of the BPF is given by

$$X(f) = \frac{A_c}{2} \sum_{n=-3}^{3} J_n(2) \delta(f - f_c - nf_m) + \delta(f + f_c + nf_m), \quad (4.157)$$

which is illustrated in Fig. 4.14.

18. (Haykin 1983) Consider a wideband PM signal produced by a sinusoidal modulating wave $m(t) = A_m \cos(2\pi f_m t)$, using a modulator with phase sensitivity k_p.

    (a) Show that if the phase deviation of the PM signal is large compared to one radian, the bandwidth of the PM signal varies linearly with the modulation frequency f_m.
    (b) Compare this bandwidth of the wideband PM signal with that of a wideband FM signal produced by $m(t)$ and frequency sensitivity k_f.

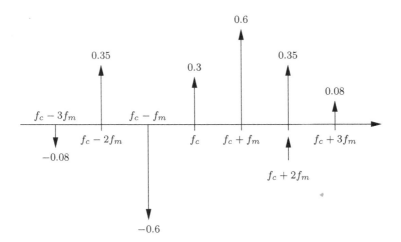

Fig. 4.14 Spectrum of the signal at the BPF output for $A_c = 2$

- *Solution*: The PM signal is given by

$$s(t) = A_c \cos(2\pi f_c t + k_p A_m \cos(2\pi f_m t)).\qquad(4.158)$$

The phase deviation is $k_p A_m$. The instantaneous frequency due to the message is

$$\frac{1}{2\pi}\frac{d}{dt}\left(k_p A_m \cos(2\pi f_m t)\right) = -k_p A_m f_m \sin(2\pi f_m t).\qquad(4.159)$$

The frequency deviation is

$$\Delta f_1 = k_p A_m f_m,\qquad(4.160)$$

which is directly proportional to f_m. Now, the PM signal in (4.158) can be considered to be an FM signal with message given by

$$m_1(t) = -(k_p/k_f)A_m f_m \sin(2\pi f_m t).\qquad(4.161)$$

Therefore, the bandwidth of the PM signal in (4.158) is

$$\begin{aligned}
B_{T,\mathrm{PM}} &= 2(\Delta f_1 + f_m)\\
&= 2 f_m (k_p A_m + 1)\\
&\approx 2 f_m k_p A_m,
\end{aligned}\qquad(4.162)$$

which varies linearly with f_m. However, in the case of FM, the frequency deviation is

$$\Delta f_2 = k_f A_m,\qquad(4.163)$$

which is independent of f_m. The bandwidth of the FM signal is

$$\begin{aligned}
B_{T,\mathrm{FM}} &= 2(\Delta f_2 + f_m)\\
&= 2(k_f A_m + f_m).
\end{aligned}\qquad(4.164)$$

19. Figure 4.15 shows the block diagram of a system. Here $s(t)$ and $h(t)$ are FM signals, given by

Fig. 4.15 Block diagram of a system

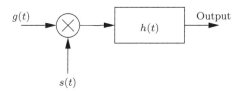

$$s(t) = \cos{(2\pi f_c t - \pi f_m t)}$$
$$h(t) = \cos{(2\pi f_c t + \pi f_m t)}. \tag{4.165}$$

Assume that $g(t)$ is a real-valued lowpass signal with bandwidth $[-W, W]$, $f_m < W$, $f_c \gg W$ and $h(t)$ is of the form

$$h(t) = 2h_c(t)\cos(2\pi f_c t) - 2h_s(t)\sin(2\pi f_c t). \tag{4.166}$$

(a) Using the method of complex envelopes, determine the output of $h(t)$.
(b) Determine the envelope of the output of $h(t)$.

• *Solution*: Let the signal at the output of the multiplier be denoted by $v_1(t)$. Then

$$v_1(t) = g(t)\cos(2\pi f_c t - \pi f_m t)$$
$$= g(t)\left[\cos(2\pi f_c t)\cos(\pi f_m t) + \sin(2\pi f_c t)\sin(\pi f_m t)\right]. \tag{4.167}$$

Comparing the above equation with that of the canonical representation of a bandpass signal, we conclude that the complex envelope of $v_1(t)$ is given by

$$\tilde{v}_1(t) = g(t)e^{-j\pi f_m t}. \tag{4.168}$$

Note that according to the representation in (4.167), $v_1(t)$ is a bandpass signal, with carrier frequency f_c. Similarly, the complex envelope of the filter is given by

$$\tilde{h}(t) = \frac{1}{2}e^{j\pi f_m t}, \tag{4.169}$$

where we have assumed that

$$h(t) = \Re\left\{2\tilde{h}(t)e^{j2\pi f_c t}\right\}. \tag{4.170}$$

Thus, the complex envelope of the filter output can be written as

$$\tilde{y}(t) = \tilde{h}(t) \star \tilde{v}_1(t), \tag{4.171}$$

where $\star$ denotes convolution. Therefore

$$\tilde{y}(t) = \int_{\tau=-\infty}^{\infty} \tilde{v}_1(\tau)\tilde{h}(t-\tau)\,d\tau$$

$$= \frac{1}{2} \int_{\tau=-\infty}^{\infty} g(\tau)e^{-j\pi f_m \tau}e^{j\pi f_m(t-\tau)}\,d\tau$$

$$= \frac{1}{2}e^{j\pi f_m t} \int_{\tau=-\infty}^{\infty} g(\tau)e^{-j2\pi f_m \tau}\,d\tau$$

$$= \frac{1}{2}e^{j\pi f_m t}G(f_m). \tag{4.172}$$

Therefore, the output of $h(t)$ is

$$y(t) = \Re\left\{\tilde{y}(t)e^{j2\pi f_c t}\right\}$$

$$= \frac{1}{2}G(f_m)\cos(2\pi(f_c - f_m/2)t). \tag{4.173}$$

Thus, the envelope of the output is given by

$$|\tilde{y}(t)| = \frac{1}{2}|G(f_m)|. \tag{4.174}$$

Hence, the envelope of the output signal is proportional to the magnitude response of $g(t)$ evaluated at $f = f_m$.

20. (Haykin 1983) Figure 4.16 shows the block diagram of the transmitter and receiver for stereophonic FM. The input signals $l(t)$ and $r(t)$ represent left-hand and right-hand audio signals. The difference signal $x_1(t) = l(t) - r(t)$ is DSB-SC modulated as shown in the figure, with $f_c = 25\,\text{kHz}$. The DSB-SC wave, $x_2(t) = l(t) + r(t)$ and the pilot carrier are summed to produce the composite signal $m(t)$. The composite signal $m(t)$ is used to frequency modulate a carrier and the resulting FM signal is transmitted. Assume that $f_2 = 20\,\text{kHz}$, $f_1 = 200\,\text{Hz}$.

    (a) Sketch the spectrum of $m(t)$. Label all the important points on the x- and y-axes. The spectrums of $l(t)$ and $r(t)$ are shown in Fig. 4.16.
    (b) Assuming that the frequency deviation of the FM signal is 90 kHz, find the transmission bandwidth of the FM signal using Carson's rule.
    (c) In the receiver block diagram determine the signal $y(t)$, the input-output characteristics of the device, and the specifications of filter1, filter2, filter3, and filter4 in the frequency domain. Assume ideal filter characteristics.

- *Solution*: Note that

$$m(t) = x_2(t) + \cos(2\pi f_c t) + x_1(t)\cos(4\pi f_c t). \tag{4.175}$$

The spectrum of $m(t)$ is depicted in Fig. 4.17.
From Carson's rule, the transmission bandwidth of the FM signal is

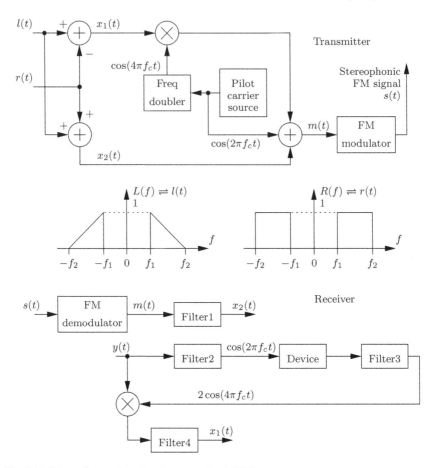

Fig. 4.16 Transmitter and receiver for stereophonic FM

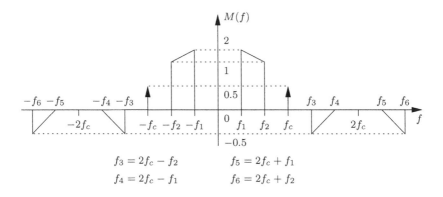

$$f_3 = 2f_c - f_2 \qquad f_5 = 2f_c + f_1$$
$$f_4 = 2f_c - f_1 \qquad f_6 = 2f_c + f_2$$

Fig. 4.17 Spectrum of $m(t)$

Fig. 4.18 Variation of phase versus time for an FM signal

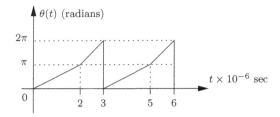

$$B_T = 2(\Delta f + W),\tag{4.176}$$

where $W = 2f_c + f_2 = 70\,\text{kHz}$ is the maximum frequency content in $m(t)$ and $\Delta f = 90\,\text{kHz}$ is the frequency deviation. Therefore $B_T = 320\,\text{kHz}$. At the receiver side $y(t) = m(t)$. The device is a squarer. Filter1 is an LPF with unity gain and bandwidth $[-f_2,\ f_2]$. Filter2 is a BPF with unity gain, center frequency f_c, and bandwidth less than 5 kHz on either side of f_c. Filter3 is a BPF with a gain of 4 with center frequency $2f_c$ and any suitable bandwidth, such that the dc component is eliminated. Filter4 is an LPF with unity gain and bandwidth $[-f_2,\ f_2]$.

21. Consider an FM signal given by

$$s(t) = A\sin(\theta(t)),$$

where $\theta(t)$ is a periodic waveform as shown in Fig. 4.18. The message signal has zero mean. Compute the carrier frequency.

- *Solution*: We know that the instantaneous frequency is given by

$$f(t) = \frac{1}{2\pi}\frac{d\theta}{dt}$$
$$= f_c + k_f m(t),\tag{4.177}$$

which is plotted in Fig. 4.19b. Observe that $f(t)$ is also periodic. Since $m(t)$ has zero mean, the mean value of $f(t)$ is f_c, where

$$f_c = \frac{25 \times 2 + 50 \times 1}{3} \times 10^4$$
$$= \frac{100}{3} \times 10^4\,\text{Hz}.\tag{4.178}$$

22. A 10 kHz periodic square wave $g_p(t)$ is applied to a first-order RL lowpass filter as shown in Fig. 4.20. It is given that $R/(2\pi L) = 10\,\text{kHz}$. The output signal $m(t)$ is FM modulated with frequency deviation equal to 75 kHz.
Determine the bandwidth of the FM signal $s(t)$, using Carson's rule. Ignore those

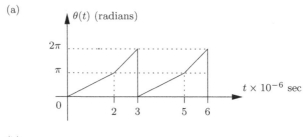

(a)

(b)

Fig. 4.19 Variation of phase versus time for an FM signal

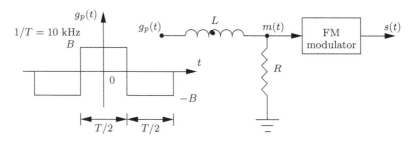

Fig. 4.20 Periodic signal applied to an RL-lowpass filter

harmonic terms in $m(t)$ whose (absolute value of the) amplitude is less than 1% of the fundamental.

- *Solution*: We know that

$$g_p(t) = \frac{4B}{\pi} \sum_{m=0}^{\infty} \frac{(-1)^m}{(2m+1)} \cos\left[2\pi(2m+1)f_0 t\right], \qquad (4.179)$$

where $f_0 = 1/T = 10\,\text{kHz}$. The transfer function of the filter is

$$H(\omega) = \frac{R}{R + j\omega L}$$

$$= \frac{1}{1 + j\omega/\omega_0}, \qquad (4.180)$$

where $\omega_0 = R/L = 2\pi f_0$ (given) is the $-3\,\mathrm{dB}$ frequency in rad/s. Therefore

$$m(t) = \frac{4B}{\pi} \sum_{m=0}^{\infty} A_m \frac{(-1)^m}{(2m+1)} \cos\left[2\pi(2m+1)f_0 t + \theta_m\right], \quad (4.181)$$

where

$$A_m = |H((2m+1)\omega_0)| = \frac{1}{\sqrt{1+(2m+1)^2}} \quad (4.182)$$

and

$$\theta_m = -\tan^{-1}(2m+1). \quad (4.183)$$

We need to ignore those harmonics that satisfy

$$\frac{4B}{\pi} \frac{1}{(2m+1)} \frac{1}{\sqrt{1+(2m+1)^2}} < 0.01 \frac{4B}{\pi} \frac{1}{\sqrt{2}}$$

$$\Rightarrow \frac{1}{(2m+1)} \frac{1}{\sqrt{1+(2m+1)^2}} < 0.00707 \quad \text{for } m > 0. \quad (4.184)$$

We find that $m \geq 6$ satisfies the inequality in (4.184). Therefore, the (one-sided) bandwidth of $m(t)$ is $(2 \times 5 + 1)f_0 = 11 f_0 = 110\,\mathrm{kHz}$. Hence, the bandwidth of $s(t)$ using Carson's rule is

$$B = 2(75 + 110) = 370\,\mathrm{kHz}. \quad (4.185)$$

23. A message signal $m(t) = A_c \cos(2\pi f_m t)$ with $f_m = 5\,\mathrm{kHz}$ is applied to a frequency modulator. The resulting FM signal has a frequency deviation of $10\,\mathrm{kHz}$. This FM signal is applied to two frequency multipliers in cascade. The first frequency multiplier has a multiplication factor of 2 and the second frequency multiplier has a multiplication factor of 3. Determine the frequency deviation and the modulation index of the FM signal obtained at the second multiplier output. What is the frequency separation between two consecutive spectral components in the spectrum of the output FM signal?

- *Solution*: Let us denote the instantaneous frequency of the input FM signal by

$$f_i(t) = f_c + \Delta f \cos(2\pi f_m t), \quad (4.186)$$

where $\Delta f = 10\,\mathrm{kHz}$. The overall multiplication factor of the system is 6, hence the instantaneous frequency of the output FM signal is

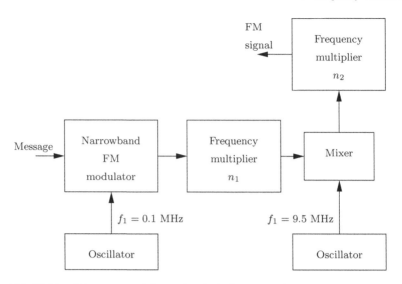

Fig. 4.21 Wideband frequency modulator using the indirect method

$$f_o(t) = 6f_c + 6\Delta f \cos(2\pi f_m t). \tag{4.187}$$

Thus, the frequency deviation of the output FM signal is 60 kHz. The modulation index is $60/5 = 12$. The frequency separation in the spectrum of the output FM signal is unchanged at 5 kHz.

24. (Haykin 1983) Figure 4.21 shows the block diagram of a wideband frequency modulator using the indirect method. Note that a mixer is essentially a multiplier, followed by a bandpass filter which allows only the difference frequency component. This transmitter is used to transmit audio signals in the range 100 Hz to 15 kHz. The narrowband frequency modulator is supplied with a carrier of frequency $f_1 = 0.1$ MHz. The second oscillator supplies a frequency of 9.5 MHz. The system specifications are as follows: carrier frequency at the transmitter output $f_c = 100$ MHz with frequency deviation, $\Delta f = 75$ kHz. Maximum modulation index at the output of the narrowband frequency modulator is 0.2 rad.

   (a) Calculate the frequency multiplication ratios n_1 and n_2.
   (b) Specify the value of the carrier frequency at the output of the first frequency multiplier.

   • *Solution*: It is given that the frequency deviation of the output FM wave is equal to 75 kHz. Note that for a tone-modulated FM signal, the frequency deviation is related to the frequency of the modulating wave as

$$\Delta f = \beta f_m. \tag{4.188}$$

The frequency deviation at the output of the narrowband FM modulator is fixed. Thus, the lowest frequency component in the message will produce the maximum modulation index, $\beta = 0.2$, modulator has a $\beta = 0.2$. Then the frequency deviation at the output of the narrowband FM modulator is

$$\Delta f_{\text{inp}} = 0.2 \times 0.1 \,\text{kHz}$$
$$= 0.02 \,\text{kHz}. \tag{4.189}$$

However, the required frequency deviation of the output FM signal is 75 kHz. Hence, we need to have

$$n_1 n_2 = \frac{75}{0.02}$$
$$= 3750. \tag{4.190}$$

The carrier frequency at the output of the first multiplier is $0.1 n_1$ MHz. The carrier frequency at the output of the second multiplier is

$$n_2(9.5 - 0.1 n_1) = 100 \,\text{MHz}. \tag{4.191}$$

Solving for n_1 and n_2 we get

$$n_1 = 75$$
$$n_2 = 50. \tag{4.192}$$

The carrier frequency at the output of the first frequency multiplier is

$$n_1 f_1 = 7.5 \,\text{MHz}. \tag{4.193}$$

25. (Haykin 1983) The equivalent circuit of the frequency-determining network of a VCO is shown in Fig. 4.22. Frequency modulation is produced by applying the modulating signal $V_m \sin(2\pi f_m t)$ plus a bias V_b to a varactor diode connected across the parallel combination of a 200-μH inductor (L) and a 100-pF capacitor

Fig. 4.22 Equivalent circuit of the frequency-determining network of a VCO

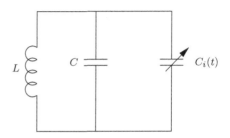

(C). The capacitance in the varactor diode is related to the voltage $V(t)$ applied across its terminals by

$$C_i(t) = 100/\sqrt{V(t)}\,\text{pF}. \tag{4.194}$$

The instantaneous frequency of oscillation is given by

$$f_i(t) = \frac{1}{2\pi} \frac{1}{\sqrt{L(C + C_i(t))}}. \tag{4.195}$$

The unmodulated frequency of oscillation is 1 MHz. The VCO output is applied to a frequency multiplier to produce an FM signal with carrier frequency of 64 MHz and a modulation index of 5.

(a) Determine the magnitude of the bias voltage V_b.
(b) Find the amplitude V_m of the modulating wave, given that $f_m = 10\,\text{kHz}$.

Assume that $V_b \gg V_m$.

- *Solution*: The instantaneous frequency of oscillation is given by

$$f_i(t) = \frac{1}{2\pi} \frac{1}{\sqrt{L(C + C_i(t))}}. \tag{4.196}$$

The unmodulated frequency of oscillation is given as 1 MHz. Thus

$$f_c = \frac{1}{2\pi} \frac{1}{\sqrt{L(C + C_0)}}$$
$$\Rightarrow C + C_0 = 126.65\,\text{pF}$$
$$\Rightarrow C_0 = 26.651\,\text{pF}$$
$$\Rightarrow 100/\sqrt{V_b} = 26.651$$
$$\Rightarrow V_b = 14.078\,\text{V}. \tag{4.197}$$

Since the final carrier frequency is 64 MHz and the frequency multiplication factor is 64. Thus, the modulation index at the VCO output is

$$\beta = \frac{5}{64} = 0.078. \tag{4.198}$$

Thus, the FM signal at the VCO output is narrowband. The instantaneous frequency of oscillation at the VCO output is

$$f_i(t) = \frac{1}{2\pi} \frac{1}{\sqrt{L(C + 100(V_b + V_m \sin(2\pi f_m t))^{-0.5})}}. \tag{4.199}$$

Since $V_b \gg V_m$ the instantaneous frequency can be approximated as

$$f_i(t) \approx \frac{1}{2\pi} \frac{1}{\sqrt{L(C + 100V_b^{-0.5}(1 - V_m/(2V_b)\sin(2\pi f_m t)))}}$$

$$= \frac{1}{2\pi} \frac{1}{\sqrt{L(C + C_0(1 - V_m/(2V_b)\sin(2\pi f_m t)))}}. \qquad (4.200)$$

Hence, the instantaneous frequency becomes

$$f_i(t) = \frac{1}{2\pi} \frac{1}{\sqrt{L(C + C_0 - C_0 V_m/(2V_b)\sin(2\pi f_m t)))}}$$

$$= \frac{1}{2\pi} \frac{1}{\sqrt{L(C + C_0)}}$$

$$\frac{1}{\sqrt{1 - C_0 V_m/(2V_b(C + C_0))\sin(2\pi f_m t)}}$$

$$\approx f_c \left(1 + \frac{C_0 V_m}{4V_b(C + C_0)} \sin(2\pi f_m t)\right). \qquad (4.201)$$

The modulation index of the narrowband FM signal is given by

$$\beta = \frac{C_0 V_m f_c}{4V_b f_m (C + C_0)} = 0.078. \qquad (4.202)$$

Substituting $f_c = 1\,\text{MHz}$, $f_m = 10\,\text{kHz}$, and $V_b = 14.078\,\text{V}$, we get

$$V_m = 0.2087\,\text{V}. \qquad (4.203)$$

26. The equivalent circuit of the frequency-determining network of a VCO is shown in Fig. 4.23. Frequency modulation is produced by applying the modulating signal $V_m \sin(2\pi f_m t)$ plus a bias V_b to a varactor diode connected across the parallel combination of a 25-μH inductor (L) and a 200-pF capacitor (C). The capacitance in the varactor diode is related to the voltage $V(t)$ applied across its terminals by

$$C_i(t) = 100/\sqrt{V(t)}\,\text{pF}. \qquad (4.204)$$

Fig. 4.23 Equivalent circuit of the frequency-determining network of a VCO

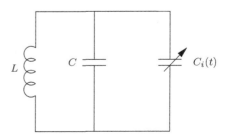

The instantaneous frequency of oscillation is given by

$$f_i(t) = \frac{1}{2\pi} \frac{1}{\sqrt{L(C + C_i(t))}}. \tag{4.205}$$

The unmodulated frequency of oscillation is 2 MHz. The VCO output is applied to a frequency multiplier to produce an FM signal with carrier frequency of 128 MHz and a modulation index of 6.

(a) Determine the magnitude of the bias voltage V_b.
(b) Find the amplitude V_m of the modulating wave, given that $f_m = 20$ kHz.

Assume that $V_b \gg V_m$.

• *Solution*: The instantaneous frequency of oscillation is given by

$$f_i(t) = \frac{1}{2\pi} \frac{1}{\sqrt{L(C + C_i(t))}}. \tag{4.206}$$

The unmodulated frequency of oscillation is given as 2 MHz. Thus

$$f_c = \frac{1}{2\pi} \frac{1}{\sqrt{L(C + C_0)}}$$
$$\Rightarrow C + C_0 = 253.303 \,\text{pF}$$
$$\Rightarrow C_0 = 53.303 \,\text{pF}$$
$$\Rightarrow 100/\sqrt{V_b} = 53.303$$
$$\Rightarrow V_b = 3.5196 \,\text{V}. \tag{4.207}$$

Since the final carrier frequency is 128 MHz, the frequency multiplication factor is $128/2 = 64$. Thus, the modulation index at the VCO output is

$$\beta = \frac{6}{64} = 0.09375. \tag{4.208}$$

Thus, the FM signal at the VCO output is narrowband. The instantaneous frequency of oscillation at the VCO output is

$$f_i(t) = \frac{1}{2\pi} \frac{1}{\sqrt{L(C + 100(V_b + V_m \sin(2\pi f_m t))^{-0.5})}}. \tag{4.209}$$

Since $V_b \gg V_m$ the instantaneous frequency can be approximated as

$$f_i(t) \approx \frac{1}{2\pi} \frac{1}{\sqrt{L(C + 100V_b^{-0.5}(1 - V_m/(2V_b)\sin(2\pi f_m t)))}}$$

$$= \frac{1}{2\pi} \frac{1}{\sqrt{L(C + C_0(1 - V_m/(2V_b)\sin(2\pi f_m t)))}}. \tag{4.210}$$

Hence, the instantaneous frequency becomes

$$f_i(t) = \frac{1}{2\pi} \frac{1}{\sqrt{L(C + C_0 - C_0 V_m/(2V_b)\sin(2\pi f_m t)))}}$$

$$= \frac{1}{2\pi} \frac{1}{\sqrt{L(C + C_0)}}$$

$$\frac{1}{\sqrt{1 - C_0 V_m/(2V_b(C + C_0))\sin(2\pi f_m t)}}$$

$$\approx f_c \left(1 + \frac{C_0 V_m}{4V_b(C + C_0)} \sin(2\pi f_m t)\right). \tag{4.211}$$

The modulation index of the narrowband FM signal is given by

$$\beta = \frac{C_0 V_m f_c}{4V_b f_m (C + C_0)} = 0.09375. \tag{4.212}$$

Substituting $f_c = 2\,\text{MHz}$, $f_m = 20\,\text{kHz}$, and $V_b = 3.5196\,\text{V}$, we get

$$V_m = 0.0627\,\text{V}. \tag{4.213}$$

27. (Haykin 1983) The FM signal

$$s(t) = A_c \cos\left[2\pi f_c t + 2\pi k_f \int_{\tau=0}^{t} m(\tau)\,d\tau\right] \tag{4.214}$$

is applied to the system shown in Fig. 4.24. Assume that the resistance R is small compared to the impedance of C for all significant frequency components of $s(t)$ and the envelope detector does not load the filter. Determine the resulting signal at the envelope detector output assuming that $k_f |m(t)| < f_c$ for all t.

• *Solution*: The transfer function of the highpass filter is

$$H(f) = \frac{R}{R + 1/(j\,2\pi f C)}$$

$$\approx j\,2\pi f RC \tag{4.215}$$

provided

Fig. 4.24 Frequency demodulation using a highpass filter

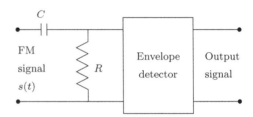

$$R \ll \frac{1}{2\pi f C}. \qquad (4.216)$$

Thus, over the range of frequencies in which (4.216) is valid, the highpass filter acts like an ideal differentiator. Hence, the output of the highpass filter is given by

$$\begin{aligned} x(t) &= RC\frac{ds(t)}{dt} \\ &= -RCA_c \left[2\pi f_c + 2\pi k_f m(t)\right] \sin\left[2\pi f_c t + 2\pi k_f \int_{\tau=0}^{t} m(\tau)\,d\tau\right]. \end{aligned}$$
$$(4.217)$$

The output of the envelope detector is given by

$$y(t) = 2\pi RCA_c \left[f_c + k_f m(t)\right]. \qquad (4.218)$$

28. Consider the FM signal

$$s(t) = A_c \cos\left(2\pi f_c t + 2\pi k_f \int_{\tau=0}^{t} m(\tau)\,d\tau\right), \qquad (4.219)$$

where k_f is 10 kHz/V and

$$m(t) = 5\cos(\omega t)\,\text{V}, \qquad (4.220)$$

where $\omega = 20{,}000\,\text{rad/s}$. Compute the frequency deviation and the modulation index.

- *Solution*: The total instantaneous frequency is

$$f_i(t) = f_c + k_f 5\cos(\omega t). \qquad (4.221)$$

Therefore, the frequency deviation is

$$5k_f = 50\,\text{kHz}. \qquad (4.222)$$

The transmitted FM signal is

$$s(t) = A_c \cos\left(2\pi f_c t + \frac{10\pi k_f}{\omega} \sin(\omega t)\right). \tag{4.223}$$

Hence, the modulation index is

$$\beta = \frac{10\pi k_f}{\omega} = 5\pi. \tag{4.224}$$

29. (Haykin 1983) Suppose that the received signal in an FM system contains some residual amplitude modulation as shown by

$$s(t) = a(t) \cos(2\pi f_c t + \phi(t)), \tag{4.225}$$

where $a(t) > 0$ and f_c is the carrier frequency. The phase $\phi(t)$ is related to the modulating signal $m(t)$ by

$$\phi(t) = 2\pi k_f \int_{\tau=0}^{t} m(\tau) \, d\tau. \tag{4.226}$$

Assume that $s(t)$ is restricted to a frequency band of width B_T centered at f_c, where B_T is the transmission bandwidth of $s(t)$ in the absence of amplitude modulation, that is, when $a(t) = A_c$. Also assume that $a(t)$ varies slowly compared to $\phi(t)$ and $f_c > k_f m(t)$. Compute the output of the ideal frequency discriminator (ideal differentiator followed by an envelope detector).

- *Solution*: The output of the ideal differentiator is

$$\begin{aligned}
s'(t) = \frac{ds(t)}{dt} &= a'(t) \cos(2\pi f_c t + \phi(t)) \\
&\quad - a(t) \sin(2\pi f_c t + \phi(t))(2\pi f_c + \phi'(t)). \tag{4.227}
\end{aligned}$$

The envelope of $s'(t)$ is given by

$$y(t) = \sqrt{(a'(t))^2 + a^2(t)(2\pi f_c + \phi'(t))^2}. \tag{4.228}$$

Since it is given that $a(t)$ varies slowly compared to $\phi(t)$, we have

$$|\phi'(t)| \gg |a'(t)| \tag{4.229}$$

$s'(t)$ can be approximated by

$$s'(t) \approx -a(t) \sin(2\pi f_c t + \phi(t))(2\pi f_c + 2\pi k_f m(t)). \tag{4.230}$$

The output of the envelope detector is

$$y(t) = 2\pi a(t)(f_c + k_f m(t)), \tag{4.231}$$

where we have assumed that $[f_c + k_f m(t)] > 0$. Thus, we see that there is distortion due to $a(t)$, at the envelope detector output.

30. (Haykin 1983) Let

$$s(t) = a(t)\cos(2\pi f_c t + \phi(t)), \tag{4.232}$$

where $a(t) > 0$, be applied to a hard limiter whose output $z(t)$ is defined by

$$\begin{aligned} z(t) &= \text{sgn}[s(t)] \\ &= \begin{cases} +1 \text{ for } s(t) > 0 \\ -1 \text{ for } s(t) < 0. \end{cases} \end{aligned} \tag{4.233}$$

(a) Show that $z(t)$ can be expressed in the form of a Fourier series as follows:

$$z(t) = \frac{4}{\pi} \sum_{n=0}^{\infty} \frac{(-1)^n}{2n+1} \cos[2\pi f_c t(2n+1) + (2n+1)\phi(t)]. \tag{4.234}$$

(b) Compute the output when $z(t)$ is applied to an ideal bandpass filter with center frequency f_c and bandwidth B_T, where B_T is the transmission bandwidth of $s(t)$ in the absence of amplitude modulation. Assume that $f_c \gg B_T$.

• *Solution*: Since $a(t) > 0$, $z(t)$ can be written as

$$z(t) = \text{sgn}[\cos(2\pi f_c t + \phi(t))]. \tag{4.235}$$

Let

$$\alpha(t) = 2\pi f_c t + \phi(t). \tag{4.236}$$

Then $z(t)$ can be rewritten as

$$z(\alpha(t)) = \text{sgn}[\cos(\alpha(t))]. \tag{4.237}$$

Note that $z(\alpha(t))$ is periodic with respect to $\alpha(t)$, that is,

$$z(\alpha(t) + 2n\pi) = z(\alpha(t)). \tag{4.238}$$

This is illustrated in Fig. 4.25. Hence, $z(t)$ can be written in the form of a Fourier series with respect to $\alpha(t)$ as follows:

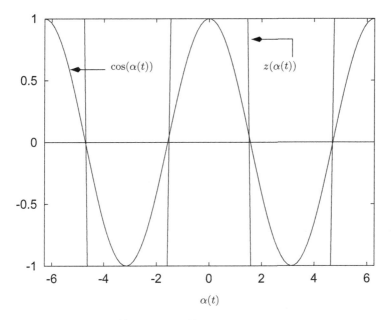

Fig. 4.25 $z(\alpha(t))$ is periodic with respect to $\alpha(t)$

$$z(\alpha(t)) = 2 \sum_{n=1}^{\infty} a_n \cos(n\alpha(t)), \tag{4.239}$$

where

$$
\begin{aligned}
a_n &= \frac{1}{2\pi} \int_{\alpha(t)=-\pi}^{\pi} z(\alpha(t)) \cos(n\alpha(t)) \, d\alpha(t) \\
&= \frac{1}{\pi} \int_{\alpha(t)=0}^{\pi} z(\alpha(t)) \cos(n\alpha(t)) \, d\alpha(t).
\end{aligned} \tag{4.240}
$$

For convenience denote $\alpha(t) = x$. Thus

$$
\begin{aligned}
a_n &= \frac{1}{\pi} \int_{x=0}^{\pi} z(x) \cos(nx) \, dx \\
&= \frac{1}{\pi} \int_{x=0}^{\pi/2} \cos(nx) \, dx - \frac{1}{\pi} \int_{x=\pi/2}^{\pi} \cos(nx) \, dx \\
&= \begin{cases} 0 & \text{for } n = 2m \\ \frac{2(-1)^m}{\pi(2m+1)} & \text{for } n = 2m + 1. \end{cases}
\end{aligned} \tag{4.241}
$$

Thus

$$z(x) = \sum_{m=0}^{\infty} \frac{4(-1)^m}{\pi(2m+1)} \cos((2m+1)x)$$

$$\Rightarrow z(\alpha(t)) = \sum_{m=0}^{\infty} \frac{4(-1)^m}{\pi(2m+1)} \cos((2m+1)\alpha(t))$$

$$= \sum_{m=0}^{\infty} \frac{4(-1)^m}{\pi(2m+1)} \cos((2m+1)2\pi f_c t + (2m+1)\phi(t)).$$

$$(4.242)$$

Observe that the mth harmonic has a carrier frequency at $(2m+1)f_c$ and bandwidth $(2m+1)B_T$, where B_T is the bandwidth of $s(t)$ with amplitude modulation removed, that is, $a(t) = A_c$. If the fundamental component at $m = 0$ is to be extracted from $z(\alpha(t))$, we require

$$f_c + \frac{B_T}{2} < (2m+1)f_c - (2m+1)\frac{B_T}{2}$$

$$\Rightarrow \frac{B_T}{2}\left(1 + \frac{1}{m}\right) < f_c \quad \text{for } m > 0. \tag{4.243}$$

The left-hand side of the second equation in (4.243) is maximum for $m = 1$. Thus, in the worst case, we require

$$f_c > B_T. \tag{4.244}$$

Now if $z(\alpha(t))$ is passed through an ideal bandpass filter with center frequency f_c and bandwidth B_T, the output is (assuming (4.244) is satisfied)

$$y(t) = \frac{4}{\pi} \cos(2\pi f_c t + \phi(t)), \tag{4.245}$$

which has no amplitude modulation.

31. The message signal

$$m(t) = \left(A\frac{\sin(tB)}{t}\right)^4 \tag{4.246}$$

is applied to an FM modulator. Compute the two-sided bandwidth (on both sides of the carrier) of the FM signal using Carson's rule. Assume frequency sensitivity of the modulator to be k_f.

- *Solution*: According to Carson's rule, the two-sided bandwidth of the FM signal is

$$B_T = 2(\Delta f + W), \tag{4.247}$$

where

$$\Delta f = \max k_f |m(t)| \tag{4.248}$$

is the frequency deviation and W is the one-sided bandwidth of the message. Clearly

$$\Delta f = k_f A^4 B^4 \tag{4.249}$$

since the maximum value of $m(t)$ is $A^4 B^4$, which occurs at $t = 0$. Next, we make use of the Fourier transform pair:

$$A \mathrm{sinc}(tB) \rightleftharpoons \frac{A}{B} \mathrm{rect}(f/B). \tag{4.250}$$

Time scaling by $1/\pi$, we obtain

$$A \mathrm{sinc}(tB/\pi) \rightleftharpoons \frac{A\pi}{B} \mathrm{rect}(f\pi/B)$$

$$\Rightarrow A \frac{\sin(tB)}{tB} \rightleftharpoons \frac{A\pi}{B} \mathrm{rect}(f\pi/B)$$

$$\Rightarrow A \frac{\sin(tB)}{t} \rightleftharpoons A\pi \, \mathrm{rect}(f\pi/B), \tag{4.251}$$

which has a one-sided bandwidth equal to $B/(2\pi)$. Therefore, $m(t)$ has a one-sided bandwidth of $W = 2B/\pi$. Hence

$$B_T = 2(k_f A^4 B^4 + 2B/\pi). \tag{4.252}$$

32. Explain the principle of operation of the PLL demodulator for FM signals. Draw the block diagram and clearly state the signal model and assumptions.

- *Solution*: Consider the block diagram in Fig. 4.26. Here $s(t)$ denotes the input FM signal (bandlimited to $[f_c - B_T/2, \ f_c + B_T/2]$; f_c is the carrier frequency, B_T is the bandwidth of $s(t)$) given by

$$s(t) = A_c \sin(2\pi f_c t + \phi_1(t)) \text{ V}, \tag{4.253}$$

where

$$\phi_1(t) = 2\pi k_f \int_{\tau=-\infty}^{t} m(\tau) \, d\tau, \tag{4.254}$$

where k_f is the frequency sensitivity of the FM modulator in Hz/V and $m(\cdot)$ is the message signal bandlimited to $[-W, \ W]$. The voltage controlled oscillator (VCO) output is

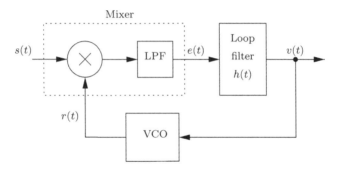

Fig. 4.26 Block diagram of the phase locked loop (PLL) demodulator for FM signals

$$r(t) = A_v \cos(2\pi f_c t + \phi_2(t)) \text{ V}, \qquad (4.255)$$

where

$$\phi_2(t) = 2\pi k_v \int_{\tau=-\infty}^{t} v(\tau) \, d\tau, \qquad (4.256)$$

where k_v is the frequency sensitivity of the VCO in Hz/V and $v(\cdot)$ is the control signal at the VCO input. The lowpass filter eliminates the sum frequency component at the multiplier output and allows only the difference frequency component. Hence

$$e(t) = A_c A_v k_m \sin(\phi_1(t) - \phi_2(t)) \text{ V}, \qquad (4.257)$$

where $k_m \left(\text{V}^{-1}\right)$ is the multiplier gain (a factor of $1/2$ is absorbed in k_m). The loop filter output is

$$v(t) = \int_{\tau=-\infty}^{\infty} e(\tau) h(t - \tau) \, d\tau \text{ V}. \qquad (4.258)$$

Note that $h(t)$ is dimensionless. Let

$$\phi_e(t) = \phi_1(t) - \phi_2(t)$$

$$\Rightarrow \phi_e(t) = \phi_1(t) - 2\pi k_v \int_{\tau=-\infty}^{t} v(\tau) \, d\tau$$

$$\Rightarrow \frac{d\phi_e(t)}{dt} = \frac{d\phi_1(t)}{dt} - 2\pi k_v v(t)$$

$$\Rightarrow \frac{d\phi_e(t)}{dt} = \frac{d\phi_1(t)}{dt} - 2\pi K_0 \int_{\tau=-\infty}^{\infty} \sin(\phi_e(\tau)) h(t - \tau) \, d\tau,$$

$$(4.259)$$

where

$$K_0 = k_m k_v A_c A_v \text{ Hz.} \tag{4.260}$$

Now, under steady state

$$\phi_e(t) \ll 1 \text{ rad} \tag{4.261}$$

for all t, therefore

$$\sin(\phi_e(t)) \approx \phi_e(t) \tag{4.262}$$

for all t. Hence, the last equation in (4.259) can be written as

$$\frac{d\phi_e(t)}{dt} = \frac{d\phi_1(t)}{dt} - 2\pi K_0 \int_{\tau=-\infty}^{\infty} \phi_e(\tau) h(t-\tau) \, d\tau. \tag{4.263}$$

Taking the Fourier transform of both sides, we get

$$j2\pi f \Phi_e(f) = j2\pi f \Phi_1(f) - 2\pi K_0 \Phi_e(f) H(f)$$
$$\Rightarrow \Phi_e(f) = \frac{jf\Phi_1(f)}{jf + K_0 H(f)}$$
$$\Rightarrow \Phi_e(f) = \frac{\Phi_1(f)}{1 + K_0 H(f)/(jf)}. \tag{4.264}$$

Now $\phi_1(t)$ is bandlimited to $[-W, \ W]$. It will be shown later that $\phi_2(t)$ is also bandlimited to $[-W, \ W]$. Therefore, both $\phi_e(t)$ and $h(t)$ are bandlimited to $[-W, \ W]$. If

$$\left| \frac{K_0 H(f)}{f} \right| \gg 1 \quad \text{for } |f| < W \tag{4.265}$$

then the last equation in (4.264) reduces to

$$\Phi_e(f) = \frac{jf\Phi_1(f)}{K_0 H(f)}. \tag{4.266}$$

Hence, the Fourier transform of (4.257), after applying (4.262), becomes

$$E(f) = A_c A_v k_m \Phi_e(f) \tag{4.267}$$

and

$$V(f) = E(f)H(f)$$
$$= A_c A_v k_m \mathrm{j}\, f\, \Phi_1(f)/K_0$$
$$= (\mathrm{j}\, f/k_v)\, \Phi_1(f). \tag{4.268}$$

The inverse Fourier transform of (4.268) gives

$$v(t) = \frac{1}{2\pi k_v} \frac{d\phi_1(t)}{dt}$$
$$= \frac{k_f}{k_v} m(t). \tag{4.269}$$

From (4.269), it is clear that $v(t)$ and $\phi_2(t)$ are bandlimited to $[-W,\ W]$.

References

Simon Haykin. *Communication Systems*. Wiley Eastern, second edition, 1983.

J. G. Proakis and M. Salehi. *Fundamentals of Communication Systems*. Pearson Education Inc., 2005.

Chapter 5
Noise in Analog Modulation

1. A DSB-SC signal of the form

$$S(t) = A_c M(t) \cos(2\pi f_c t + \Theta) \qquad (5.1)$$

is transmitted over a channel that adds additive Gaussian noise with psd shown in Fig. 5.1. The message spectrum extends over $[-4, 4]$ kHz and the carrier frequency is 200 kHz. Assuming that the average power of $S(t)$ is 10 W and coherent detection, determine the output SNR of the receiver.

Assume that the IF filter is ideal with unity gain in the passband and zero for other frequencies, and the narrowband representation of noise at the IF filter output is

$$N(t) = N_c(t) \cos(2\pi f_c t) - N_s(t) \sin(2\pi f_c t), \qquad (5.2)$$

where $N_c(t)$ and $N_s(t)$ are both independent of Θ.

- *Solution*: The received signal at the output of the IF filter is

$$\begin{aligned} X(t) &= A_c M(t) \cos(2\pi f_c t + \Theta) + N(t) \\ &= A_c M(t) \cos(2\pi f_c t + \Theta) + N_c(t) \cos(2\pi f_c t) - N_s(t) \sin(2\pi f_c t) \end{aligned}$$
$$, \qquad (5.3)$$

where $M(t)$ denotes the random process corresponding to the message, Θ is uniformly distributed in $[0, 2\pi]$ and $N(t)$ is a narrowband noise process. The psd of $N(t)$ is illustrated in Fig. 5.2. If the local oscillator (LO) at the receiver supplies $2 \cos(2\pi f_c t + \Theta)$, then the LPF output is

$$Y(t) = A_c M(t) + N_c(t) \cos(\Theta) + N_s(t) \sin(\Theta). \qquad (5.4)$$

K. Vasudevan, *Analog Communications*, https://doi.org/10.1007/978-3-030-50337-6_5

Fig. 5.1 Noise psd

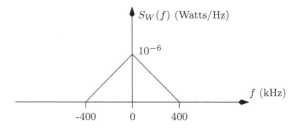

Fig. 5.2 Psd of narrowband noise at the IF filter output

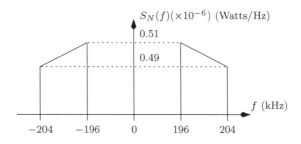

Hence the noise power at the LPF output is

$$E\left[N_c^2(t)\cos^2(\Theta)\right] + E\left[N_s^2(t)\sin^2(\Theta)\right] = E\left[N^2(t)\right]$$
$$= \int_{f=-\infty}^{\infty} S_N(f)\,df$$
$$= 8000 \times 10^{-6}\,\text{W}.$$

$$(5.5)$$

Note that

$$E\left[N_c(t)N_s(t)\right] = 0$$
$$E\left[\sin(\Theta)\cos(\Theta)\right] = 0 \qquad\qquad (5.6)$$

where we have used the fact that the cross spectral density $S_{N_cN_s}(f)$ is an odd function, therefore $R_{N_cN_s}(0) = 0$.

The power of the modulated message signal is

$$A_c^2 P/2 = 10\,\text{W}. \qquad\qquad (5.7)$$

Thus, the power of the demodulated message in (5.4) is $A_c^2 P = 20\,\text{W}$. Hence

$$\text{SNR}_O = 2.5 \times 10^3 \equiv 33.98\,\text{dB}. \qquad\qquad (5.8)$$

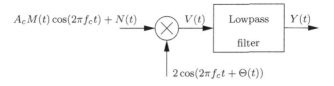

Fig. 5.3 DSB-SC demodulator having a phase error $\Theta(t)$

2. (Haykin 1983) In a DSB-SC receiver, the sinusoidal wave generated by the local oscillator suffers from a phase error $\Theta(t)$ with respect to the input carrier wave $\cos(2\pi f_c t)$, as illustrated in Fig. 5.3. Assuming that $\Theta(t)$ is a zero-mean Gaussian process of variance σ_Θ^2, and that most of the time $|\Theta(t)|$ is small compared to unity, find the mean squared error at the receiver output for DSB-SC modulation.

The mean squared error is defined as the expected value of the squared difference between the receiver output for $\Theta(t) \neq 0$ and the message signal component of the receiver output when $\Theta(t) = 0$.

Assume that the message $M(t)$, the carrier phase $\Theta(t)$, and noise $N(t)$ are statistically independent of each other. Also assume that $N(t)$ is a zero-mean narrowband noise process with psd $N_0/2$ extending over $f_c - W \leq |f| \leq f_c + W$ and having the representation:

$$N(t) = N_c(t) \cos(2\pi f_c t) - N_s(t) \sin(2\pi f_c t). \tag{5.9}$$

The psd of $M(t)$ extends over $[-W, W]$ with power P. The LPF is ideal with unity gain in $[-W, W]$.

- *Solution*: The output of the multiplier is

$$V(t) = 2 [A_c M(t) \cos(2\pi f_c t) + N(t)] \cos(2\pi f_c t + \Theta(t)), \tag{5.10}$$

where $N(t)$ is narrowband noise. The output of the lowpass filter is

$$Y(t) = A_c M(t) \cos(\Theta(t)) + N_c(t) \cos(\Theta(t)) + N_s(t) \sin(\Theta(t)). \tag{5.11}$$

When $\Theta(t) = 0$ the signal component is

$$Y_0(t) = A_c M(t). \tag{5.12}$$

Thus, the mean squared error is

$$\begin{aligned}
E \left[(Y_0(t) - Y(t))^2 \right] \\
= E [(A_c M(t)(1 - \cos(\Theta(t))) \\
+ N_c(t) \cos(\Theta(t)) + N_s(t) \sin(\Theta(t)))^2]
\end{aligned}$$

$$= E\left[A_c^2 M^2(t)(1 - \cos(\Theta(t)))^2\right] + E\left[N_c^2(t) \cos^2(\Theta(t))\right]$$
$$+ E\left[N_s^2(t) \sin^2(\Theta(t))\right]. \tag{5.13}$$

We now use the following relations (assuming $|\Theta(t)| \ll 1$ for all t):

$$(1 - \cos(\Theta(t)))^2 \approx \frac{\Theta^4(t)}{4}$$
$$E\left[N_c^2(t)\right] = 2N_0 W$$
$$E\left[N_s^2(t)\right] = 2N_0 W$$
$$E\left[M^2(t)\right] = P. \tag{5.14}$$

Thus

$$E\left[(Y_0(t) - Y(t))^2\right]$$
$$= \frac{A_c^2 P}{4} E\left[\Theta^4(t)\right] + 2N_0 W E\left[\cos^2(\Theta(t)) + \sin^2(\Theta(t))\right]$$
$$= \frac{3A_c^2 P \sigma_\Theta^4}{4} + 2N_0 W, \tag{5.15}$$

where we have used the fact that if X is a zero-mean Gaussian random variable with variance σ^2

$$E\left[X^{2n}\right] = 1 \times 3 \times \cdots \times (2n-1)\sigma^{2n}. \tag{5.16}$$

3. (Haykin 1983) Let the message $M(t)$ be transmitted using SSB modulation. The psd of $M(t)$ is

$$S_M(f) = \begin{cases} \frac{a|f|}{W} & \text{for } |f| < W \\ 0 & \text{elsewhere}, \end{cases} \tag{5.17}$$

where a and W are constants. Let the transmitted signal be of the form:

$$S(t) = A_c M(t) \cos(2\pi f_c t + \theta) + A_c \hat{M}(t) \sin(2\pi f_c t + \theta), \tag{5.18}$$

where θ is a uniformly distributed random variable in $[0, 2\pi)$ and independent of $M(t)$. White Gaussian noise of zero mean and psd $N_0/2$ is added to the SSB signal at the receiver input. Compute SNR_O assuming coherent detection and an appropriate unity gain IF filter in the receiver front-end.

- *Solution*: The average message power is

$$P = \int_{f=-W}^{W} S_M(f)\,df$$
$$= aW. \tag{5.19}$$

The received SSB signal at the output of the IF filter is given by

$$x(t) = A_c M(t)\cos(2\pi f_c t + \theta) + A_c \hat{M}(t)\sin(2\pi f_c t + \theta)$$
$$+N_c(t)\cos(2\pi f_c t) - N_s(t)\sin(2\pi f_c t). \tag{5.20}$$

Assuming coherent demodulation by $2\cos(2\pi f_c t + \theta)$, the LPF output is given by

$$Y(t) = A_c M(t) + N_c(t)\cos(\theta) + N_s(t)\sin(\theta). \tag{5.21}$$

The noise power is

$$E\left[(N_c(t)\cos(\theta) + N_s(t)\sin(\theta))^2\right] = E\left[N_c^2(t)\right] E\left[\cos^2(\theta)\right]$$
$$= +E\left[N_s^2(t)\right] E\left[\sin^2(\theta)\right]$$
$$= E\left[N^2(t)\right]$$
$$= N_0 W, \tag{5.22}$$

since

$$E\left[N_c^2(t)\right] = E\left[N_s^2(t)\right] = E\left[N^2(t)\right]$$
$$= N_0 W$$
$$E\left[\cos^2(\theta) + \sin^2(\theta)\right] = 1. \tag{5.23}$$

Therefore, the output SNR is

$$\text{SNR}_O = \frac{A_c^2 aW}{N_0 W} = \frac{aA_c^2}{N_0}. \tag{5.24}$$

4. An unmodulated carrier of amplitude A_c and frequency f_c and bandlimited white noise are summed and passed through an ideal envelope detector. Assume that the noise psd to be of height $N_0/2$ and bandwidth $2W$ centered at f_c. Determine the output SNR when the input carrier-to-noise ratio is high.

- *Solution*: The input to the envelope detector is

$$x(t) = A_c \cos(2\pi f_c t) + n(t)$$
$$= A_c \cos(2\pi f_c t)$$
$$+n_c(t)\cos(2\pi f_c t) - n_s(t)\sin(2\pi f_c t). \tag{5.25}$$

The output of the envelope detector is

$$y(t) = \sqrt{(A_c + n_c(t))^2 + n_s^2(t)}. \tag{5.26}$$

It is given that the carrier-to-noise ratio is high. Hence

$$y(t) \approx A_c + n_c(t). \tag{5.27}$$

The output signal power is A_c^2 and the output noise power is $2N_0 W$. Hence

$$\text{SNR}_O = \frac{A_c^2}{2N_0 W}. \tag{5.28}$$

5. (Haykin 1983) A frequency division multiplexing (FDM) system uses SSB modulation to combine 12 independent voice channels and then uses frequency modulation to transmit the composite signal. Each voice signal has an average power P and occupies the frequency band $[-4, 4]$ kHz. Only the lower sideband is transmitted. The modulated voice signals used for the first stage of modulation are defined by

$$S_k(t) = A_k M_k(t) \cos(2\pi k f_0 t + \theta) + A_k \hat{M}_k(t) \sin(2\pi k f_0 t + \theta), \tag{5.29}$$

for $1 \le k \le 12$, where $f_0 = 4$ kHz. Note that $E[M_k^2(t)] = P$. The received signal consists of the transmitted FM signal plus zero-mean white Gaussian noise of psd $N_0/2$.
Assume that the output of the FM receiver is given by

$$Y(t) = k_f S(t) + N_o(t), \tag{5.30}$$

where

$$S(t) = \sum_{k=1}^{12} S_k(t). \tag{5.31}$$

The psd of $N_o(t)$ is

$$S_{N_o}(f) = \begin{cases} N_0 f^2 / A_c^2 & \text{for } |f| < 48 \text{ kHz} \\ 0 & \text{otherwise}, \end{cases} \tag{5.32}$$

where A_c is the amplitude of the transmitted FM signal.
Find the relationship between the subcarrier amplitudes A_k so that the modulated voice signals have equal SNRs at the FM receiver output.

- *Solution*: The power in the kth voice signal is

$$E\left[S_k^2(t)\right] = A_k^2 P. \tag{5.33}$$

The overall signal at the output of the first stage of modulation is given by

$$S(t) = \sum_{k=1}^{12} S_k(t). \tag{5.34}$$

The bandwidth of $S(t)$ extends over $[-48, 48]$ kHz. The signal $S(t)$ is given as input to the second stage, which is a frequency modulator.
The output of the FM receiver is given by

$$Y(t) = k_f S(t) + N_o(t). \tag{5.35}$$

The psd of $N_o(t)$ is

$$S_{N_o}(f) = \begin{cases} N_0 f^2 / A_c^2 & \text{for } |f| < 48 \text{ kHz} \\ 0 & \text{otherwise}, \end{cases} \tag{5.36}$$

where A_c is the amplitude of the transmitted FM signal. The noise power in the kth received voice band is

$$\begin{aligned}
P_{N_k} &= 2 \int_{(k-1)B}^{kB} S_{N_o}(f)\, df \\
&= \frac{2N_0 B^3}{3A_c^2} \left[3k^2 - 3k + 1\right],
\end{aligned} \tag{5.37}$$

where $B = 4$ kHz. The power in the kth received voice signal is

$$P_k = k_f^2 A_k^2 P. \tag{5.38}$$

The output SNR for the kth SSB modulated voice signal is

$$\text{SNR}_{O,k} = \frac{3k_f^2 A_k^2 P A_c^2}{2N_0 B^3 (3k^2 - 3k + 1)}. \tag{5.39}$$

We require $\text{SNR}_{O,k}$ to be independent of k. Hence the required condition on A_k is

$$A_k^2 = C(3k^2 - 3k + 1), \tag{5.40}$$

where C is a constant.

Fig. 5.4 Multiplying a
random process with a sine
wave

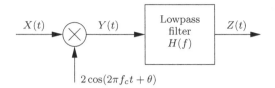

6. Consider the system shown in Fig. 5.4. Here, $H(f)$ is an ideal LPF with unity
   gain in the band $[-W, W]$. *Carefully* follow the two procedures outlined below:

   (a) Let

   $$X(t) = M(t)\cos(2\pi f_c t + \theta), \tag{5.41}$$

   where θ is a uniformly distributed random variable in $[0, 2\pi)$, $M(t)$ is a
   random process with psd $S_M(f)$ in the range $[-W, W]$. Assume that $M(t)$
   and θ are independent.
   Compute the autocorrelation and the psd of $Y(t)$. Hence compute the psd of
   $Z(t)$.

   (b) Let $X(t)$ be any random process with autocorrelation $R_X(\tau)$ and psd $S_X(f)$.
   Compute the psd of $Z(t)$ in terms of $S_X(f)$.
   Now assuming that $X(t)$ is given by (5.41), compute $S_X(f)$. Substitute this
   expression for $S_X(f)$ into the expression for the psd of $Z(t)$.

   (c) Explain the result for the psd of $Z(t)$ obtained using procedure (a) and procedure (b).

- *Solution*: In the case of procedure (a)

$$Y(t) = M(t)[1 + \cos(4\pi f_c t + 2\theta)]. \tag{5.42}$$

Therefore

$$R_Y(\tau) = E[Y(t)Y(t - \tau)]$$
$$= R_M(\tau)\left[1 + \frac{1}{2}\cos(4\pi f_c \tau)\right]. \tag{5.43}$$

Hence

$$S_Y(f) = S_M(f) + \frac{1}{4}[S_M(f - 2f_c) + S_M(f + 2f_c)]. \tag{5.44}$$

The psd of $Z(t)$ is

$$S_Z(f) = S_Y(f)|H(f)|^2$$
$$= S_M(f). \tag{5.45}$$

Using procedure (b) we have

$$Y(t) = 2X(t)\cos(2\pi f_c t + \theta). \tag{5.46}$$

Hence

$$R_Y(\tau) = 2R_X(\tau)\cos(2\pi f_c \tau)$$
$$\Rightarrow S_Y(f) = S_X(f - f_c) + S_X(f + f_c). \tag{5.47}$$

Therefore

$$S_Z(f) = S_Y(f)|H(f)|^2$$
$$= [S_X(f - f_c) + S_X(f + f_c)]|H(f)|^2. \tag{5.48}$$

Now, assuming that

$$X(t) = M(t)\cos(2\pi f_c t + \theta), \tag{5.49}$$

we get

$$R_X(\tau) = \frac{1}{2}R_M(\tau)\cos(2\pi f_c \tau)$$
$$\Rightarrow S_X(f) = \frac{1}{4}[S_M(f - f_c) + S_M(f + f_c)]. \tag{5.50}$$

Substituting the above value of $S_X(f)$ into (5.48) we get

$$S_Z(f) = \frac{1}{4}[S_M(f - 2f_c) + S_M(f) + S_M(f) + S_M(f + 2f_c)]|H(f)|^2$$
$$= \frac{1}{2}S_M(f). \tag{5.51}$$

The reason for the difference in the psd using the two procedures is that in the first case we are doing coherent demodulation.
However, in the second case coherent demodulation is not assumed. In fact, the local oscillator supplying any arbitrary phase α would have given the result in (5.51).

7. Consider the communication system shown in Fig. 5.5. The message is assumed to be a random process $X(t)$ given by

$$X(t) = \sum_{k=-\infty}^{\infty} S_k p(t - kT - \alpha), \tag{5.52}$$

Fig. 5.5 Block diagram of a
communication system

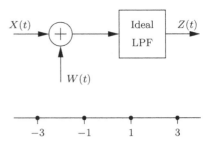

where $1/T$ denotes the symbol-rate, α is a random variable uniformly distributed
in $[0, T]$, and

$$p(t) = \text{sinc}\,(t/T). \tag{5.53}$$

The symbols S_k are drawn from a 4-ary constellation as indicated in Fig. 5.5.
The symbols are independent, that is

$$E[S_k S_{k+n}] = E[S_k]E[S_{k+n}] \quad \text{for } n \neq 0 \tag{5.54}$$

and equally likely, that is

$$P(-3) = P(-1) = P(1) = P(3). \tag{5.55}$$

Also assume that S_k and α are independent. The term $W(t)$ denotes an additive
white noise process with psd $N_0/2$. The LPF is ideal with unity gain in the
bandwidth $[-W, W]$. Assume that the LPF does not distort the message.

(a) Derive the expression for the autocorrelation and the psd of $X(t)$.
(b) Compute the signal-to-noise ratio at the LPF output.
(c) What should be the value of W so that the SNR at the LPF output is maximized
     without distorting the message?

- *Solution*: The autocorrelation of $X(t)$ is given by

$$
\begin{aligned}
R_X(\tau) &= E[X(t)X(t - \tau)] \\
&= E\left[\sum_{i=-\infty}^{\infty} S_i\, p(t - iT - \alpha) \sum_{j=-\infty}^{\infty} S_j\, p(t - \tau - jT - \alpha) \right] \\
&= \sum_{i=-\infty}^{\infty} \sum_{j=-\infty}^{\infty} E[S_i S_j]E[p(t - iT - \alpha)p(t - \tau - jT - \alpha)].
\end{aligned}
$$
$$\tag{5.56}$$

Now

$$E[S_i S_j] = 5\delta_K(i - j), \tag{5.57}$$

where

$$\delta_K(i - j) = \begin{cases} 1 \text{ for } i = j \\ 0 \text{ for } i \neq j \end{cases} \tag{5.58}$$

Similarly

$$E\ [p(t - iT - \alpha)p(t - \tau - jT - \alpha)]$$
$$= \frac{1}{T} \int_{\alpha=0}^{T} p(t - iT - \alpha)p(t - \tau - jT - \alpha)\, d\alpha. \tag{5.59}$$

Let

$$t - iT - \alpha = z \tag{5.60}$$

Thus

$$R_X(\tau) = \frac{5}{T} \sum_{i=-\infty}^{\infty} \sum_{j=-\infty}^{\infty} \delta_K(i - j)$$
$$\int_{z=t-iT-T}^{t-iT} p(z)p(z + iT - \tau - jT)\, dz$$
$$= \frac{5}{T} \sum_{i=-\infty}^{\infty} \int_{z=t-iT-T}^{t-iT} p(z)p(z - \tau)\, dz$$
$$= \frac{5}{T} \int_{z=-\infty}^{\infty} p(z)p(z - \tau)\, dz$$
$$= \frac{5}{T} R_{pp}(\tau), \tag{5.61}$$

where $R_{pp}(\tau)$ is the autocorrelation of $p(t)$. The power spectral density of $X(t)$ is given by

$$S_X(f) = \frac{5}{T}|P(f)|^2$$
$$= 5T \text{ rect}\, (fT)$$
$$\rightleftharpoons 5 \text{ sinc}\, (t/T). \tag{5.62}$$

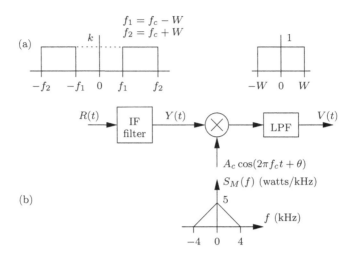

Fig. 5.6 Block diagram of a DSB-SC receiver

Therefore

$$R_X(\tau) = 5 \operatorname{sinc}(t/T). \tag{5.63}$$

The power in $X(t)$ is

$$R_X(0) = \int_{f=-\infty}^{\infty} S_X(f)\,df$$
$$= 5. \tag{5.64}$$

The noise power at the LPF output is

$$P_N = \frac{N_0}{2} \times 2W = N_0 W. \tag{5.65}$$

Therefore

$$\mathrm{SNR}_O = \frac{5}{N_0 W}. \tag{5.66}$$

Clearly, SNR_O is maximized when $W = 1/(2T)$.

8. Consider the coherent DSB-SC receiver shown in Fig. 5.6a. The signal $R(t)$ is given by

$$R(t) = S(t) + W(t)$$
$$= A_c M(t) \cos(2\pi f_c t + \theta) + W(t), \tag{5.67}$$

where $M(t)$ has the psd as shown in Fig. 5.6b, θ is a uniformly distributed random variable in $[0, 2\pi)$, and $W(t)$ is a zero-mean random process with psd

$$S_W(f) = af^2 \text{ W/kHz} \quad \text{for } -\infty < f < \infty. \tag{5.68}$$

The power of $S(t)$ is 160 W. The representation of narrowband noise is

$$N(t) = N_c(t) \cos(2\pi f_c t) - N_s(t) \sin(2\pi f_c t). \tag{5.69}$$

Assume that $M(t), \theta, N(t)$ are independent of each other.

(a) Compute A_c.
(b) Derive the expression for the signal power at the receiver output.
(c) Derive the expression for the noise power at the receiver output.
(d) Compute the SNR in dB at the receiver output, if $a = 10^{-3}$, $f_c = 8$ kHz, and $W = 4$ kHz.

• *Solution*: It is given that the power of $S(t)$ is 160 W. Thus

$$E\left[S^2(t)\right] = A_c^2 E\left[M^2(t)\right] E\left[\cos^2(2\pi f_c t + \theta)\right] = \frac{A_c^2 P}{2} = 160, \tag{5.70}$$

where

$$P = E\left[M^2(t)\right] = \int_{f=-4}^{4} S_M(f)\,df = 20 \text{ W}. \tag{5.71}$$

Therefore

$$A_c = 4. \tag{5.72}$$

The IF filter output is

$$Y(t) = kA_c M(t) \cos(2\pi f_c t + \theta) + N(t). \tag{5.73}$$

The psd of $N(t)$ is

$$S_N(f) = ak^2 f^2 \quad \text{for } f_c - W \le |f| \le f_c + W. \tag{5.74}$$

The receiver output is

$$V(t) = \frac{kA_c^2 M(t)}{2} + \frac{A_c}{2} N_c(t) \cos(\theta) + \frac{A_c}{2} N_s(t) \sin(\theta). \tag{5.75}$$

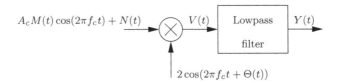

Fig. 5.7 DSB-SC demodulator having a phase error $\Theta(t)$

The signal power at the receiver output is

$$k^2 A_c^4 P/4 = 1280k^2. \tag{5.76}$$

The noise power at the receiver output is

$$\frac{A_c^2}{4} \left\{ E\left[N_c^2(t)\right] E\left[\cos^2(\theta)\right] + E\left[N_s^2(t)\right] E\left[\sin^2(\theta)\right] \right\} = \frac{A_c^2}{4} E\left[N^2(t)\right]. \tag{5.77}$$

Now

$$E\left[N^2(t)\right] = 2ak^2 \int_{f=f_c-W}^{f_c+W} f^2 \, df$$

$$= \frac{4ak^2}{3} \left[3f_c^2 W + W^3\right]. \tag{5.78}$$

Therefore, the noise power at the receiver output becomes

$$\frac{ak^2 A_c^2}{3} \left[3f_c^2 W + W^3\right] = 4.4373k^2. \tag{5.79}$$

The SNR at the receiver output in dB is

$$\text{SNR} = 10 \log(1280/4.4373) = 24.6 \, \text{dB}. \tag{5.80}$$

9. In a DSB-SC receiver, the sinusoidal wave generated by the local oscillator suffers from a phase error $\Theta(t)$ with respect to the input carrier wave $\cos(2\pi f_c t)$, as illustrated in Fig. 5.7. Assuming that $\Theta(t)$ is uniformly distributed in $[-\delta, \ \delta]$, where $\delta \ll 1$, find the mean squared error at the receiver output.

   The mean squared error is defined as the expected value of the squared difference between the receiver output for $\Theta(t) \neq 0$ and the message signal component of the receiver output when $\Theta(t) = 0$.

   Assume that the message $M(t)$, the carrier phase $\Theta(t)$, and noise $N(t)$ are statistically independent of each other. Also assume that $N(t)$ is a zero-mean narrowband noise process with psd $N_0/2$ extending over $f_c - W \leq |f| \leq f_c + W$ and having the representation:

$$N(t) = N_c(t)\cos(2\pi f_c t) - N_s(t)\sin(2\pi f_c t). \tag{5.81}$$

The psd of $M(t)$ extends over $[-W, W]$ with power P. The LPF is ideal with unity gain in $[-W, W]$.

- *Solution*: The output of the multiplier is

$$V(t) = 2[A_c M(t)\cos(2\pi f_c t) + N(t)]\cos(2\pi f_c t + \Theta(t)), \tag{5.82}$$

where $N(t)$ is narrowband noise. The output of the lowpass filter is

$$Y(t) = A_c M(t)\cos(\Theta(t)) + N_c(t)\cos(\Theta(t)) + N_s(t)\sin(\Theta(t)). \tag{5.83}$$

When $\Theta(t) = 0$ the signal component is

$$Y_0(t) = A_c M(t). \tag{5.84}$$

Thus, the mean squared error is

$$
\begin{aligned}
E &\left[(Y_0(t) - Y(t))^2\right] \\
&= E\left[(A_c M(t)(1 - \cos(\Theta(t))) \right. \\
&\quad \left. + N_c(t)\cos(\Theta(t)) + N_s(t)\sin(\Theta(t)))^2\right] \\
&= E\left[A_c^2 M^2(t)(1 - \cos(\Theta(t)))^2\right] + E\left[N_c^2(t)\cos^2(\Theta(t))\right] \\
&\quad + E\left[N_s^2(t)\sin^2(\Theta(t))\right].
\end{aligned} \tag{5.85}
$$

We now use the following relations (assuming $|\Theta(t)| \ll 1$ for all t):

$$
\begin{aligned}
(1 - \cos(\Theta(t)))^2 &\approx \frac{\Theta^4(t)}{4} \\
E\left[N_c^2(t)\right] &= 2N_0 W \\
E\left[N_s^2(t)\right] &= 2N_0 W \\
E\left[M^2(t)\right] &= P.
\end{aligned} \tag{5.86}
$$

Thus

$$
\begin{aligned}
E &\left[(Y_0(t) - Y(t))^2\right] \\
&= \frac{A_c^2 P}{4} E\left[\Theta^4(t)\right] + 2N_0 W E\left[\cos^2(\Theta(t)) + \sin^2(\Theta(t))\right] \\
&= \frac{A_c^2 P \delta^4}{20} + 2N_0 W,
\end{aligned} \tag{5.87}
$$

where we have used the fact that

$$E\left[\Theta^4(t)\right] = \frac{1}{2\delta}\int_{\alpha=-\delta}^{\delta} \alpha^4 \, d\alpha$$

$$= \frac{\delta^4}{5}. \tag{5.88}$$

10. (Haykin 1983) Consider a phase modulation (PM) system, with the received signal at the output of the IF filter given by

$$x(t) = A_c \cos(2\pi f_c t + k_p m(t)) + n(t), \tag{5.89}$$

where

$$n(t) = n_c(t) \cos(2\pi f_c t) - n_s(t) \sin(2\pi f_c t). \tag{5.90}$$

Assume that the carrier-to-noise ratio of $x(t)$ to be high, the message power to be P, the message bandwidth to extend over $[-W, W]$, and the transmission bandwidth of the PM signal to be B_T. The psd of $n(t)$ is equal to $N_0/2$ for $f_c - B_T/2 \le |f| \le f_c + B_T/2$ and zero elsewhere.

(a) Find the output SNR. Show all the steps.
(b) Determine the figure-of-merit of the system.
(c) If the PM system uses a pair of pre-emphasis and de-emphasis filters defined by

$$H_{\text{pe}}(f) = 1 + \text{j}\, f/f_0 = 1/H_{\text{de}}(f) \tag{5.91}$$

determine the improvement in the output SNR.

• *Solution*: Let $\phi(t) = k_p m(t)$ and

$$n(t) = r(t) \cos(2\pi f_c t + \psi(t)). \tag{5.92}$$

Then the received signal $x(t)$ can be written as (see Fig. 5.8):

$$x(t) = a(t) \cos(2\pi f_c t + \theta(t)), \tag{5.93}$$

Fig. 5.8 Phasor diagram for $x(t)$

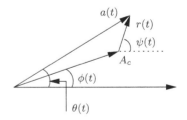

where

$$\theta(t) = \phi(t) + \tan^{-1}\left(\frac{r(t)\sin(\psi(t) - \phi(t))}{A_c + r(t)\cos(\psi(t) - \phi(t))}\right). \qquad (5.94)$$

The phase detector consists of the cascade of a hard limiter, bandpass filter, frequency discriminator, and an integrator. The hard limiter and bandpass filter remove envelope variations in $x(t)$ to yield

$$x_1(t) = \cos(2\pi f_c t + \theta(t)). \qquad (5.95)$$

The frequency discriminator (cascade of differentiator, envelope Detector, and dc blocking capacitor) output is

$$x_2(t) = d\theta(t)/dt, \qquad (5.96)$$

assuming that

$$2\pi f_c + d\theta(t)/dt > 0, \qquad (5.97)$$

for all t. The output of the integrator is

$$x_3(t) = \theta(t)$$
$$\approx \phi(t) + \frac{r(t)\sin(\psi(t) - \phi(t))}{A_c}$$
$$= \phi(t) + n'_s(t)/A_c, \qquad (5.98)$$

where we have assumed that $A_c \gg r(t)$ (high carrier-to-noise ratio). It is reasonable to assume that the psd of $n'_s(t)$ to be identical to $n_s(t)$ and is equal to

$$S_{N'_s}(f) = \begin{cases} N_0 & \text{for } f_c - B_T/2 < |f| < f_c + B_T/2 \\ 0 & \text{elsewhere.} \end{cases} \qquad (5.99)$$

The noise power at the output of the postdetection (baseband) lowpass filter is $2N_0 W/A_c^2$. The output SNR is

$$\text{SNR}_O = \frac{k_p^2 P A_c^2}{2N_0 W}. \qquad (5.100)$$

The average power of the PM signal is $A_c^2/2$ and the average noise power in the message bandwidth is $N_0 W$. Therefore, the channel signal-to-noise ratio is

$$\text{SNR}_C = \frac{A_c^2}{2N_0 W}. \qquad (5.101)$$

Therefore, the figure-of-merit of the receiver is

$$\frac{\text{SNR}_O}{\text{SNR}_C} = k_p^2 P. \tag{5.102}$$

When pre-emphasis and de-emphasis is used, the noise psd at the output of the de-emphasis filter is

$$S_{N_o}(f) = \frac{N_0}{A_c^2(1 + (f/f_0)^2)} \qquad |f| < W. \tag{5.103}$$

Therefore, the noise power at the output of the de-emphasis filter is

$$\int_{f=-W}^{W} S_{N_o}(f)\,df = 2N_0 f_0 \tan^{-1}(W/f_0)/A_c^2. \tag{5.104}$$

Therefore, the improvement in the output SNR is

$$D = \frac{W/f_0}{\tan^{-1}(W/f_0)}. \tag{5.105}$$

11. (Haykin 1983) Suppose that the transfer functions of the pre-emphasis and de-emphasis filters of an FM system are scaled as follows:

$$H_{pe}(f) = k\left(1 + \frac{j\,f}{f_0}\right)$$

$$H_{de}(f) = \frac{1}{k}\left(\frac{1}{1 + j\,f/f_0}\right). \tag{5.106}$$

The scaling factor k is chosen so that the average power of the emphasized signal is the same as the original message $M(t)$.

(a) Find the value of k that satisfies this requirement for the case when the psd of the message is

$$S_M(f) = \begin{cases} 1/\left(1 + (f/f_0)^2\right) & \text{for } -W \le f \le W \\ 0 & \text{otherwise.} \end{cases} \tag{5.107}$$

(b) What is the corresponding value of the improvement factor obtained by using this pair of pre-emphasis and de-emphasis filters.

• *Solution*: The message power is

$$P = \int_{f=-W}^{W} S_M(f)\,df$$

$$= 2f_0 \tan^{-1}\left(\frac{W}{f_0}\right). \tag{5.108}$$

The message psd at the output of the pre-emphasis filter is

$$S'_M(f) = \begin{cases} \frac{k^2(1+(f/f_0)^2)}{1+(f/f_0)^2} & \text{for } -W \le f \le W \\ 0 & \text{otherwise} \end{cases}$$

$$= \begin{cases} k^2 & \text{for } -W \le f \le W \\ 0 & \text{otherwise.} \end{cases} \tag{5.109}$$

The message power at the output of the pre-emphasis filter is

$$P' = \int_{f=-W}^{W} S'_M(f)\,df = k^2 2W. \tag{5.110}$$

Solving for k we get

$$k = \left[\frac{f_0}{W} \tan^{-1}\left(\frac{W}{f_0}\right)\right]^{1/2}. \tag{5.111}$$

The improvement factor due to this pair of pre-emphasis and de-emphasis combination is

$$D = \frac{k^2(W/f_0)^3}{3[(W/f_0) - \tan^{-1}(W/f_0)]}$$

$$= \frac{(W/f_0)^2 \tan^{-1}(W/f_0)}{3[(W/f_0) - \tan^{-1}(W/f_0)]}. \tag{5.112}$$

12. Consider an AM receiver using a square-law detector whose output is equal to the square of the input as indicated in Fig. 5.9. The AM signal is defined by

$$s(t) = A_c[1 + \mu \cos(2\pi f_m t)] \cos(2\pi f_c t) \tag{5.113}$$

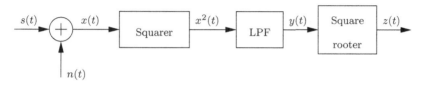

Fig. 5.9 Square-law detector in the presence of noise

and $n(t)$ is narrowband noise given by

$$n(t) = n_c(t) \cos(2\pi f_c t) - n_s(t) \sin(2\pi f_c t). \tag{5.114}$$

Assume that

(a) The bandwidth of the LPF is large enough so as to reject only those components centered at $2f_c$.
(b) The channel signal-to-noise ratio at the receiver input is high.
(c) The psd of $n(t)$ is flat with a height of $N_0/2$ in the range $f_c - W \leq |f| \leq f_c + W$.
Compute the output SNR.

- *Solution*: The squarer output is

$$
\begin{aligned}
x^2(t) &= s^2(t) + n^2(t) + 2s(t)n(t) \\
&= A_c^2[1 + \mu \cos(2\pi f_m t)]^2 \frac{(1 + \cos(4\pi f_c t))}{2} \\
&\quad + n_c^2(t) \frac{(1 + \cos(4\pi f_c t))}{2} + n_s^2(t) \frac{(1 - \cos(4\pi f_c t))}{2} \\
&\quad - 2n_c(t)n_s(t) \cos(2\pi f_c t) \sin(2\pi f_c t) \\
&\quad + 2A_c n_c(t)[1 + \mu \cos(2\pi f_m t)] \frac{(1 + \cos(4\pi f_c t))}{2} \\
&\quad - 2A_c n_s(t)[1 + \mu \cos(2\pi f_m t)] \cos(2\pi f_c t) \sin(2\pi f_c t).
\end{aligned}
\tag{5.115}
$$

The output of the lowpass filter is

$$
\begin{aligned}
y(t) &= \frac{A_c^2}{2}[1 + \mu \cos(2\pi f_m t)]^2 + \frac{n_c^2(t)}{2} + \frac{n_s^2(t)}{2} \\
&\quad + A_c n_c(t)[1 + \mu \cos(2\pi f_m t)] \\
&\approx \frac{A_c^2}{2}[1 + \mu \cos(2\pi f_m t)]^2 + A_c n_c(t)[1 + \mu \cos(2\pi f_m t)], \tag{5.116}
\end{aligned}
$$

where we have made the high channel SNR approximation. The output of the square rooter is

$$
\begin{aligned}
z(t) &= \sqrt{\frac{A_c^2}{2}[1 + \mu \cos(2\pi f_m t)]^2 + A_c n_c(t)[1 + \mu \cos(2\pi f_m t)]} \\
&= \frac{A_c}{\sqrt{2}}[1 + \mu \cos(2\pi f_m t)] \sqrt{1 + \frac{2n_c(t)}{A_c(1 + \mu \cos(2\pi f_m t))}} \\
&\approx \frac{A_c}{\sqrt{2}}[1 + \mu \cos(2\pi f_m t)] + \frac{n_c(t)}{\sqrt{2}}. \tag{5.117}
\end{aligned}
$$

The signal power is

$$P = \frac{A_c^2 \mu^2}{4}. \tag{5.118}$$

The noise power is

$$P_N = \frac{2N_0 W}{2}. \tag{5.119}$$

The output SNR is

$$\text{SNR}_O = \frac{A_c^2 \mu^2}{4N_0 W}. \tag{5.120}$$

13. (Haykin 1983) Consider the random process

$$X(t) = A + W(t), \tag{5.121}$$

where A is a constant and $W(t)$ is a zero-mean WSS random process with psd $N_0/2$. The signal $X(t)$ is passed through a first-order RC-lowpass filter. Find the expression for the output SNR, with the dc component at the LPF output regarded as the signal of interest.

- *Solution*: The signal component at the output of the lowpass filter is A. Assuming that $1/f_0 = 2\pi RC$, the noise power at the LPF output is

$$P_N = \frac{N_0}{2} \int_{f=-\infty}^{\infty} \frac{1}{1+(f/f_0)^2} \, df = \frac{N_0}{4RC}. \tag{5.122}$$

Therefore the output SNR is:

$$\text{SNR}_O = \frac{4RCA^2}{N_0}. \tag{5.123}$$

14. Consider the modified receiver for detecting DSB-SC signals, as illustrated in Fig. 5.10. Note that in this receiver configuration, the IF filter is absent, hence $w(t)$ is zero-mean AWGN with psd $N_0/2$. The DSB-SC signal $s(t)$ is given by

Fig. 5.10 Modified receiver for DSB-SC signals

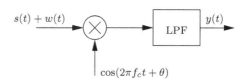

$$s(t) = A_c m(t) \cos(2\pi f_c t + \theta), \qquad (5.124)$$

where θ is a uniformly distributed random variable in $[0, 2\pi]$. The message $m(t)$ extends over the band $[-W, W]$ and has a power P. Assume the LPF to be ideal, with unity gain over $[-W, W]$. Assume that θ and $w(t)$ are statistically independent.

Compute the figure-of-merit of this receiver. Does this mean that the IF filter is redundant? Justify your answer.

- *Solution*: The noise autocorrelation at the output of the multiplier is

$$E[w(t)w(t - \tau) \cos(2\pi f_c t + \theta) \cos(2\pi f_c (t - \tau) + \theta)]$$
$$= \frac{N_0}{2}\delta(\tau)\frac{\cos(2\pi f_c \tau)}{2}$$
$$= \frac{N_0}{4}\delta(\tau). \qquad (5.125)$$

The average noise power at the LPF output is

$$P_N = \frac{N_0}{4}2W = \frac{N_0 W}{2}. \qquad (5.126)$$

The signal component at the LPF output is

$$s_1(t) = A_c m(t)/2. \qquad (5.127)$$

The average signal power at the LPF output is

$$E\left[s_1^2(t)\right] = A_c^2 E\left[m^2(t)\right]/4 = A_c^2 P/4. \qquad (5.128)$$

Therefore, SNR at the LPF output is

$$\text{SNR}_O = \frac{A_c^2 P}{2 N_0 W}. \qquad (5.129)$$

The average power of the modulated signal is

$$E\left[s^2(t)\right] = \frac{A_c^2 P}{2}. \qquad (5.130)$$

The average noise power in the message bandwidth for *baseband transmission* is

$$P_{N_1} = \frac{N_0}{2}(2W) = N_0 W. \qquad (5.131)$$

Thus, the channel SNR is

$$\text{SNR}_C = \frac{A_c^2 P}{2 N_0 W}. \tag{5.132}$$

Hence the figure-of-merit of the receiver is 1.

The above result does not imply that the IF filter is redundant. In fact, the main purpose of the IF filter in a superheterodyne receiver is to reject undesired *adjacent* stations. If we had used a bandpass filter in the RF section for this purpose, the Q-factor of the BPF would have to be very high, since the adjacent stations are spaced very closely. In fact, the required Q-factor of the RF BPF to reject adjacent stations would have been

$$Q \approx \frac{f_{\text{RF, max}}(= 1605\,\text{kHz})}{\text{bandwidth of message }(=\ 10\,\text{kHz})}$$
$$= 160.5, \tag{5.133}$$

which is very high.

However, the Q-factor requirement of the IF BPF is

$$Q \approx \frac{f_{\text{IF}}(=\ 455\ \text{kHz})}{\text{bandwidth of message }(=\ 10\,\text{kHz})}$$
$$= 45.5, \tag{5.134}$$

which is reasonable.

Note that the IF filter *cannot* reject *image* stations. The image stations are rejected by the RF bandpass filter. The frequency spacing between the image stations is $2 f_{\text{IF}}$, which is quite large, hence the Q-factor requirement of the RF bandpass filter is not very high.

15. Consider an FM demodulator in the presence of noise. The input signal to the demodulator is given by

$$X(t) = A_c \cos\left(2\pi f_c t + 2\pi k_f \int_{\tau=0}^{t} M(\tau)\,d\tau\right) + N(t) \tag{5.135}$$

where $N(t)$ denotes narrowband noise process with psd as illustrated in Fig. 5.11, and B_T is the bandwidth of the FM signal. The psd of the message is

$$S_M(f) = \begin{cases} af^2 & \text{for } |f| < W \\ 0 & \text{otherwise.} \end{cases} \tag{5.136}$$

(a) Write down the expression for the signal at the output of the FM discriminator (cascade of a differentiator, envelope detector, dc blocking capacitor, and a gain of $1/(2\pi)$). No derivation is required.

(b) Compute the SNR at the output of the FM demodulator.

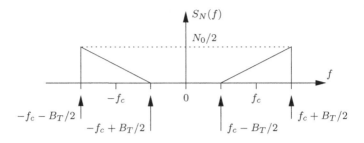

Fig. 5.11 PSD of $N(t)$

- *Solution*: The output of the FM discriminator is

$$V(t) = k_f M(t) + N_d(t)$$

$$= k_f M(t) + \frac{1}{2\pi A_c} \frac{dN_s'(t)}{dt} \tag{5.137}$$

where $N_s'(t)$ has the same statistical properties as $N_s(t)$. Here $N_s(t)$ denotes the quadrature component of $N(t)$, as given by

$$N(t) = N_c(t) \cos(2\pi f_c t) - N_s(t) \sin(2\pi f_c t). \tag{5.138}$$

Thus the psd of $N_s'(t)$ is identical to that of $N_s(t)$ and is given by

$$S_{N_s'}(f) = \begin{cases} S_N(f - f_c) + S_N(f + f_c) & \text{for } |f| < B_T/2 \\ 0 & \text{otherwise} \end{cases}$$

$$= \begin{cases} N_0/2 & \text{for } |f| < B_T/2 \\ 0 & \text{otherwise.} \end{cases} \tag{5.139}$$

The noise psd at the demodulator output is

$$S_{N_o}(f) = \begin{cases} \dfrac{f^2}{A_c^2} S_{N_s'}(f) & \text{for } |f| < W \\ 0 & \text{otherwise} \end{cases}$$

$$= \begin{cases} \dfrac{f^2 N_0}{2 A_c^2} & \text{for } |f| < W \\ 0 & \text{otherwise.} \end{cases} \tag{5.140}$$

The noise power at the demodulator output is

$$\int_{f=-W}^{W} S_{N_o}(f) \, df = \frac{N_0 W^3}{3 A_c^2}. \tag{5.141}$$

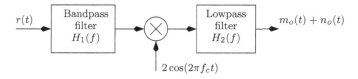

Fig. 5.12 Demodulation of an AM signal using nonideal filters

The signal power at the demodulator output is

$$\int_{f=-W}^{W} S_M(f)\,df = \frac{2k_f^2 a W^3}{3}. \tag{5.142}$$

Therefore, the output SNR is

$$\text{SNR}_O = \frac{2a A_c^2 k_f^2}{N_0}. \tag{5.143}$$

16. (Haykin 1983) Consider the AM receiver shown in Fig. 5.12 where

$$r(t) = [1 + m(t)] \cos(2\pi f_c t) + w(t), \tag{5.144}$$

where $w(t)$ has zero mean and power spectral density $N_0/2$ and $m(t)$ is WSS with psd $S_M(f)$. Note that $H_1(f)$ and $H_2(f)$ in Fig. 5.12 are nonideal filters, $h_1(t)$ and $h_2(t)$ are real-valued, and

$$h_1(t) = 2\Re \left\{ \tilde{h}_1(t)\, e^{j 2\pi f_c t} \right\}. \tag{5.145}$$

It is also given that $H_1(f)$ has conjugate symmetry about f_c, that is

$$H_1(f_c + f) = H_1^*(f_c - f) \quad \text{for } 0 \le f \le B. \tag{5.146}$$

The two-sided bandwidth of $H_1(f)$ about f_c is $2B$. The one-sided bandwidth of $H_2(f)$ about zero frequency is B. The one-sided bandwidth of $S_M(f)$ may be greater than B. Assume that $m(t)$ and $m_o(t)$ do not contain any dc component and any other dc component at the receiver output is removed by a capacitor. As a measure of distortion introduced by $H_1(f)$, $H_2(f)$ and noise, we define the mean squared error as

$$\mathscr{E} = E \left[(m_o(t) + n_o(t) - m(t))^2 \right]. \tag{5.147}$$

Note that $m_o(t)$, $n_o(t)$ and $m(t)$ are real-valued. All symbols have their usual meaning.

(a) Show that

$$\mathscr{E} = \int_{f=-\infty}^{\infty} \left[|1 - H(f)|^2 \, S_M(f) + N_0 \, |H(f)|^2 \right] df$$

, (5.148)

where

$$H(f) = \tilde{H}_1(f) H_2(f).$$ (5.149)

(b) Given that

$$S_M(f) = \frac{S_0}{1 + (f/f_0)^2} \qquad |f| < \infty, \, f_c \gg f_0$$ (5.150)

and

$$H(f) = \begin{cases} 1 \text{ for } |f| \le B, \, f_c \gg B \\ 0 \text{ otherwise} \end{cases}$$ (5.151)

find B such that $\mathscr{E}$ is minimized. Ignore the effects of aliasing due to the $2f_c$ component during demodulation.

• *Solution*: Given that

(a)

$$h_1(t) = 2\Re \left\{ \tilde{h}_1(t) \, e^{j 2\pi f_c t} \right\}.$$ (5.152)

(b)

$$H_1(f_c + f) = H_1^*(f_c - f) \qquad \text{for } 0 \le f \le B.$$ (5.153)

(c)

$$S_M(f) = \frac{S_0}{1 + (f/f_0)^2} \qquad |f| < \infty, \, f_c \gg f_0.$$ (5.154)

(d)

$$H(f) = \begin{cases} 1 \text{ for } |f| \le B, \, f_c \gg B \\ 0 \text{ otherwise.} \end{cases}$$ (5.155)

The Fourier transform of (5.152) is

$$H_1(f) = \tilde{H}_1(f - f_c) + \tilde{H}_1^*(-f - f_c),$$ (5.156)

where $\tilde{H}_1(f)$ denotes the Fourier transform of the complex envelope, $\tilde{h}_1(t)$. Substituting (5.156) in (5.153) we obtain

$$\tilde{H}_1(f) = \tilde{H}_1^*(-f) \quad \text{for } 0 \leq f \leq B$$
$$\Rightarrow \tilde{h}_1(t) = h_{1,c}(t) \tag{5.157}$$

that is, $\tilde{h}_1(t)$ is real-valued. The signal component at the bandpass filter output is

$$y(t) = \Re\left\{\tilde{y}(t)\, e^{j2\pi f_c t}\right\}, \tag{5.158}$$

where

$$\begin{aligned}
\tilde{y}(t) &= (1 + m(t)) \star \tilde{h}_1(t) \\
&= (1 + m(t)) \star h_{1,c}(t) \\
&= y_c(t), \tag{5.159}
\end{aligned}$$

where "$\star$" denotes convolution. Substituting (5.159) in (5.158) we get

$$y(t) = y_c(t) \cos(2\pi f_c t). \tag{5.160}$$

The signal component at the lowpass filter output is

$$\begin{aligned}
y_c(t) \star h_2(t) &= (1 + m(t)) \star h_{1,c}(t) \star h_2(t) \\
&= (1 + m(t)) \star h(t), \tag{5.161}
\end{aligned}$$

where

$$h(t) = h_{1,c}(t) \star h_2(t)$$
$$\Rightarrow H(f) = \tilde{H}_1(f) H_2(f). \tag{5.162}$$

The dc term $(1 \star h(t))$ in (5.161) is removed by a capacitor. Therefore, the message component at the lowpass filter output is

$$m_o(t) = m(t) \star h(t). \tag{5.163}$$

Therefore, the message psd at the lowpass filter output is

$$S_{M_o}(f) = S_M(f)|H(f)|^2. \tag{5.164}$$

Let us now turn our attention to the noise component. The noise psd at the bandpass filter output is

$$S_N(f) = \frac{N_0}{2}|H_1(f)|^2. \tag{5.165}$$

The narrowband noise at the bandpass filter output is given by

$$N(t) = N_c(t)\cos(2\pi f_c t) - N_s(t)\sin(2\pi f_c t). \tag{5.166}$$

The noise component at the output of the lowpass filter is

$$n_o(t) = N_c(t) \star h_2(t). \tag{5.167}$$

The noise psd at the lowpass filter output is

$$S_{N_o}(f) = S_{N_c}(f)|H_2(f)|^2, \tag{5.168}$$

where $S_{N_c}(f)$ is the psd of $N_c(t)$ and is given by

$$
\begin{aligned}
S_{N_c}(f) &= \begin{cases} S_N(f - f_c) + S_N(f + f_c) & \text{for } |f| < B \\ 0 & \text{otherwise} \end{cases} \\
&= \begin{cases} (N_0/2)\left[|H_1(f - f_c)|^2 + |H_1(f + f_c)|^2\right] & \text{for } |f| < B \\ 0 & \text{otherwise} \end{cases} \\
&= \begin{cases} (N_0/2)\left[\left|\tilde{H}_1^*(-f)\right|^2 + \left|\tilde{H}_1(f)\right|^2\right] & \text{for } |f| < B \\ 0 & \text{otherwise} \end{cases} \\
&= \begin{cases} N_0\left|\tilde{H}_1(f)\right|^2 & \text{for } |f| < B \\ 0 & \text{otherwise,} \end{cases}
\end{aligned}
\tag{5.169}
$$

where we have used (5.165), (5.156), and (5.157). Therefore (5.168) becomes

$$
\begin{aligned}
S_{N_o}(f) &= N_0|H_2(f)|^2\left|\tilde{H}_1(f)\right|^2 \\
&= N_0|H(f)|^2, \tag{5.170}
\end{aligned}
$$

where we have used (5.162).
Now

$$
\begin{aligned}
\mathcal{E} &= E\left[(m_o(t) + n_o(t) - m(t))^2\right] \\
&= E\left[m_o^2(t) + n_o^2(t) + m^2(t) - 2m(t)m_o(t)\right], \tag{5.171}
\end{aligned}
$$

where we have assumed that the noise and message are independent, and that the noise has zero mean. Simplifying (5.171) we obtain

$$\mathscr{E} = \int_{f=-\infty}^{\infty} \left[S_{M_o}(f) + S_{N_o}(f) + S_M(f) \right.$$
$$\left. - S_M(f)H(f) - S_M(f)H^*(f) \right] df$$
$$= \int_{f=-\infty}^{\infty} \left[|1 - H(f)|^2 \, S_M(f) + N_0 \, |H(f)|^2 \right] df, \qquad (5.172)$$

where we have used the following relations

$$E\left[m_o^2(t) \, dt \right] = \int_{f=-\infty}^{\infty} S_{M_o}(f) \, df$$

$$E\left[n_o^2(t) \, dt \right] = \int_{f=-\infty}^{\infty} S_{N_o}(f) \, df$$

$$E\left[m^2(t) \, dt \right] = \int_{f=-\infty}^{\infty} S_M(f) \, df$$

$$E\left[m(t)m_o(t) \, dt \right] = E\left[m(t) \int_{\alpha=-\infty}^{\infty} h(\alpha)m(t-\alpha) \, d\alpha \right]$$
$$= \int_{\alpha=-\infty}^{\infty} h(\alpha) E\left[m(t) \, m(t-\alpha) \right] d\alpha$$
$$= \int_{\alpha=-\infty}^{\infty} h(\alpha) R_M(\alpha) \, d\alpha$$
$$= \int_{f=-\infty}^{\infty} H(f) S_M(f) \, df$$
$$= \int_{f=-\infty}^{\infty} H^*(f) S_M(f) \, df. \qquad (5.173)$$

The last two equations in (5.173) are obtained as follows

$$\int_{\alpha=-\infty}^{\infty} h(\alpha) R_M(\alpha) \, d\alpha = h(-t) \star R_M(t)|_{t=0}$$
$$= h(t) \star R_M(-t)|_{t=0}, \qquad (5.174)$$

which is also equal to the inverse Fourier transform evaluated at $t = 0$.
The second part may be solved by substituting $S_M(f)$ in (5.154) and $H(f)$ in (5.155) into (5.172). Therefore (5.172) becomes

$$\mathscr{E} = 2 \int_{f=B}^{\infty} \frac{S_0}{1 + (f/f_0)^2} \, df + 2N_0 B$$
$$= 2S_0 f_0 \left[\pi/2 - \tan^{-1}(B/f_0) \right] + 2N_0 B. \qquad (5.175)$$

In order to minimize $\mathscr{E}$, we differentiate with respect to B and set the result to zero. Thus we obtain

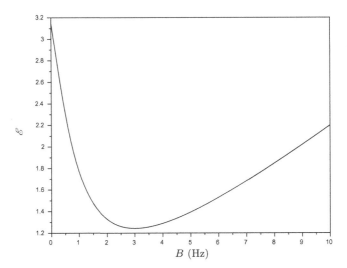

Fig. 5.13 Plot of $\mathscr{E}$ versus B

Fig. 5.14 Block diagram of
a system

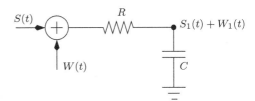

$$\frac{d\mathscr{E}}{dB} = -2S_0 \frac{1}{1 + (B/f_0)^2} + 2N_0$$
$$= 0$$
$$\Rightarrow B = f_0 \sqrt{\frac{S_0}{N_0} - 1}. \qquad (5.176)$$

The plot of $\mathscr{E}$ versus B is shown in Fig. 5.13, for $S_0 = 1.0\,\text{W/Hz}$, $N_0 = 0.1\,\text{W/Hz}$, and $f_0 = 1.0\,\text{Hz}$.

17. Consider the block diagram in Fig. 5.14. Here $S(t)$ is a narrowband FM signal given by

$$S(t) = A_c \cos(2\pi f_c t + \theta) - \beta A_c \sin(2\pi f_c t + \theta) \sin(2\pi f_m t + \alpha), \quad (5.177)$$

where θ, α are independent random variables, uniformly distributed in $[0, 2\pi)$. The psd of $W(t)$ is $N_0/2$. Compute $E\left[S_1^2(t)\right]/E\left[W_1^2(t)\right]$.

- *Solution*: Now $S(t)$ can be written as

$$S(t) = A_c \cos(2\pi f_c t + \theta)$$
$$-\frac{\beta A_c}{2} \cos(2\pi f_1 t + \theta - \alpha)$$
$$+\frac{\beta A_c}{2} \cos(2\pi f_2 t + \theta + \alpha), \qquad (5.178)$$

where

$$f_1 = f_c - f_m$$
$$f_2 = f_c + f_m. \qquad (5.179)$$

The autocorrelation of $S(t)$ is

$$R_S(\tau) = E\left[S(t)S(t - \tau)\right]$$
$$= \frac{A_c^2}{2} \cos(2\pi f_c \tau)$$
$$+\frac{\beta^2 A_c^2}{8} \cos(2\pi f_1 \tau)$$
$$+\frac{\beta^2 A_c^2}{8} \cos(2\pi f_2 \tau). \qquad (5.180)$$

The psd of $S(t)$ is the Fourier transform of $R_S(\tau)$ in (5.180), and is given by

$$S_S(f) = \frac{A_c^2}{4} \left[\delta(f - f_c) + \delta(f + f_c)\right]$$
$$+\frac{\beta^2 A_c^2}{16} \left[\delta(f - f_1) + \delta(f + f_1)\right]$$
$$+\frac{\beta^2 A_c^2}{16} \left[\delta(f - f_2) + \delta(f + f_2)\right]. \qquad (5.181)$$

The psd of $S_1(t)$ is

$$S_{S_1}(f) = \frac{A_c^2}{4} |H(f_c)|^2 \left[\delta(f - f_c) + \delta(f + f_c)\right]$$
$$+\frac{\beta^2 A_c^2}{16} |H(f_1)|^2 \left[\delta(f - f_1) + \delta(f + f_1)\right]$$
$$+\frac{\beta^2 A_c^2}{16} |H(f_2)|^2 \left[\delta(f - f_2) + \delta(f + f_2)\right], \qquad (5.182)$$

where

$$|H(f)|^2 = \frac{1}{1 + (2\pi f RC)^2}$$
$$\Rightarrow |H(f)|^2 = |H(-f)|^2. \qquad (5.183)$$

The power of $S_1(t)$ is

$$
\begin{aligned}
E\left[S_1^2(t)\right] &= \int_{f=-\infty}^{\infty} S_{S_1}(f)\,df \\
&= \frac{A_c^2}{2}\,|H(f_c)|^2 + \frac{\beta^2 A_c^2}{8}\,|H(f_1)|^2 + \frac{\beta^2 A_c^2}{8}\,|H(f_2)|^2\,.
\end{aligned}
$$

(5.184)

Finally, the power of $W_1(t)$ is

$$
\begin{aligned}
E\left[W_1^2(t)\right] &= \frac{N_0}{2} \int_{f=-\infty}^{\infty} \frac{1}{1 + (2\pi f RC)^2}\,df \\
&= \frac{N_0}{4\pi RC} \int_{x=-\infty}^{\infty} \frac{1}{1 + x^2}\,dx \\
&= \frac{N_0}{2\pi RC} \int_{x=0}^{\infty} \frac{1}{1 + x^2}\,dx \\
&= \frac{N_0}{2\pi RC} \left[\tan^{-1}(x)\right]_{x=0}^{\infty} \\
&= \frac{N_0}{4RC}.
\end{aligned}
$$

(5.185)

Reference

Simon Haykin. *Communication Systems*. Wiley Eastern, second edition, 1983.

Chapter 6
Pulse Code Modulation

1. A speech signal has a total duration of $10\,$s. It is sampled at a rate of $8\,$kHz and then encoded. The SNR_Q must be greater than $40\,$dB. Calculate the minimum storage capacity required to accommodate this digitized speech signal. Assume that the speech signal has a Gaussian pdf with zero mean and variance σ^2 and the overload factor is 5. Assume that the quantizer is uniform and of the mid-rise type.

   • *Solution*: We know that

   $$\text{SNR}_Q = \frac{12\sigma^2}{\Delta^2}, \tag{6.1}$$

   where the step-size Δ is given by

   $$\Delta = \frac{2x_{\max}}{2^n}. \tag{6.2}$$

   Here $x_{\max} = 5\sigma$ is the maximum input the quantizer can handle and n is the number of bits used to encode any representation level. Substituting for Δ, the SNR_Q becomes

   $$\text{SNR}_Q = \frac{12 \times 2^{2n}}{100} = 10^4$$
   $$\Rightarrow n = 8.17. \tag{6.3}$$

   Since n has to be an integer and SNR_Q must be greater than $40\,$dB, we take $n = 9$.
   Now, the number of samples obtained in $10\,$s is 8×10^4. The number of bits obtained is $9 \times 8 \times 10^4$, which is the storage capacity required for the speech signal.

K. Vasudevan, *Analog Communications*,
https://doi.org/10.1007/978-3-030-50337-6_6

2. A PCM system uses a uniform quantizer followed by a 7-bit binary encoder. The bit-rate of the system is 56 Mbps.

   (a) What is the maximum message bandwidth for which the system operates satisfactorily.
   (b) Determine the SNR_Q when a full-load sinusoidal signal of frequency 1 MHz is applied to the input.

   • *Solution*: The sampling-rate is

   $$\frac{56}{7} = 8\,\text{MHz}. \tag{6.4}$$

   Hence the maximum message bandwidth is 4 MHz.
   For a sinusoidal signal of amplitude A, the power is $A^2/2$. Since the sinusoid is stated to be at full-load, $x_{max} = A$. The step-size is

   $$\Delta = \frac{2A}{2^7}. \tag{6.5}$$

   The mean-square quantization error is

   $$\sigma_Q^2 = \frac{\Delta^2}{12}. \tag{6.6}$$

   Hence the SNR_Q is

   $$SNR_Q = \frac{6 \times 128^2}{4} \equiv 43.91\,\text{dB}. \tag{6.7}$$

3. Twenty-four voice signals are sampled uniformly and then time division multi-plexed (TDM). The sampling operation uses flat-top samples with 1 μs duration. The multiplexing operation includes provision for synchronization by adding an extra pulse also of 1 μs duration. The highest frequency component of each voice signal is 3.4 kHz.

   (a) Assuming a sampling-rate of 8 kHz and uniform spacing between pulses, calculate the spacing (the time gap between the ending of a pulse and the starting of the next pulse) between successive pulses of the multiplexed signal.
   (b) Repeat your calculation using Nyquist-rate sampling.

   • *Solution*: Since the sampling-rate is 8 kHz, the time interval between two con-secutive samples of the same message is $10^6/8000 = 125$ μs. Thus, we can visualize a "frame" of duration 125 μs containing samples of the 24 voice signals plus the extra synchronization pulse. Thus the spacing between the starting points of 2 consecutive pulses is $125/25 = 5$ μs. Since the pulse-width is 1 μs, the spacing between consecutive pulses is 4 μs.

When Nyquist-rate sampling is used, the frame duration is $10^6/6800 = 147.059\,\mu s$. Therefore the spacing between the starting points of two consecutive pulses is $147.059/25 = 5.88\,\mu s$. Hence the spacing between the pulses is $4.88\,\mu s$.

4. Twelve different message signals, each with a bandwidth of $[-W, W]$ where $W = 10\,kHz$, are to be multiplexed and transmitted. Determine the minimum bandwidth required for each method if the modulation/multiplexing method used is

(a) SSB, FDM.
(b) Pulse amplitude modulation (PAM), TDM with Dirac-delta samples followed by Nyquist pulse shaping.

- *Solution*: Bandwidth required for SSB/FDM is $12 \times 10 = 120\,kHz$.
  The sampling rate required for each message is $20\,kHz$. Thus, the output of a PAM transmitter for a single message is 20 ksamples/s. The samples are weighted Dirac-delta functions. The output of the TDM is 240 ksamples/s. Using Nyquist pulse shaping, the transmission bandwidth required is $120\,kHz$.

5. (Vasudevan 2010) Compute the power spectrum of NRZ unipolar signals as shown in Fig. 6.1. Assume that the symbols are equally likely, statistically independent and WSS.

- *Solution*: Consider the system model in Fig. 6.2. Here the input $X(t)$ is given by

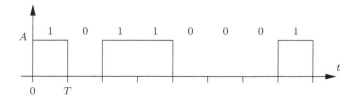

Fig. 6.1 NRZ unipolar signals

Fig. 6.2 System model for computing the psd of linearly modulated digital signals

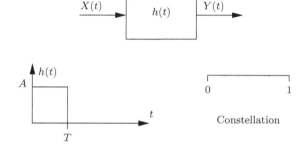

$$X(t) = \sum_{k=-\infty}^{\infty} S_k \delta(t - kT - \alpha), \tag{6.8}$$

where S_k denotes a symbol occurring at time kT, drawn from an M-ary PAM constellation, $1/T$ denotes the symbol-rate and α is a uniformly distributed RV in $[0, T]$. Since the symbols are independent and WSS

$$E[S_k S_j] = E[S_k]E[S_j] = m_S^2, \tag{6.9}$$

where m_S denotes the mean value of the symbols in the constellation and is equal to

$$m_S = \sum_{i=1}^{M} P(S_i) S_i, \tag{6.10}$$

where S_i denotes the ith symbol in the constellation and $P(S_i)$ denotes the probability of occurrence of S_i, and M is the number of symbols in the constellation. In the given problem $M = 2$, $P(S_i) = 0.5$, and $S_1 = 0$ and $S_2 = 1$, hence $m_S = 0.5$.

The output $Y(t)$ is given by

$$Y(t) = \sum_{k=-\infty}^{\infty} S_k h(t - kT - \alpha), \tag{6.11}$$

where $h(t)$ is the impulse response of the transmit filter. We know that the psd of $Y(t)$ is given by

$$S_Y(f) = \frac{P - m_S^2}{T} |H(f)|^2 + \frac{m_S^2}{T^2} \sum_{n=-\infty}^{\infty} |H(n/T)|^2 \delta(f - n/T), \tag{6.12}$$

where $H(f)$ is the Fourier transform of $h(t)$ and

$$P = \sum_{i=1}^{M} P(S_i) S_i^2 = 0.5 \tag{6.13}$$

is the average power of the constellation. Here $H(f)$ is equal to

$$H(f) = e^{-j\pi f T} AT \operatorname{sinc}(fT)$$
$$\Rightarrow |H(f)|^2 = A^2 T^2 \operatorname{sinc}^2(fT). \tag{6.14}$$

Therefore

Table 6.1 The 15-segment μ-law companding characteristic

Segment number	Step-size	Projections of the segment end points onto the horizontal axis	Representation levels
0	2	±31	$0, \ldots, \pm 15$
1a, 1b	4	±95	$\pm 16, \ldots, \pm 31$
2a, 2b	8	±223	$\pm 32, \ldots, \pm 47$
3a, 3b	16	±479	●
4a, 4b	32	±991	●
5a, 5b	64	±2015	●
6a, 6b	128	±4063	
7a, 7b	256	±8159	

$$S_Y(f) = \frac{A^2 T^2 \operatorname{sinc}^2(fT)}{4T} + \frac{A^2}{4} \sum_{k=-\infty}^{\infty} \operatorname{sinc}^2(nT/T)\delta(f - n/T). \quad (6.15)$$

Since

$$\operatorname{sinc}(fT) = \begin{cases} 1 \text{ for } f = 0 \\ 0 \text{ for } f = n/T, n \neq 0 \end{cases} \quad (6.16)$$

and $\delta(f - n/T)$ is not defined for $f = n/T$, it is assumed that the product $\operatorname{sinc}(n)\delta(f - n/T)$ is zero for $n \neq 0$. The reason is as follows. Consider an analogy in the time-domain. If $a\delta(t)$ is input to a filter with impulse response $h(t)$, the output is $ah(t)$. This implies that if $a = 0$, then the output is also zero, which in turn is equivalent to zero input. Thus we conclude that $0\delta(f)$ is equivalent to zero.
Thus

$$S_Y(f) = \frac{A^2 T \operatorname{sinc}^2(fT)}{4} + \frac{A^2}{4}\delta(f). \quad (6.17)$$

6. Consider the 15-segment piecewise linear characteristic used for μ-law companding ($\mu = 255$), shown in Table 6.1. Assume that the input signal lies in the range $[-8159, 8159]$ mV. Compute the representation level corresponding to an input of 500 mV. Note that in Table 6.1 the representation levels are not properly scaled, hence $c(x_{\max}) \neq x_{\max}$.

• *Solution*: From Table 6.1 we see that 500 mV lies in segment 4a. The first representation level corresponding to segment 4a is 64. The step-size for 4a is 32. Hence the end point of the first uniform segment in 4a is $479 + 32 = 511$. Thus the representation level is 64.

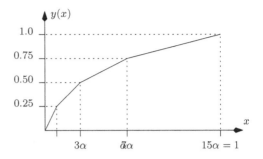

7. Consider an 8-segment piecewise linear characteristic which approximates the
   μ-law for companding. Four segments are in the first quadrant and the other four
   are in the third quadrant (odd symmetry). The μ-law is given by

$$c(x) = \frac{\ln(1 + \mu x)}{\ln(1 + \mu)} \quad \text{for } 0 \le x \le 1. \tag{6.18}$$

The projections of all segments along the y-axis are spaced uniformly. For the
segments in the first quadrant, the projections of the segments along the x-axis
are such that the length of the projection is double that of the previous segment.

(a) Compute μ.
(b) Let $e(x)$ denote the difference between the values obtained by the μ-law
    and that obtained by the 8-segment approximation. Determine $e(x)$ for the
    second segment in the first quadrant.

- *Solution*: Let the projection of the first segment in the first quadrant along
  the x-axis be denoted by α. Then the projection of the second segment in the
  first quadrant along the x-axis is 2α and so on. The projection of each of the
  segments along the y-axis is of length $1/4$ since $c(1) = 1$. This is illustrated
  in Fig. 6.3. Thus

$$\alpha + 2\alpha + 4\alpha + 8\alpha = 1$$
$$\Rightarrow \alpha = 1/15. \tag{6.19}$$

Also

$$\frac{\ln(1 + \mu\alpha)}{\ln(1 + \mu)} = 1/4$$
$$\frac{\ln(1 + \mu 3\alpha)}{\ln(1 + \mu)} = 1/2$$
$$\Rightarrow \mu = 1/\alpha$$
$$= 15. \tag{6.20}$$

Fig. 6.4 Message pdf

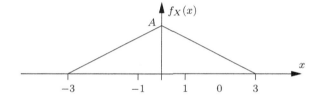

The expression for the second segment in the first quadrant is

$$y(x) = 15x/8 + 1/8 \quad \text{for } 1/15 \le x \le 3/15. \tag{6.21}$$

Hence

$$e(x) = c(x) - y(x) \quad \text{for } 1/15 \le x \le 3/15. \tag{6.22}$$

8. A DPCM system uses a second-order predictor of the form

$$\hat{x}(n) = p_1 x(n-1) + p_2 x(n-2). \tag{6.23}$$

The autocorrelation of the input is given by: $R_X(0) = 1$, $R_X(1) = 0.8$, $R_X(2) = 0.6$.
Compute the optimum forward prediction coefficients and the prediction gain.

- *Solution*: We know that

$$\begin{bmatrix} R_X(0) & R_X(1) \\ R_X(1) & R_X(0) \end{bmatrix} \begin{bmatrix} p_1 \\ p_2 \end{bmatrix} = \begin{bmatrix} R_X(1) \\ R_X(2) \end{bmatrix}. \tag{6.24}$$

Solving for p_1 and p_2 we get $p_1 = 8/9$, $p_2 = -1/9$.
The prediction error is

$$\sigma_E^2 = \sigma_X^2 - p_1 R_X(1) - p_2 R_X(2) = 3.2/9 = 0.355. \tag{6.25}$$

The prediction gain is

$$\frac{\sigma_X^2}{\sigma_E^2} = \frac{1}{\sigma_E^2} = 2.8125. \tag{6.26}$$

9. Consider the message pdf shown in Fig. 6.4. The decision thresholds of a non-uniform quantizer are at 0, ± 1, and ± 3.
Compute SNR_Q at the output of the expander.

- *Solution*: Since the area under the pdf must be unity, we must have $A = 1/3$.
Moreover

Fig. 6.5 Message pdf

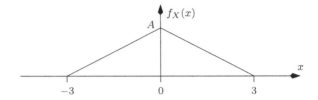

$$f_X(x) = -x/9 + 1/3. \tag{6.27}$$

The representation levels at the output of the expander are at ± 0.5 and ± 2. Here $f_X(x)$ cannot be considered a constant in any decision region, therefore

$$
\begin{aligned}
\sigma_Q^2 &= 2 \int_{x=0}^{1} (x - 1/2)^2 f_X(x)\, dx + 2 \int_{x=1}^{3} (x - 2)^2 f_X(x)\, dx \\
&= 2 \left[\frac{-x^4}{36} + \frac{4x^3}{27} - \frac{13x^2}{72} + \frac{x}{12} \right]\Big|_{x=0}^{1} \\
&\quad + 2 \left[\frac{-x^4}{36} + \frac{7x^3}{27} - \frac{16x^2}{18} + \frac{4x}{3} \right]\Big|_{x=1}^{3} \\
&= 0.19444. \tag{6.28}
\end{aligned}
$$

10. Consider the message pdf shown in Fig. 6.5.

   (a) Compute the representation levels of the optimum 2-level Lloyd-Max quantizer.
   (b) What is the corresponding value of σ_Q^2?

   • *Solution*: Since the area under the pdf is unity, we must have $A = 1/3$. Due to symmetry of the pdf, one of the decision thresholds is $x_1 = 0$. The other two decision thresholds are $x_0 = -3$ and $x_2 = 3$. The corresponding representation levels are y_1 and y_2 which are related by $y_1 = -y_2$. Thus the variance of the quantization error is

$$\sigma_Q^2 = 2 \int_{x=0}^{3} (x - y_2)^2 f_X(x)\, dx, \tag{6.29}$$

   which can be minimized by differentiating wrt y_2. Thus

$$y_2 = \frac{\int_{x=0}^{3} x f_X(x)\, dx}{\int_{x=0}^{3} f_X(x)\, dx}. \tag{6.30}$$

   The numerator of (6.30) is

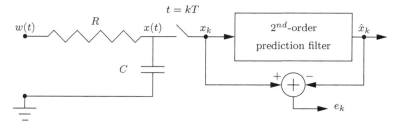

Fig. 6.6 Block diagram of a system using a prediction filter

$$\int_{x=0}^{3} x f_X(x)\, dx = \int_{x=0}^{3} x(-x/9 + 1/3)\, dx$$
$$= (-x^3/27 + x^2/6)\big|_{x=0}^{3}$$
$$= 1/2. \tag{6.31}$$

The denominator of (6.30) is

$$\int_{x=0}^{3} f_X(x)\, dx = \frac{1}{2} \times 3 \times \frac{1}{3} = \frac{1}{2}. \tag{6.32}$$

Thus $y_2 = 1$.

The minimum variance of the quantization error is

$$\sigma^2_{Q,\,min} = 2 \int_{x=0}^{3} (x-1)^2 f_X(x)\, dx$$
$$= 2 \int_{x=0}^{3} (x-1)^2(-x/9 + 1/3)\, dx$$
$$= 0.5. \tag{6.33}$$

11. Consider the block diagram in Fig. 6.6. Assume that $w(t)$ is zero-mean AWGN with psd $N_0/2 = 0.5 \times 10^{-4}$ W/Hz. Assume that $RC = 10^{-4}$ s and $T = RC/4$ s. Let

$$\hat{x}_k = p_1 x_{k-1} + p_2 x_{k-2}. \tag{6.34}$$

(a) Compute p_1 and p_2 so that the prediction error variance, $E[e_k^2]$, is minimized.

(b) What is the minimum value of the prediction error variance?

- *Solution*: We know that

$$S_X(f) = \frac{N_0}{2}|H(f)|^2$$

$$\frac{N_0/2}{1+(2\pi f RC)^2}. \tag{6.35}$$

Now

$$H(f) = \frac{1}{1+j2\pi f RC} \rightleftharpoons \frac{1}{RC}e^{-t/RC}u(t). \tag{6.36}$$

We also use the following relation:

$$|H(f)|^2 \rightleftharpoons h(t) \star h(-t). \tag{6.37}$$

Thus the autocorrelation function of $x(t)$ is

$$E[x(t)x(t-\tau)] = R_X(\tau) = \frac{N_0}{2}(h(\tau) \star h(-\tau))$$

$$= \frac{N_0}{2}\int_{t=-\infty}^{\infty} h(t)h(t-\tau)\,dt$$

$$= \frac{N_0}{2(RC)^2}\int_{t=-\infty}^{\infty} e^{-t/RC}u(t)$$

$$\times e^{(\tau-t)/RC}u(t-\tau)\,dt. \tag{6.38}$$

When $\tau > 0$

$$E[x(t)x(t-\tau)] = \frac{N_0}{2(RC)^2}\int_{t=\tau}^{\infty} e^{-t/RC}e^{(\tau-t)/RC}\,dt$$

$$= \frac{N_0}{4RC}\exp\left(-\frac{\tau}{RC}\right). \tag{6.39}$$

When $\tau < 0$

$$E[x(t)x(t-\tau)] = \frac{N_0}{2(RC)^2}\int_{t=0}^{\infty} e^{-t/RC}e^{(\tau-t)/RC}\,dt$$

$$= \frac{N_0}{4RC}\exp\left(\frac{\tau}{RC}\right). \tag{6.40}$$

Thus $R_X(\tau)$ is equal to

$$E[x(t)x(t-\tau)] = R_X(\tau) = \frac{N_0}{4RC}\exp\left(-\frac{|\tau|}{RC}\right). \tag{6.41}$$

The autocorrelation of the samples of $x(t)$ is

$$E[x(kT)x(kT - mT)] = R_X(mT) = \frac{N_0}{4RC} \exp\left(-\frac{|mT|}{RC}\right)$$

$$\Rightarrow R_X(m) = \frac{1}{4} \exp\left(-|m|/4\right)$$

(6.42)

The relevant autocorrelation values are

$$R_X(0) = 0.25$$
$$R_X(1) = 0.25\,e^{-1/4} = 0.1947$$
$$R_X(2) = 0.25\,e^{-1/2} = 0.1516.$$

(6.43)

We know that

$$\begin{bmatrix} R_X(0) & R_X(1) \\ R_X(1) & R_X(0) \end{bmatrix} \begin{bmatrix} p_1 \\ p_2 \end{bmatrix} = \begin{bmatrix} R_X(1) \\ R_X(2) \end{bmatrix}.$$

(6.44)

Solving for p_1 and p_2 we get $p_1 = 0.7788$, $p_2 = 0$.
The minimum prediction error is

$$\sigma_E^2 = \sigma_X^2 - p_1 R_X(1) - p_2 R_X(2) = 0.25(1 - e^{-1/2}) = 0.09837. \quad (6.45)$$

12. A delta modulator is designed to operate on speech signals limited to 3.4 kHz. The specifications of the modulator are

(a) Sampling-rate is ten times the Nyquist-rate of the speech signal.
(b) Step-size $\Delta = 100\,$mV.
    The modulator is tested with a 1 kHz sinusoidal signal. Determine the maximum amplitude of this test signal required to avoid slope overload.

- *Solution*

$$A_m < \frac{0.1\,f_s}{2\pi\,f_c}$$
$$= \frac{0.1 \times 68}{2\pi}$$
$$= 1.08\,\text{V}.$$

(6.46)

13. Consider a message pdf given by

$$f_X(x) = a\,e^{-|x|} \qquad \text{for } -\infty < x < \infty.$$

(6.47)

(a) Compute the representation levels of the optimum 2-level Lloyd-Max quantizer.
(b) What is the minimum value of σ_Q^2?

- *Solution*: Since the area under the pdf is unity, we must have

$$2a \int_{x=0}^{\infty} e^{-x} \, dx = 1$$

$$\Rightarrow a = 1/2. \tag{6.48}$$

Due to symmetry of the pdf, one of the decision thresholds is $x_1 = 0$. The other two decision thresholds are $x_0 = -\infty$ and $x_2 = \infty$. The corresponding representation levels are y_1 and y_2 which are related by $y_1 = -y_2$. Thus the variance of the quantization error is

$$\sigma_Q^2 = 2 \int_{x=0}^{\infty} (x - y_2)^2 f_X(x) \, dx \tag{6.49}$$

which can be minimized by differentiating wrt y_2. Thus

$$y_2 = \frac{\int_{x=0}^{\infty} x f_X(x) \, dx}{\int_{x=0}^{\infty} f_X(x) \, dx}. \tag{6.50}$$

The numerator of (6.50) is

$$\int_{x=0}^{\infty} x f_X(x) \, dx = \int_{x=0}^{\infty} (x/2) e^{-x} \, dx$$

$$= 1/2. \tag{6.51}$$

The denominator of (6.50) is

$$\int_{x=0}^{\infty} f_X(x) \, dx = \frac{1}{2}. \tag{6.52}$$

Thus $y_2 = 1$.
The minimum variance of the quantization error is

$$\sigma_{Q,\,min}^2 = 2 \int_{x=0}^{\infty} (x - 1)^2 f_X(x) \, dx$$

$$= 1. \tag{6.53}$$

14. A message signal having a pdf shown in Fig. 6.7 is applied to a μ-law compressor. The compressor characteristic is given by

$$c(|x|) = \frac{x_{\max}}{\ln(1 + \mu)} \ln\left(1 + \mu |x|/x_{\max}\right) \qquad \text{for } 0 \le |x| \le x_{\max}. \tag{6.54}$$

Derive the expression for the signal-to-quantization noise ratio at the expander output.

Fig. 6.7 Message pdf

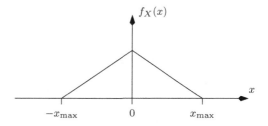

Assume that the number of representation levels is L and the overload level is x_{max}.

- *Solution*: We know that

$$\text{SNR}_Q = \frac{3L^2 \sigma_X^2}{x_{\text{max}}^2 \int_{x=-x_{\text{max}}}^{x_{\text{max}}} f_X(x)(dc/dx)^{-2}\, dx}. \tag{6.55}$$

The pdf of the message is

$$f_X(x) = \frac{1}{x_{\text{max}}}\left(1 - \frac{|x|}{x_{\text{max}}}\right). \tag{6.56}$$

Since

$$c(x) = \begin{cases} B\ln\left(1 + \mu x/x_{\text{max}}\right) & \text{for } 0 \le x \le x_{\text{max}} \\ -B\ln\left(1 - \mu x/x_{\text{max}}\right) & \text{for } -x_{\text{max}} \le x \le 0, \end{cases} \tag{6.57}$$

where

$$B = \frac{x_{\text{max}}}{\ln(1 + \mu)} \tag{6.58}$$

we have

$$\frac{dc}{dx} = \frac{\mu}{\ln(1 + \mu)}\left[\frac{1}{1 + \mu|x|/x_{\text{max}}}\right] \qquad \text{for } 0 \le |x| \le x_{\text{max}}. \tag{6.59}$$

Since both $f_X(x)$ and dc/dx are even functions of x we have

$$\int_{x=-x_{\text{max}}}^{x_{\text{max}}} f_X(x)(dc/dx)^{-2}\, dx = 2\int_{x=0}^{x_{\text{max}}} f_X(x)(dc/dx)^{-2}\, dx$$
$$= I \quad \text{(say)}. \tag{6.60}$$

Substituting for $f_X(x)$ and dc/dx we get

$$I = K \int_{x=0}^{x_{\max}} (1 - x/x_{\max})(1 + \mu x/x_{\max})^2 \, dx$$

$$= K \int_{x=0}^{x_{\max}} \left[1 + \frac{\mu^2 x^2}{x_{\max}^2} + \frac{2\mu x}{x_{\max}} + \frac{x}{x_{\max}} - \frac{\mu^2 x^3}{x_{\max}^3} - \frac{2\mu x^2}{x_{\max}^2} \right] dx$$

$$= K \left[x + \frac{\mu^2 x^3}{3x_{\max}^2} + \frac{2\mu x^2}{2x_{\max}} - \frac{x^2}{2x_{\max}} - \frac{\mu^2 x^4}{4x_{\max}^3} - \frac{2\mu x^3}{3x_{\max}^2} \right]_{x=0}^{x_{\max}}$$

$$= C \left[1 + \frac{\mu^2}{3} + \mu - 0.5 - \frac{\mu^2}{4} - \frac{2\mu}{3} \right]$$

$$= C \left[0.5 + \frac{\mu^2}{12} + \frac{\mu}{3} \right], \tag{6.61}$$

where

$$K = \frac{2}{x_{\max}} \left[\frac{\ln(1+\mu)}{\mu} \right]^2$$

$$C = K x_{\max}. \tag{6.62}$$

The signal power is given by

$$\sigma_X^2 = \int_{x=-x_{\max}}^{x_{\max}} x^2 f_X(x) \, dx$$

$$= \frac{2}{x_{\max}} \int_{x=0}^{x_{\max}} x^2 \left[1 - \frac{x}{x_{\max}} \right] dx$$

$$= \frac{2}{x_{\max}} \left[\frac{x^3}{3} - \frac{x^4}{4x_{\max}} \right]_{x=0}^{x_{\max}}$$

$$= \frac{x_{\max}^2}{6}. \tag{6.63}$$

Therefore

$$\text{SNR}_Q = \frac{L^2}{2I}. \tag{6.64}$$

15. A message signal having a pdf shown in Fig. 6.8 is applied to a uniform mid-step quantizer also shown in the same figure. Compute a and the pdf of the quantizer output Y. Assume that the overload level of the quantizer is ± 3.

• *Solution*: Firstly we have

$$2a \int_{x=0}^{3} x^2 \, dx = 1$$

$$\Rightarrow a = 1/18. \tag{6.65}$$

Fig. 6.8 Message pdf

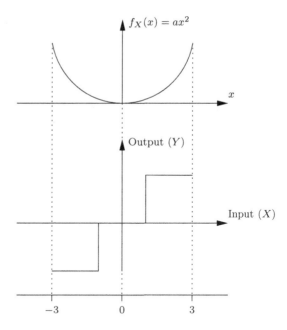

Moreover, Y is a discrete random variable which takes only three values ($L = 3$). The step-size of the quantizer is

$$\Delta = \frac{2x_{\max}}{L} = \frac{6}{3} = 2. \tag{6.66}$$

The various quantizer levels are indicated in Fig. 6.9. Thus

$$P(Y = 0) = P(-1 < X < 1)$$
$$= \frac{2}{18} \int_{x=0}^{1} x^2 \, dx$$
$$= \frac{1}{27}$$
$$P(Y = 2) = P(1 < X < 3)$$
$$= \frac{1}{18} \int_{x=1}^{3} x^2 \, dx$$
$$= \frac{13}{27}$$
$$= P(Y = -2). \tag{6.67}$$

Therefore

Fig. 6.9 Message pdf and
quantizer levels

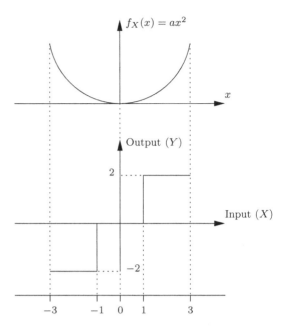

Fig. 6.10 Message pdf

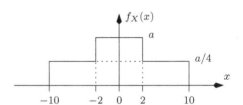

$$f_Y(y) = \frac{13}{27}\delta(y+2) + \frac{1}{27}\delta(y) + \frac{13}{27}\delta(y-2). \qquad (6.68)$$

16. The probability density function $f_X(x)$ of a message signal is given in Fig. 6.10.
    The message is quantized by a 3-bit quantizer such that all the eight reconstruc-
    tion levels occur with equal probability.

    (a) Determine a.
    (b) Determine the power of the message signal.
    (c) Compute the decision thresholds.
    (d) For each partition cell, compute the reconstruction level such that the quan-
        tization noise power is minimized.
    (e) Determine the overall quantization noise power.

    • *Solution*: Since the area under the pdf is unity we must have

$$2\left[2a + 8\frac{a}{4}\right] = 1$$

$$\Rightarrow a = \frac{1}{8}. \tag{6.69}$$

The power in the message is

$$2\int_{x=0}^{10} x^2 f_X(x)\, dx = 21.33. \tag{6.70}$$

Since the message pdf is symmetric, we expect the decision thresholds also to be symmetric about the origin. Let us denote the decision thresholds on the positive x-axis as x_0, x_1, x_2, x_3, and x_4. The decision thresholds along the negative x-axis are $-x_1$, $-x_2$, $-x_3$, and $-x_4$. Note that $x_4 = 10$ and $x_0 = 0$. Let us first compute x_3. Since each of the reconstruction levels occur with probability equal to $1/8$, we have

$$P(x_3 \le x \le 10) = 1/8$$
$$(a/4)(10 - x_3) = 1/8$$
$$\Rightarrow x_3 = 6. \tag{6.71}$$

Similarly we get $x_2 = 2$ and $x_1 = 1$.

Let us denote the reconstruction levels along the positive x-axis as y_1, y_2, y_3, and y_4. Due to symmetry, the reconstruction levels along the negative x-axis are $-y_1$, $-y_2$, $-y_3$ and $-y_4$. The quantization noise power for the ith partition cell is

$$Q_i = \int_{x=x_{i-1}}^{x_i} (x - y_i)^2 f_X(x)\, dx \qquad \text{for } 1 \le i \le 4. \tag{6.72}$$

The overall quantization power is

$$Q = 2\sum_{i=1}^{4} Q_i. \tag{6.73}$$

In order to minimize the overall quantization noise power, we need to minimize Q_i. For minimizing Q_i in (6.72) we differentiate wrt y_i and set the result to zero. Thus we get

$$\frac{Q_i}{d y_i} = 2\int_{x=x_{i-1}}^{x_i} (x - y_i) f_X(x)\, dx = 0$$

$$\Rightarrow y_i = \frac{\int_{x=x_{i-1}}^{x_i} x f_X(x)\, dx}{\int_{x=x_{i-1}}^{x_i} f_X(x)\, dx} \tag{6.74}$$

Fig. 6.11 Message pdf

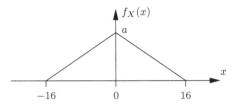

which is the centroid of each partition cell. For the given problem $f_X(x)$ is constant for each partition cell, therefore

$$y_i = \frac{x_{i-1} + x_i}{2}. \tag{6.75}$$

Therefore the reconstruction levels are at $y_1 = 1/2$, $y_2 = 3/2$, $y_3 = 4$, and $y_4 = 8$.
The overall quantization noise power is

$$\begin{aligned}
Q &= 2[Q_1 + Q_2 + Q_3 + Q_4] \\
&= 2[2/96 + 2/6] \\
&= 17/24 \\
&= 0.70833. \tag{6.76}
\end{aligned}$$

17. The probability density function $f_X(x)$ of a message signal is given in Fig. 6.11. The message is quantized by a 3-bit quantizer such that all the eight reconstruction levels occur with equal probability.

   (a) Determine a.
   (b) Determine the power of the message signal.
   (c) Compute the decision thresholds.
   (d) For each partition cell, compute the reconstruction level such that the quantization noise power is minimized.
   (e) Determine the overall quantization noise power.

   • *Solution*: Consider Fig. 6.12. Since the area under the pdf is unity we must have

$$2 \times \frac{1}{2} \times 16a = 1$$

$$\Rightarrow a = \frac{1}{16}. \tag{6.77}$$

   The power in the message is

Fig. 6.12 Message pdf

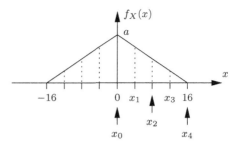

$$2 \int_{x=0}^{16} x^2 f_X(x)\, dx = 2 \int_{x=0}^{16} x^2 \left(\frac{1}{16} - \frac{x}{256} \right) dx$$
$$= \frac{128}{3}. \tag{6.78}$$

Since the message pdf is symmetric, we expect the decision thresholds also to be symmetric about the origin. Let us denote the decision thresholds on the positive x-axis as x_0, x_1, x_2, x_3, and x_4. The decision thresholds along the negative x-axis are $-x_1$, $-x_2$, $-x_3$, and $-x_4$. Note that $x_4 = 16$ and $x_0 = 0$. Let us first compute x_3. Since each of the reconstruction levels occur with probability equal to $1/8$, we have

$$P(x_3 \le x \le 16) = 1/8$$
$$\Rightarrow \frac{1}{2} \left(\frac{1}{16} - \frac{x_3}{256} \right) (16 - x_3) = 1/8$$
$$\Rightarrow x_3 = 8. \tag{6.79}$$

Similarly we get

$$P(x_2 \le x \le x_3) = 1/8$$
$$\Rightarrow \frac{1}{2} \left(\frac{1}{32} + \frac{1}{16} - \frac{x_2}{256} \right) (x_3 - x_2) = 1/8$$
$$\Rightarrow x_2 = 4.6862915 \tag{6.80}$$

and

$$P(0 \le x \le x_1) = 1/8$$
$$\Rightarrow \frac{1}{2} \left(\frac{1}{16} + \frac{1}{16} - \frac{x_1}{256} \right) x_1 = 1/8$$
$$\Rightarrow x_1 = 2.1435935. \tag{6.81}$$

Let us denote the reconstruction levels along the positive x-axis as y_1, y_2, y_3, and y_4. Due to symmetry, the reconstruction levels along the negative x-

axis are $-y_1$, $-y_2$, $-y_3$, and $-y_4$. The quantization noise power for the ith partition cell is

$$Q_i = \int_{x=x_{i-1}}^{x_i} (x - y_i)^2 f_X(x)\, dx \qquad \text{for } 1 \le i \le 4. \qquad (6.82)$$

The overall quantization power is

$$Q = 2 \sum_{i=1}^{4} Q_i. \qquad (6.83)$$

In order to minimize the overall quantization noise power, we need to minimize Q_i. For minimizing Q_i in (6.82) we differentiate wrt y_i and set the result to zero. Thus we get

$$\frac{Q_i}{dy_i} = 2 \int_{x=x_{i-1}}^{x_i} (x - y_i) f_X(x)\, dx = 0$$

$$\Rightarrow y_i = \frac{\int_{x=x_{i-1}}^{x_i} x f_X(x)\, dx}{\int_{x=x_{i-1}}^{x_i} f_X(x)\, dx}$$

$$= 8 \int_{x=x_{i-1}}^{x_i} x \left(\frac{1}{16} - \frac{x}{256} \right) dx$$

$$= 8 \left[\frac{x^2}{32} - \frac{x^3}{768} \right]_{x_{i-1}}^{x_i} \qquad \text{for } 1 \le i \le 4 \qquad (6.84)$$

which is the centroid of each partition cell. Therefore the reconstruction levels are at

$$y_1 = 1.0461463$$
$$y_2 = 3.3721317$$
$$y_3 = 6.2483887$$
$$y_4 = 10.666667. \qquad (6.85)$$

From (6.82), the quantization noise in the ith partition cell is

$$Q_i = \int_{x=x_{i-1}}^{x_i} (x - y_i)^2 \left(\frac{1}{16} - \frac{x}{256} \right) dx \qquad \text{for } 1 \le i \le 4$$

$$= \left[\frac{(x - y_i)^3}{48} - \frac{1}{256} \left(\frac{x^4}{4} + \frac{y_i^2 x^2}{2} - \frac{2 y_i x^3}{3} \right) \right]_{x=x_{i-1}}^{x_i} \qquad (6.86)$$

which evaluates to

$$Q_1 = 0.0477823$$
$$Q_2 = 0.0671179$$
$$Q_3 = 0.1132596$$
$$Q_4 = 0.4444444. \tag{6.87}$$

Therefore

$$Q = 2(Q_1 + Q_2 + Q_3 + Q_4)$$
$$= 1.3452084. \tag{6.88}$$

18. A speech signal has a total duration of 20 s. It is sampled at a rate of 8 kHz and then encoded. The SNR_Q must be greater than 50 dB. Calculate the minimum storage capacity required to accommodate this digitized speech signal. Assume that the speech signal has a Gaussian pdf with zero mean and variance σ^2 and the overload factor is 6. Assume that the quantizer is uniform and of the mid-rise type.

- *Solution*: We know that

$$\text{SNR}_Q = \frac{12\sigma^2}{\Delta^2}, \tag{6.89}$$

where the step-size Δ is given by

$$\Delta = \frac{2x_{\text{max}}}{2^n}. \tag{6.90}$$

Here, $x_{\text{max}} = 6\sigma$ is the maximum input the quantizer can handle and n is the number of bits used to encode any representation level. Substituting for Δ, the SNR_Q becomes

$$\text{SNR}_Q = \frac{12 \times 2^{2n}}{144} = 10^5$$
$$\Rightarrow n = 10.097. \tag{6.91}$$

Since n has to be an integer and SNR_Q must be greater than 50 dB, we take $n = 11$.
Now, the number of samples obtained in 20 s is 16×10^4. The number of bits obtained is $11 \times 16 \times 10^4 = 176 \times 10^4$, which is the storage capacity required for the speech signal.

19. A DPCM system uses a second-order predictor of the form

$$\hat{x}(n) = p_1 x(n-1) + p_2 x(n-2). \tag{6.92}$$

The autocorrelation of the input is given by: $R_X(0) = 2$, $R_X(1) = 1.8$, $R_X(2) = 1.6$.

Compute the optimum prediction coefficients and the prediction gain.

• *Solution*: We know that

$$\begin{bmatrix} R_X(0) & R_X(1) \\ R_X(1) & R_X(0) \end{bmatrix} \begin{bmatrix} p_1 \\ p_2 \end{bmatrix} = \begin{bmatrix} R_X(1) \\ R_X(2) \end{bmatrix}. \tag{6.93}$$

Solving for p_1 and p_2 we get $p_1 = 0.9473684$, $p_2 = -0.0526316$.
The prediction error is

$$\sigma_E^2 = \sigma_X^2 - p_1 R_X(1) - p_2 R_X(2) = 0.3789474. \tag{6.94}$$

The prediction gain is

$$\frac{\sigma_X^2}{\sigma_E^2} = \frac{2}{\sigma_E^2} = 5.28. \tag{6.95}$$

20. A random process $X(t)$ is defined by

$$X(t) = Y \cos(2\pi f_c t + \theta), \tag{6.96}$$

where Y and θ are independent random variables. Y is uniformly distributed in the range $[-3, 3]$ and θ is uniformly distributed in the range $[0, 2\pi)$.

(a) Compute the autocorrelation and psd of $X(t)$.
(b) $X(t)$ is fed to a uniform mid-rise type quantizer. Compute the number of bits per sample required so that the SQNR is at least 40 dB with no overload distortion.

• *Solution*: Clearly

$$E[X(t)] = E[Y]E[\cos(2\pi f_c t + \theta)]$$
$$= 0. \tag{6.97}$$

The autocorrelation of $X(t)$ is given by

$$\begin{aligned} E[X(t)X(t - \tau)] &= R_X(\tau) \\ &= E[Y \cos(2\pi f_c t + \theta)Y \cos(2\pi f_c(t - \tau) + \theta)] \\ &= E[Y^2] E[\cos(2\pi f_c t + \theta) \cos(2\pi f_c(t - \tau) + \theta)] \\ &= \frac{3}{2} \cos(2\pi f_0 \tau). \end{aligned} \tag{6.98}$$

The psd is

$$S_X(f) = \frac{3}{4}[\delta(f - f_0) + \delta(f + f_0)]. \tag{6.99}$$

The signal power is $3/2$. The quantization noise power is

$$\sigma_Q^2 = \frac{\Delta^2}{12}, \tag{6.100}$$

where

$$\Delta = \frac{2m_{\max}}{2^n} = \frac{6}{2^n}, \tag{6.101}$$

where $m_{\max} = 3$ denotes the maximum amplitude of the random process and n is the number of bits per sample.
Therefore the SQNR is

$$\begin{aligned}
\text{SQNR} &= \frac{3}{2} \times \frac{2^{2n}}{3} \\
&= 2^{2n-1} \\
\Rightarrow 10\log_{10}(2^{2n-1}) &\geq 40 \\
\Rightarrow n &\geq 7.14 \\
\Rightarrow n &= 8.
\end{aligned} \tag{6.102}$$

21. The pdf of a full-load message signal is uniformly distributed in $[-6, 6]$. Compute the SNR_Q in dB, due to a 2-bit, uniform mid-rise quantizer.

- *Solution*: The signal power is

$$\begin{aligned}
\sigma_X^2 &= \int_{x=-6}^{6} x^2 f_X(x)\, dx \\
&= \frac{2}{12} \int_{x=0}^{6} x^2\, dx \\
&= 12.
\end{aligned} \tag{6.103}$$

Since a 2-bit quantizer has four representation levels, the step-size is

$$\begin{aligned}
\Delta &= \frac{2x_{\max}}{4} \\
&= \frac{2 \times 6}{4} \\
&= 3.
\end{aligned} \tag{6.104}$$

Therefore the decision thresholds are at $x_0 = -6$, $x_1 = -3$, $x_2 = 0$, $x_3 = 3$, $x_4 = 6$ and the representation levels are at the mid points of two consecutive decision thresholds, that is, $y_1 = -9/2$, $y_2 = -3/2$, $y_3 = 3/2$, and $y_4 = 9/2$. The quantization noise power is

$$
\begin{aligned}
\sigma_Q^2 &= \sum_{i=1}^{4} \int_{x=x_{i-1}}^{x_i} (x - y_i)^2 \, f_X(x) \, dx \\
&= \frac{2}{12} \int_{x=0}^{3} (x - 3/2)^2 \, dx + \frac{2}{12} \int_{x=3}^{6} (x - 9/2)^2 \, dx \\
&= 3/4.
\end{aligned}
\tag{6.105}
$$

Hence we get

$$
\begin{aligned}
\mathrm{SNR}_Q &= \frac{\sigma_X^2}{\sigma_Q^2} \\
&= 16 \\
\Rightarrow \mathrm{SNR}_Q \ (\text{in dB}) &= 10 \log_{10}(16) \\
&= 12.0412 \, \text{dB}.
\end{aligned}
\tag{6.106}
$$

22. The μ-law compressor characteristic is given by

$$
c(|x|) = \frac{x_{\max}}{\ln(1 + \mu)} \ln\left(1 + \mu|x|/x_{\max}\right) \qquad \text{for } 0 \le |x| \le x_{\max}. \tag{6.107}
$$

Derive and compute the companding gain for $\mu = 255$.

- *Solution*: The companding gain is defined as

$$
G_c = \frac{\Delta_u}{\Delta_{r,\,\min}}, \tag{6.108}
$$

where Δ_u is the step-size of the uniform quantizer and $\Delta_{r,\,\min}$ is the minimum step-size of the robust quantizer having the same overload level ($x_{\max}$) and representation levels (L). Now

$$
\Delta_u = \frac{2x_{\max}}{L}. \tag{6.109}
$$

Similarly, for the kth segment of the robust quantizer we have

$$
\left.\frac{dc(x)}{dx}\right|_k = \frac{2x_{\max}}{L\Delta_k}, \tag{6.110}
$$

where Δ_k is the length of the kth segment along the x-axis (input). Note that the length of each segment along the y-axis is identical and equal to $2x_{\max}/L$. From (6.110) it is clear that Δ_k is minimum when dc/dx is maximum, which occurs at the origin. Substituting (6.109) and (6.110) in (6.108) we get

$$
\begin{aligned}
G_c &= \left.\frac{dc(x)}{dx}\right|_{x=0} \\
&= \frac{\mu}{\ln(1+\mu)} \\
&= 45.985904
\end{aligned}
\tag{6.111}
$$

for $\mu = 255$.

23. The A-law compressor characteristic is given by

$$
\frac{c(|x|)}{x_{max}} = \begin{cases} A|x|/(Kx_{max}) & \text{for } 0 \leq |x|/x_{max} \leq 1/A \\ (1+\ln(A|x|/x_{max}))/K & \text{for } 1/A \leq |x|/x_{max} \leq 1, \end{cases}
\tag{6.112}
$$

where $K = 1 + \ln(A)$. Derive and compute the companding gain for $A = 87.56$.

- *Solution*: The companding gain is defined as

$$
G_c = \frac{\Delta_u}{\Delta_{r,\,min}},
\tag{6.113}
$$

where Δ_u is the step-size of the uniform quantizer and $\Delta_{r,\,min}$ is the minimum step-size of the robust quantizer having the same overload level (x_{max}) and representation levels (L). Now

$$
\Delta_u = \frac{2x_{max}}{L}.
\tag{6.114}
$$

Similarly, for the kth segment of the robust quantizer we have

$$
\left.\frac{dc(x)}{dx}\right|_k = \frac{2x_{max}}{L\Delta_k},
\tag{6.115}
$$

where Δ_k is the length of the kth segment along the x-axis (input). Note that the length of each segment along the y-axis is identical and equal to $2x_{max}/L$. From (6.115) it is clear that Δ_k is minimum when dc/dx is maximum, which occurs at the origin. Substituting (6.114) and (6.115) in (6.113) we get

$$
\begin{aligned}
G_c &= \left.\frac{dc(x)}{dx}\right|_{x=0} \\
&= \frac{A}{(1+\ln(A))} \\
&= 16.000514
\end{aligned}
\tag{6.116}
$$

for $A = 87.56$.

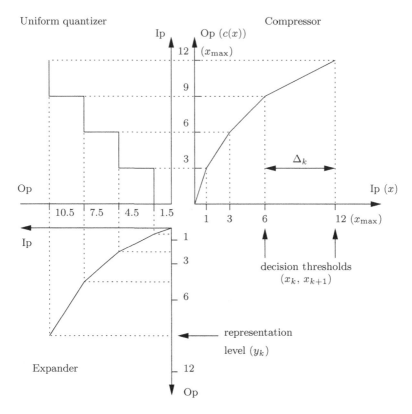

Fig. 6.13 Illustrating the transfer characteristics of the compressor and expander

24. (Haykin 1988) derived the ideal compressor characteristic. Clearly state the objective of the compressor. Can the ideal compressor be used in practice? Give reasons.

   • *Solution*: Recall that in the case of a uniform quantizer, the SNR_Q is directly proportional to the input signal power. Therefore, if the input signal power decreases, the SNR_Q also decreases. However, in the case of a robust quantizer, the SNR_Q remains constant over a wide range of input signal power. In order to achieve this feature, the robust quantizer uses a compressor. This is shown in Fig. 6.13. For ease of illustration, the compressor characteristic is assumed to be piecewise-linear. The compressor and uniform quantizer are located at the transmitter, whereas the expander is located at the receiver. Note that in the case of the compressor, both input and output amplitudes are continuous. However, in the case of the expander, both input and output amplitudes are discrete.

   The representation level y_k is related to the decision thresholds x_k and x_{k+1} as follows:

$$y_k = \frac{1}{2}(x_k + x_{k+1}).$$ (6.117)

The quantization error for the kth decision region is defined as

$$q_k = x - y_k,$$ (6.118)

where x denotes the compressor input. The compressor characteristic must satisfy the following property:

$$c(-x) = -c(x).$$ (6.119)

It is also convenient to have

$$c(x) = \begin{cases} x_{\max} & \text{for } x = x_{\max} \\ 0 & \text{for } x = 0 \\ -x_{\max} & \text{for } x = -x_{\max}. \end{cases}$$ (6.120)

Let L denote the number of decision regions or representation levels. Let Δ_k, for $0 \leq k \leq L - 1$, denote the length of the kth decision region at the compressor input. We also observe from Fig. 6.13 that the compressor output is divided into equal intervals. Hence

$$\left(\frac{dc(x)}{dx}\right)_k = \frac{2x_{\max}}{L\Delta_k}$$

$$\Rightarrow \Delta_k = \frac{2x_{\max}}{L(dc(x)/dx)_k}.$$ (6.121)

We now make the following assumptions regarding the input x.

(a) The pdf of x, $f_X(x)$, is an even function of x. This ensures that x has zero-mean.
(b) In each interval Δ_k, $f_X(x)$, is approximately a constant. Hence

$$f_X(x) \approx f_X(y_k) \qquad \text{for } x_k \leq x \leq x_{k+1}.$$ (6.122)

This ensures that the quantization error is uniformly distributed in $[-\Delta_k/2, \Delta_k/2]$.
(c) The input does not overload the compressor.

Now, the variance of the quantization error in the kth interval is

$$\sigma_{Q,k}^2 = \Delta_k^2/12.$$ (6.123)

The overall quantization noise power is

$$\sigma_Q^2 = \sum_{k=0}^{L-1} P_k \sigma_{Q,k}^2$$

$$= \frac{1}{12} \sum_{k=0}^{L-1} P_k \Delta_k^2, \tag{6.124}$$

where

$$P_k = f_X(y_k) \Delta_k \tag{6.125}$$

denotes the probability that x lies in the region Δ_k. Substituting (6.121) in (6.124), we get

$$\sigma_Q^2 = \frac{1}{12} \sum_{k=0}^{L-1} P_k \frac{4x_{\max}^2}{L^2 (dc(x)/dx)_k^2}. \tag{6.126}$$

Now, for a given $x_{\max}$, as L becomes large, we obtain

$$P_k \rightarrow f_X(x)\, dx$$
$$\left(\frac{dc(x)}{dx}\right)_k \rightarrow \frac{dc(x)}{dx} \tag{6.127}$$

and the summation in (6.126) can be replaced by an integral, as follows:

$$\sigma_Q^2 = \frac{x_{\max}^2}{3L^2} \int_{x=-x_{\max}}^{x_{\max}} \frac{f_X(x)}{(dc(x)/dx)^2}\, dx. \tag{6.128}$$

Therefore

$$\mathrm{SNR}_Q = \frac{\sigma_X^2}{\sigma_Q^2}$$

$$= \frac{1}{\sigma_Q^2} \int_{x=-x_{\max}}^{x_{\max}} x^2 f_X(x)\, dx \tag{6.129}$$

is independent of the input signal power when

$$\frac{dc(x)}{dx} = \frac{K}{x}$$
$$\Rightarrow c(x) = K \ln(x) + C_0 \qquad \text{for } x > 0, \tag{6.130}$$

where K and C_0 are constants. Using (6.120) we obtain

$$C_0 = x_{max} - K \ln(x_{max}). \tag{6.131}$$

Hence the ideal compressor characteristic is

$$c(x) = K \ln \left(\frac{x}{x_{max}} \right) + x_{max} \quad \text{for } x > 0. \tag{6.132}$$

Note that when $x < 0$, (6.119) needs to be applied. The ideal $c(x)$ is not used in practice, since $c(x) \to -\infty$ as $x \to 0^+$.

25. The probability density function of a message signal is

$$f_X(x) = ae^{-3|x|} \quad \text{for } -\infty < x < \infty. \tag{6.133}$$

The message is applied to a 4-representation level Lloyd-Max quantizer. The initial set of representation levels are given by: $y_{1,0} = -4$, $y_{2,0} = -1$, $y_{3,0} = 1$, $y_{4,0} = 4$, where the second subscript denotes the 0th iteration. Compute the next set of decision thresholds and representation levels (in the 1st iteration).

- *Solution*: Since the area under the pdf is unity we must have

$$\int_{x=-\infty}^{\infty} ae^{-3|x|}\, dx = \int_{x=0}^{\infty} 2ae^{-3x}\, dx$$
$$= 1$$
$$\Rightarrow a = 3/2. \tag{6.134}$$

Let the decision thresholds in the 1st iteration be given by $x_{0,1}, \ldots, x_{4,1}$. We have

$$x_{0,1} = -\infty$$
$$x_{1,1} = \frac{y_{1,0} + y_{2,0}}{2}$$
$$= -2.5$$
$$x_{2,1} = \frac{y_{2,0} + y_{3,0}}{2}$$
$$= 0$$
$$x_{3,1} = \frac{y_{3,0} + y_{4,0}}{2}$$
$$= 2.5$$
$$x_{4,1} = \infty. \tag{6.135}$$

The representation levels in the 1st iteration are given by

$$y_{k,1} = \frac{\int_{x=x_{k-1,1}}^{x_{k,1}} x f_X(x)\, dx}{\int_{x=x_{k-1,1}}^{x_{k,1}} f_X(x)\, dx} \quad \text{for } 1 \le k \le 4. \tag{6.136}$$

Solving (6.136) we get

$$
\begin{aligned}
y_{1,1} &= -2.83 \\
&= -y_{4,1} \\
y_{2,1} &= -0.33194 \\
&= -y_{3,1}.
\end{aligned}
\tag{6.137}
$$

References

Simon Haykin. *Digital Communications*. John Wiley & Sons, first edition, 1988.

K. Vasudevan. *Digital Communications and Signal Processing, Second edition (CDROM included)*. Universities Press (India), Hyderabad, www.universitiespress.com, 2010.

Chapter 7
Signaling Through AWGN Channel

1. For the transmit filter with impulse response given in Fig. 7.1, draw the output of the matched filter. Assume that the matched filter has an impulse response $p(-t)$.

   - *Solution*: The matched filter output is plotted in Fig. 7.2.

2. Consider a communication system shown in Fig. 7.3. Here, an input bit $b_k = 1$ is represented by the signal $x_1(t)$ and the bit $b_k = 0$ is represented by the signal $x_2(t)$. Note that $x_1(t)$ and $x_2(t)$ are orthogonal, that is

$$\int_{t=0}^{T} x_1(t)x_2(t)\, dt = 0. \tag{7.1}$$

   (a) It is given that the impulse response of the transmit filter is

$$p(t) = \begin{cases} 1 & \text{for } 0 \le t \le T/4 \\ 0 & \text{otherwise,} \end{cases} \tag{7.2}$$

   where $1/T$ denotes the bit-rate of the input data stream b_k.
   The "bit manipulator" converts the input bit $b_k = 0$ into a sequence of Dirac-delta functions weighted by symbols from a binary constellation. Similarly for input bit $b_k = 1$. Thus the output of the bit manipulator is given by

$$y_1(t) = \sum_{k=-\infty}^{\infty} a_k \delta(t - kT/4), \tag{7.3}$$

   where a_k denotes symbols from a binary constellation. The final PCM signal $y(t)$ is

© The Editor(s) (if applicable) and The Author(s), under exclusive license to Springer Nature Switzerland AG 2021
K. Vasudevan, *Analog Communications*,
https://doi.org/10.1007/978-3-030-50337-6_7

357

Fig. 7.1 Impulse response
of the transmit filter

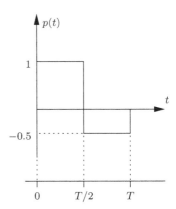

Fig. 7.2 Matched filter
output

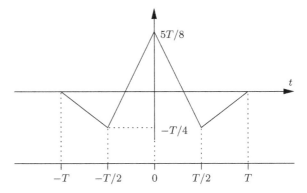

$$y(t) = \sum_{k=-\infty}^{\infty} a_k p(t - kT/4). \qquad (7.4)$$

Note that the symbol-rate of a_k is $4/T$, that is four times the input bit-rate.
Draw the constellation and write down the sequence of symbols that are
generated corresponding to an input bit $b_k = 0$ and an input bit $b_k = 1$.

(b) The received signal is given by

$$u(t) = y(t) + w(t), \qquad (7.5)$$

where $y(t)$ is the transmitted signal and $w(t)$ is a sample function of a zero-
mean AWGN process with psd $N_0/2$. Compute the mean and variance of
z_1 and z_2, given that 1 ($x_1(t)$) was transmitted in the interval $[0,\ T]$. Also
compute $\mathrm{cov}(z_1,\ z_2)$.

(c) Derive the detection rule for the optimal detector.

(d) Derive the average probability of error.

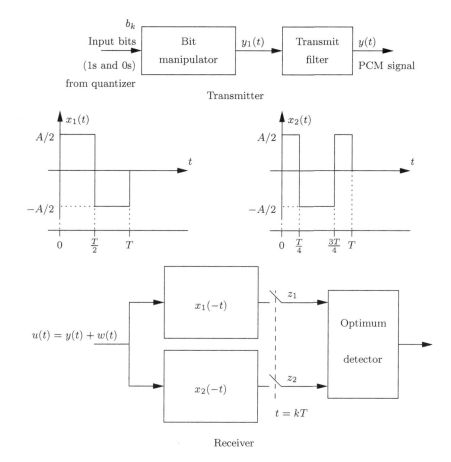

Fig. 7.3 Block diagram of a communication system

- *Solution*: The impulse response of the transmit filter is

$$p(t) = \begin{cases} 1 \text{ for } 0 \leq t \leq T/4 \\ 0 \text{ otherwise.} \end{cases} \qquad (7.6)$$

The input bit 0 gets converted to the symbol sequence:

$$\{A/2, \ -A/2, \ -A/2, \ A/2\}. \qquad (7.7)$$

The input bit 1 gets converted to the symbol sequence:

$$\{A/2, \ A/2, \ -A/2, \ -A/2\}. \qquad (7.8)$$

Given that 1 ($x_1(t)$) has been transmitted in the interval $[0, T]$, the received signal can be written as

$$u(t) = x_1(t) + w(t). \tag{7.9}$$

The output of the upper MF is

$$\begin{aligned}
z_1 &= u(t) \star x_1(-t)|_{t=0} \\
&= \int_{\tau=-\infty}^{\infty} u(\tau)x_1(-t+\tau)\,d\tau \Big|_{t=0} \\
&= \int_{t=0}^{T} u(t)x_1(t)\,dt \\
&= \frac{A^2 T}{4} + w_1,
\end{aligned} \tag{7.10}$$

where

$$w_1 = \int_{t=0}^{T} w(t)x_1(t)\,dt. \tag{7.11}$$

Similarly

$$\begin{aligned}
z_2 &= u(t) \star x_2(-t)|_{t=0} \\
&= \int_{t=0}^{T} u(t)x_2(t)\,dt \\
&= w_2,
\end{aligned} \tag{7.12}$$

where we have used the fact that $x_1(t)$ and $x_2(t)$ are orthogonal and

$$w_2 = \int_{t=0}^{T} w(t)x_2(t)\,dt. \tag{7.13}$$

Since $w(t)$ is zero mean, w_1 and w_2 are also zero mean. Since $x_1(t)$ and $x_2(t)$ are LTI filters, w_1 and w_2 are Gaussian distributed RVs. Moreover

$$\begin{aligned}
E[w_1 w_2] &= E\left[\int_{t=0}^{T} w(t)x_1(t)\,dt \int_{\tau=0}^{T} w(\tau)x_2(\tau)\,d\tau\right] \\
&= \int_{t=0}^{T}\int_{\tau=0}^{T} x_1(t)x_2(\tau)\frac{N_0}{2}\delta(t-\tau)\,dt\,d\tau \\
&= \int_{t=0}^{T} x_1(t)x_2(t)\frac{N_0}{2}\,dt \\
&= 0.
\end{aligned} \tag{7.14}$$

Thus w_1 and w_2 are uncorrelated, and being Gaussian, they are also statistically independent. This implies that z_1 and z_2 are also statistically independent. The variance of w_1 and w_2 is

$$
\begin{aligned}
E\left[w_1^2\right] &= E\left[\int_{t=0}^{T} w(t)x_1(t)\, dt \int_{\tau=0}^{T} w(\tau)x_1(\tau)\, d\tau\right] \\
&= \int_{t=0}^{T}\int_{\tau=0}^{T} x_1^2(t)\frac{N_0}{2}\delta(t-\tau)\, dt\, d\tau \\
&= \int_{t=0}^{T} x_1^2(t)\frac{N_0}{2}\, dt \\
&= \frac{N_0 A^2 T}{8} \\
&= E[w_2^2] \\
&= \operatorname{var}(z_1) = \operatorname{var}(z_2) = \sigma^2 \quad (\text{say}).
\end{aligned}
\tag{7.15}
$$

The mean values of z_1 and z_2 are (given that 1 was transmitted)

$$
\begin{aligned}
E[z_1] &= \frac{A^2 T}{4} = m_{1,1} \quad (\text{say}) \\
E[z_2] &= 0 = m_{2,1} \quad (\text{say}),
\end{aligned}
\tag{7.16}
$$

where $m_{1,1}$ and $m_{2,1}$ denote the means of z_1 and z_2, respectively, when 1 is transmitted. Similarly it can be shown that when 0 is transmitted

$$
\begin{aligned}
E[z_1] &= 0 = m_{1,0} \quad (\text{say}) \\
E[z_2] &= \frac{A^2 T}{4} = m_{2,0} \quad (\text{say}).
\end{aligned}
\tag{7.17}
$$

Let

$$
\mathbf{z} = \begin{bmatrix} z_1 & z_2 \end{bmatrix}^T.
\tag{7.18}
$$

The maximum likelihood (ML) detector computes the probabilities:

$$
P(j|\mathbf{z}) \quad \text{for } j = 0, 1
\tag{7.19}
$$

and decides in favor of that bit for which the probability is maximum. Using Bayes' rule we have

$$
P(j|\mathbf{z}) = \frac{f_{\mathbf{z}}(\mathbf{z}|j)P(j)}{f_{\mathbf{z}}(\mathbf{z})},
\tag{7.20}
$$

where $P(j)$ denotes the probability that bit j ($j = 0, 1$) was transmitted. Since $P(j) = 0.5$ and $f_{\mathbf{Z}}(\mathbf{z})$ are independent of j they can be ignored and computing the maximum of $P(j|\mathbf{z})$ is equivalent to computing the maximum of $f_{\mathbf{Z}}(\mathbf{z}|j)$ which can be conveniently written as

$$\max_j \, f_{\mathbf{Z}}(\mathbf{z}|j) \quad \text{for } j = 0, 1. \tag{7.21}$$

Since z_1 and z_2 are statistically independent (7.21) can be written as

$$\max_j \, f_{Z_1}(z_1|j) f_{Z_2}(z_2|j)$$

$$\Rightarrow \max_j \, \frac{1}{2\pi\sigma^2} \exp\left(-\frac{(z_1 - m_{1,j})^2 + (z_2 - m_{2,j})^2}{2\sigma^2}\right). \tag{7.22}$$

Taking the natural logarithm and ignoring constants (7.22) reduces to

$$\min_j (z_1 - m_{1,j})^2 + (z_2 - m_{2,j})^2 \quad \text{for } j = 0, 1. \tag{7.23}$$

To compute the average probability of error we first compute the probability of detecting 0 when 1 is transmitted ($P(0|1)$). This happens when

$$(z_1 - m_{1,0})^2 + (z_2 - m_{2,0})^2 < (z_1 - m_{1,1})^2 + (z_2 - m_{2,1})^2. \tag{7.24}$$

Let

$$e_1 = m_{1,1} - m_{1,0}$$
$$e_2 = m_{2,1} - m_{2,0}$$
$$e_1^2 + e_2^2 = d^2. \tag{7.25}$$

Thus the detector makes an error when

$$2e_1 w_1 + 2e_2 w_2 < -d^2. \tag{7.26}$$

Let

$$Z = 2e_1 w_1 + 2e_2 w_2. \tag{7.27}$$

Then

$$E[Z] = 2e_1 E[w_1] + 2e_2 E[w_2]$$
$$= 0$$
$$E[Z^2] = 4e_1^2 \sigma^2 + 4e_2^2 \sigma^2$$
$$= 4d^2 \sigma^2$$

$$= \sigma_Z^2 \quad \text{(say)}. \tag{7.28}$$

Thus the probability of detecting 0 when 1 is transmitted is

$$P(Z < -d^2) = \frac{1}{\sigma_Z \sqrt{2\pi}} \int_{-\infty}^{-d^2} \exp\left(-\frac{Z^2}{2\sigma_Z^2}\right) dz$$

$$= \frac{1}{2}\text{erfc}\left(\frac{d^2}{\sigma_Z \sqrt{2}}\right)$$

$$= \frac{1}{2}\text{erfc}\left(\sqrt{\frac{d^2}{8\sigma^2}}\right). \tag{7.29}$$

3. Consider the passband PAM system in Fig. 7.4. The bits 1 and 0 from the quantizer are equally likely. The signal $b(t)$ is given by

$$b(t) = \sum_{k=-\infty}^{\infty} a_k \delta(t - kT), \tag{7.30}$$

where a_k denotes a symbol at time kT taken from a binary constellation shown in Fig. 7.4 and $\delta(t)$ is the Dirac-Delta function. The mapping of the bits to symbols is also shown. The transmit filter is a pulse corresponding to the root-raised cosine (RRC) spectrum. The roll-off factor of the raised cosine (RC) spectrum

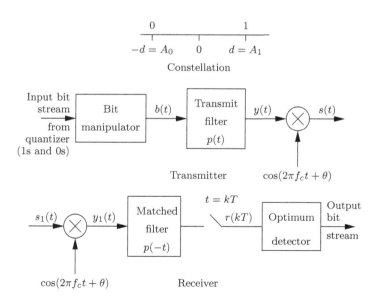

Fig. 7.4 Block diagram of a passband PAM system

(from which the root-raised cosine spectrum is obtained) is $\alpha = 0.5$. Assume that the bit-rate $1/T = 1\,\mathrm{kbps}$, the energy of $p(t) = 2$, θ is a uniformly distributed random variable in $[0,\ 2\pi]$ and θ and $w(t)$ are statistically independent. The received signal is given by

$$s_1(t) = s(t) + w(t), \tag{7.31}$$

where $w(t)$ is zero-mean AWGN with psd $N_0/2$.

(a) Compute the two-sided bandwidth (bandwidth on either side of the carrier) of the psd of the transmitted signal $s(t)$.
(b) Derive the expression for $r(kT)$.
(c) Derive the ML detection rule. Start from the MAP rule

$$\max_j P(A_j|r_k) \qquad \text{for } j = 0,\ 1, \tag{7.32}$$

where for brevity $r_k = r(kT)$.
(d) Assume that the transmitted bit at time kT is 1. Compute the probability of making an erroneous decision, in terms of d and N_0.

4. *Solution*: The two-sided bandwidth is

$$B = 2 \times 500(1 + \alpha)$$
$$= 1500\,\mathrm{Hz}. \tag{7.33}$$

The signal at the multiplier output is given by

$$s_1(t)\cos(2\pi f_c t) = \frac{y(t)}{2}(1 + \cos(4\pi f_c t)) + w_1(t), \tag{7.34}$$

where $w_1(t)$ is

$$w_1(t) = w(t)\cos(2\pi f_c t). \tag{7.35}$$

The autocorrelation of $w_1(t)$ is

$$\begin{aligned}
E[w_1(t)w_1(t-\tau)] &= E[w(t)w(t-\tau)] \\
&\quad E[\cos(2\pi f_c t + \theta)\cos(2\pi f_c(t-\tau) + \theta)] \\
&= \frac{N_0}{2}\delta(\tau)\frac{1}{2}\cos(2\pi f_c \tau) \\
&= \frac{N_0}{4}\delta(\tau). \tag{7.36}
\end{aligned}$$

Since the MF eliminates the signal at $2f_c$, the effective signal at the output of the multiplier is

$$y_1(t) = \frac{1}{2} \sum_{i=-\infty}^{\infty} a_i p(t - kT) + w_1(t). \tag{7.37}$$

The MF output is

$$r(t) = y_1(t) \star p(-t)$$
$$= \frac{1}{2} \sum_{i=-\infty}^{\infty} a_i g(t - iT) + z(t), \tag{7.38}$$

where $\star$ denotes convolution and

$$g(t) = p(t) \star p(-t)$$
$$z(t) = w_1(t) \star p(-t). \tag{7.39}$$

Note that $g(t)$ is a pulse corresponding to the raised cosine spectrum, hence it satisfies the Nyquist criterion for zero intersymbol interference (ISI). The MF output sampled at time kT is

$$r(kT) = (1/2)a_k g(0) + z(kT)$$
$$= a_k + z(kT). \tag{7.40}$$

The above equation can be written more concisely as

$$r_k = a_k + z_k. \tag{7.41}$$

Note that z_k is a zero-mean Gaussian random variable with variance

$$E[z_k^2] = \frac{N_0}{4}g(0) = \frac{N_0}{2} = \sigma^2 \quad \text{(say).} \tag{7.42}$$

At time kT given that $a_k = A_j$ was transmitted, the mean value of r_k is

$$E[r_k|A_j] = A_j. \tag{7.43}$$

Note that

$$E[r_k] = \sum_{j=0}^{1} E[r_k|A_j]P(A_j) = 0. \tag{7.44}$$

In other words, the unconditional mean is zero. At time kT, the MAP detector computes the probabilities $P(A_j|r_k)$ for $j = 0, 1$ and decides in favor of that symbol for which the probability is maximum. Using Bayes' rule, the MAP detector can be re-written as

$$\max_{j} \frac{f_{R_k|A_j}(r_k|A_j)P(A_j)}{f_{R_k}(r_k)} \quad \text{for } j = 0, 1, \tag{7.45}$$

where $f_{R_k|A_j}(r_k|A_j)$ is the conditional pdf of r_k given that A_j was transmitted. Since the symbols are equally likely $P(A_j) = 1/2$, and is independent of j. Moreover $f_{R_k}(r_k)$ is also independent of j. Thus the MAP detector reduces to an ML detector given by

$$\max_{j} f_{R_k|A_j}(r_k|A_j) \quad \text{for } j = 0, 1. \tag{7.46}$$

Substituting for the conditional pdf we get

$$\max_{j} \frac{1}{\sigma\sqrt{2\pi}} e^{-(r_k - A_j)^2/(2\sigma^2)} \quad \text{for } j = 0, 1. \tag{7.47}$$

Ignoring constants and noting that maximizing e^x is equivalent to maximizing x, the ML detection rule simplifies to

$$\max_{j} -(r_k - A_j)^2 \quad \text{for } j = 0, 1. \tag{7.48}$$

However for $x > 0$, maximizing $-x$ is equivalent to minimizing x. Hence the ML detection rule can be rewritten as

$$\min_{j} (r_k - A_j)^2 \quad \text{for } j = 0, 1. \tag{7.49}$$

Given that 1 was transmitted at time kT, the ML detector makes an error when

$$
\begin{aligned}
(r_k - A_0)^2 &< (r_k - A_1)^2 \\
\Rightarrow (z_k + d_1)^2 &< z_k^2 \\
\Rightarrow z_k^2 + d_1^2 + 2z_k d_1 &< z_k^2 \\
\Rightarrow 2z_k d_1 &< -d_1^2 \\
\Rightarrow z_k &< -d_1/2,
\end{aligned}
\tag{7.50}
$$

where

$$d_1 = (A_1 - A_0) = 2d. \tag{7.51}$$

Thus the probability of detecting 0 when 1 is transmitted is

Fig. 7.5 Noise samples at the output of a root-raised cosine filter

$$P(z_k < -d_1/2) = \frac{1}{\sigma\sqrt{2\pi}} \int_{z_k=-\infty}^{-d_1/2} e^{-z_k^2/(2\sigma^2)} \, dz_k$$

$$= \frac{1}{2}\text{erfc}\left(\sqrt{\frac{d_1^2}{8\sigma^2}}\right)$$

$$= \frac{1}{2}\text{erfc}\left(\sqrt{\frac{d^2}{N_0}}\right). \tag{7.52}$$

5. Consider the block diagram in Fig. 7.5. Here $w(t)$ is a sample function of a zero-mean AWGN process with psd $N_0/2$. Let $p(t)$ denote the pulse corresponding to the root-raised cosine spectrum. Let $P(f)$ denote the Fourier transform of $p(t)$ and let the energy of $p(t)$ be 2. Assume that $w(t)$ is WSS. Compute $E[z(kT)z(kT - mT)]$.

  • *Solution*: The psd of $z(t)$ is

$$S_Z(f) = \frac{N_0}{2}|P(f)|^2, \tag{7.53}$$

where $|P(f)|^2$ has a raised-cosine spectrum. Note that since $w(t)$ is WSS, $z(t)$ is also WSS. This implies that the autocorrelation of $z(t)$ is

$$E[z(t)z(t - \tau)] = R_Z(\tau) = \frac{N_0}{2}g(\tau), \tag{7.54}$$

where $g(\tau)$ is the inverse Fourier transform of the raised cosine spectrum. Therefore

$$E[z(kT)z(kT - mT)] = R_Z(mT) = \frac{N_0}{2}g(mT)$$

$$= \begin{cases} N_0 & \text{for } m = 0 \\ 0 & \text{otherwise} \end{cases} \tag{7.55}$$

Index

© The Editor(s) (if applicable) and The Author(s), under exclusive license
to Springer Nature Switzerland AG 2021
K. Vasudevan, *Analog Communications*,
https://doi.org/10.1007/978-3-030-50337-6

Printed in the United States
by Baker & Taylor Publisher Services